STUDY GUIDE

KENNETH PINZKE

EARTH SCIENCE

Eighth Edition

TARBUCK ◆ LUTGENS

PRENTICE HALL, UPPER SADDLE RIVER, NJ 07458

Supplement Acquisitions Editor: *Wendy Rivers*
Production Editor: *Robert C. Walters*
Cover Designer: *Paul Gourhan*
Special Projects Manager: *Barbara A. Murray*
Manufacturing Buyer: *Benjamin Smith*

Printed in the United States of America

10 9 8 7 6 5 4 3 2

ISBN 0-13-572744-8

Prentice-Hall International (UK) Limited, *London*
Prentice-Hall of Australia Pty. Limited, *Sydney*
Prentice-Hall Canada Inc., *Toronto*
Prentice-Hall Hispanoamericana, S.A., *Mexico*
Prentice-Hall of India Private Limited, *New Delhi*
Prentice-Hall of Japan, Inc., *Tokyo*
Simon & Schuster Asia Pte. Ltd., *Singapore*
Editoria Prentice-Hall do Brasil, *Ltda., Rio de Janero*

CONTENTS

Keys to Successful Learning

Learning involves much more than reading a textbook and attending classes. It is a process that requires continuous preparation, practice, and review in a dedicated attempt to understand a topic through study, instruction, and experience. Furthermore, most learning experts agree that successful learning is based upon association, repetition, and visualization.

Students often ask questions such as "What is the best way to learn Earth science?" or "How should I study?" Although there are no set answers to these questions, many successful students use similar study techniques and utilize many of the same study habits. For you to become a successful learner, you need to develop certain skills that include, among others,

1. **Carefully observing the world around you.** Pay special attention to the relations between natural phenomena. For example, when studying weather, you might observe and note the direction that storms move in your area, or the relation between air temperature and wind direction. Or when studying minerals and rocks, you may want to list, locate, and identify those that are found in your area.

2. **Being curious and inquisitive.** Ask questions of your instructor, your classmates, and yourself, especially about concepts or ideas you do not fully understand.

3. **Developing critical thinking skills.** Carefully analyze all available information before forming a conclusion, and be prepared to modify your conclusions when necessary on the basis of new evidence.

4. **Visualizing.** Take time to create mental images of what you are studying, especially those concepts that involve change through time. If you can "picture" it, you probably understand it.

5. **Quantifying.** Accurate measurement is the foundation of scientific inquiry. To fully comprehend scientific literature you must be familiar with the common systems and units of measurement, in particular the metric system. If you are not already comfortable with the metric system, take time to study and review it. Practice "thinking metric" by estimating everyday measurements such as the distance to the store or the volume of water in a can or pail using metric units.

6. **Being persistent.** Be prepared to repeat experiments, observations, and questions until you understand and are comfortable with the concepts in your own mind.

Remember that an important component of learning is identifying what you *should* know and understand and what you *do* know and understand about the material presented in the textbook and during class. With this in mind, the following steps may help you become an active, successful learner.

1. **Prepare.** Quickly scan the pages of the textbook chapter and accompanying section of this study guide. Try to visualize the organization of the chapter in simple, broad concepts. Read the chapter overview and learning objectives provided for each chapter in the Study Guide. These will help you focus on the most important concepts the authors present in the chapter.

2. **Read the Assignment.** Read each assignment before attending class. Taking written notes as you read can be an effective means of helping you remember major points presented in the chapter. As you read the assignment, you may also find the following suggestions helpful.

   - To help you visualize concepts, look at all the illustrations carefully and read the captions. Also try to develop mental images of the material you are reading.

   - Pace yourself as you read. Often a single chapter contains more subject matter than can be assimilated in one long reading. Take frequent short breaks to help keep your mind fresh.

   - If you find a section confusing, stop and read the passage again. Make sure you understand the material before proceeding.

   - As you read, write down questions you may want to ask your instructor during class time.

3. **Attend and Participate.** To get the most from any subject, you should attend all lecture classes, study sessions, review sessions, and (if required) laboratory sessions.

   - Bring any questions concerning the subject matter to these sessions.

   - Communicate with your instructor and the other students in the class.

   - Learn how to become a good listener and how to take useful class notes. Remember that taking a lot of notes may not be as effective as taking the right notes.

   - Come fully prepared for each session.

4. **Review.** Review your class notes and the chapter material by examining the section headings, illustrations, and chapter notes. Furthermore,

   - Identify the key facts, concepts, and principles in the chapter by highlighting or underlining them.

   - Examine the chapter summary statements presented in the Study Guide.

   - Write out the answers to the review questions presented at the end of the textbook chapter.

5. **Study.** Studying is one of the most important keys to successful learning. Very few individuals have the ability to retain vast amounts of detailed information after only one reading. In fact, most of us require seeing and thinking about a concept many times, perhaps over several days, before we fully comprehend it. Consider the following as you study and prepare for exams.

   - As you study, keep "putting the pieces together" by integrating the key concepts and facts as you go along. The chapter outlines presented in the Study Guide may prove useful for reviewing and organizing the major points of each chapter.

   - As you review your textbook notes, lecture notes, and chapter highlights, keep in mind that most introductory science courses emphasize terms and definitions because scientists must have consistent, precise meanings for purposes of communication. To quiz your knowledge of each chapter's key terms, complete the vocabulary review section of the appropriate chapter in the Study Guide. Review your answers often prior to an examination.

- Complete the comprehensive review section of the appropriate chapter in the Study Guide by writing out complete answers to the questions. By actually writing the answers, you will be able to retain the material with greater detail than simply answering the questions in your mind or copying the answers from the answer key.

- Several days before an exam, test your overall retention of the chapter material by taking the practice test for the appropriate chapter(s) in the Study Guide. Thoroughly research and review any of the questions you miss.

Using the Study Guide Effectively

This study guide has been written to accompany the textbook *Earth Science*, Eighth Edition, by Edward J. Tarbuck and Frederick K. Lutgens. For each chapter of the textbook there is a corresponding Study Guide chapter. For example, Chapter 1 of the Study Guide corresponds to Chapter 1 of the textbook. Each chapter in the Study Guide contains nine parts. They are

1. *Chapter Overview.* The overview provides a brief summary of the topics presented in the chapter. The overview should be read before reading the textbook chapter.

2. *Learning Objectives.* The learning objectives are a list of what you should know after you have read and studied the textbook chapter and completed the corresponding chapter in the Study Guide. You should examine the objectives before and after you read and study the chapter.

3. *Chapter Summary.* This section provides a brief summary of the most important concepts presented in the chapter. The statements are useful for a quick review of the major chapter concepts as well as preparing for examinations.

4. *Chapter Outline.* The chapter outline presents, in outline form, the major topics of the chapter. It is helpful for both reviewing chapter concepts and preparing for examinations.

5. *Key Terms.* The key terms list includes the new vocabulary terms introduced in the textbook chapter. The page number shown in parentheses ( ) after each term identifies the textbook page where the term first appears.

6. *Vocabulary Review.* This section consists of fill-in-the-blank statements that are intended to test your knowledge of the new terms presented in the textbook chapter. You should make sure you have a good understanding of the terms before completing the remaining sections of the Study Guide chapter. Answers to the vocabulary review are listed in the answer keys located at the back of the Study Guide.

7. *Comprehensive Review.* The comprehensive review contains several questions that test your knowledge of the basic facts and concepts presented in the chapter. In this section you will be asked to write out answers, label diagrams, and list key facts. Answers to the comprehensive review questions are listed in the answer keys located at the back of the Study Guide.

8. *Practice Test.* This section of each Study Guide chapter will test your total knowledge and understanding of the textbook chapter. Each practice test consists of multiple choice, true/false, word choice, and written questions. Taking the practice test is one way for you to determine how ready you are for an actual exam. You may find it helpful to write your answers to the practice test questions on a separate sheet of paper and leave the test in the Study Guide unmarked so you can review it several times while preparing for a unit exam, mid-term exam, and/or final exam. Answers to the practice test questions are listed in the answer keys located at the back of the Study Guide.

9. *Answer Key.* Chapter answer keys are located at the back of the Study Guide.

To effectively use the Study Guide, it is recommended that you

- Quickly glance over the appropriate chapter in the Study Guide before reading the corresponding textbook chapter.

- When required, write out complete answers to the Study Guide questions.

- Check your answers with those provided at the back of the Study Guide.

- Research the answers you don't know by looking them up in your textbook and/or class notes. You will learn very little by simply copying the answers from the answer keys.

- Test yourself by taking the practice test provided at the end of each Study Guide chapter.

- Constantly review your answers to all the Study Guide questions.

ACKNOWLEDGMENTS

My sincere thanks to each of the many individuals who assisted in the preparation of this edition of the *Study Guide* for *Earth Science*. Furthermore, I would like to acknowledge all my students, past and present, whose comments and questions have helped me focus on the nature, meaning, and essentials of Earth science. A special debt of gratitude goes to Wendy Rivers, Dan Kaveney, and their Prentice Hall colleagues who skillfully guided this manuscript from its inception to completion. They are true professionals with whom I feel fortunate to be associated.

Thanks,
Ken Pinzke

I hear and I forget.

I see and I remember.

I do and I understand.

- Ancient Proverb

Introduction

The *Introduction* opens with a brief discussion of Earth's four "spheres." A definition of *Earth science* is followed by a discussion of the four areas that are traditionally included—geology, oceanography, meteorology, and astronomy. The importance of understanding basic Earth science principles when examining resources and environmental issues is also presented. The *Introduction* closes with a discussion of the nature of scientific inquiry.

Learning Objectives

After reading, studying, and discussing the *Introduction*, you should be able to:

- Describe Earth's four "spheres."
- List the sciences traditionally included in Earth science.
- Discuss some resource and environmental issues.
- Describe the nature of scientific inquiry.

Introduction Summary

- The four "spheres" of Earth are 1) the *hydrosphere*, the dynamic mass of water that is continually on the move; 2) the *atmosphere*, the gaseous envelope surrounding Earth; 3) the *biosphere*, which includes all life; and 4) the *lithosphere*, which refers to the rigid outer layer of Earth. The solid Earth is divided into the *core, mantle,* and *crust.* The lithosphere includes the crust and part of the upper mantle.

- *Earth science* is the name for all the sciences that collectively seek to understand Earth and its neighbors in space. It includes *geology, oceanography, meteorology,* and *astronomy.*

- *Environment* refers to everything that surrounds and influences an organism. These influences can be biological, social, or physical. *Resources* are an important environmental concern. The two broad categories of resources are 1) *renewable*, which means that they can be replenished over relatively short time spans, and 2) *nonrenewable.*

- Environmental problems can be local, regional, or global. Human-induced problems include urban air pollution, acid rain, ozone depletion, and global warming. Natural hazards imposed by the physical environment include earthquakes, landslides, floods, and hurricanes. As world population continues to grow, pressures on the environment increase as well.

- All science is based on the assumption that the natural world behaves in a consistent and predictable manner. The process through which scientists gather facts through observations and formulate scientific *hypotheses, theories,* and *laws,* is called the *scientific method.* To help determine what is occurring in

1

the natural world, scientists often 1) collect facts, 2) develop a scientific hypothesis, 3) construct experiments to validate the hypothesis, and 4) accept, modify, or reject the hypothesis on the basis of extensive testing. Other discoveries represent purely theoretical ideas which have stood up to extensive examination. Still other scientific advancements have been made when a totally unexpected happening occurred during an experiment.

Chapter Outline

I. Earth's four "spheres"
   A. Hydrosphere
      1. Ocean—the most prominent feature of the hydrosphere
         a. Nearly 71 percent of Earth's surface
         b. About 97 percent of Earth's water
      2. Also includes fresh water
   B. Atmosphere
   C. Biosphere
      1. Includes all life
      2. Influences other three spheres
   D. Lithosphere
      1. Earth's rigid outer layer
      2. Solid Earth consists of
         a. Core
         b. Mantle
         c. Crust
      3. Lithosphere includes
         a. Crust
         b. Part of the upper mantle
      4. Divisions of Earth's surface
         a. Continents
         b. Ocean Basins
II. Earth science
   A. Encompasses all sciences that seek to understand
      1. Earth
      2. Earth's neighbors in space
   B. Includes
      1. Geology
         a. Physical geology examines the materials composing Earth
         b. Historical geology is the study of the origin and development of Earth
      2. Oceanography
         a. Not a separate and distinct science
         b. Oceanography integrates
            1. Chemistry
            2. Physics
            3. Geology
            4. Biology
      3. Meteorology

      4. Astronomy
III. Resources and environmental issues
   A. Environment
      1. Surrounds and influences organisms
      2. Influences on organisms
         a. Biological
         b. Social
         c. Nonliving (physical environment)
            1. Water
            2. Air
            3. Soil
            4. Rock
   B. Resources
      1. Important environmental concern
      2. Include
         a. Water
         b. Soil
         c. Minerals
         d. Energy
      3. Two broad categories
         a. Renewable
            1. Can be replenished
            2. Examples
               a. Plants
               b. Wind energy
         b. Nonrenewable
            1. Fixed quantities
            2. Examples
               a. Metals
               b. Fuels
   C. Environmental problems
      1. Local, regional, and global
      2. Human-induced and accentuated
         a. Urban air pollution
         b. Acid rain
         c. Ozone depletion
         d. Global warming
      3. Natural hazards
         a. Earthquakes
         b. Landslides
         c. Floods
         d. Hurricanes
      4. World population pressures

IV. Scientific inquiry
    A. Science assumes the natural world is
        1. Consistent
        2. Predictable
    B. Goal of science
        1. To discover patterns in nature, and
        2. To use the knowledge to predict
    C. An idea can become a
        1. Hypothesis (untested)
        2. Theory (tested and confirmed)
        3. Law (no known deviation)
    D. Scientific method
        1. Gather facts through observation
        2. Formulate

        a. Hypotheses
        b. Theories
        c. Laws
    E. Scientific knowledge is gained through
        1. Following systematic steps
            a. Collecting facts
            b. Developing a hypothesis
            c. Conduct experiments
            d. Reexamine the hypothesis and
                1. Accept
                2. Modify
                3. Reject
        2. Theories that withstand examination
        3. Totally unexpected occurrences

Key Terms

Page numbers shown in ( ) refer to the textbook page where the term first appears.

astronomy (p. 6)	hydrosphere (p. 2)	nonrenewable resource (p. 10)
atmosphere (p. 3)	hypothesis (p. 12)	oceanography (p. 5)
biosphere (p. 3)	law (p. 13)	renewable resource (p. 10)
core (p. 4)	lithosphere (p. 4)	theory (p. 13)
crust (p. 4)	mantle (p. 4)	
geology (p. 5)	meteorology (p. 6)	

Vocabulary Review

Choosing from the list of key terms, furnish the most appropriate response for the following statements.

1. The solid Earth is divided into three principal units: the very dense _____; the less dense _____; and the _____.

2. In scientific inquiry, a preliminary untested explanation is a scientific _____.

3. The science of _____ is traditionally divided into two parts—physical and historical.

4. _____ is the study of the atmosphere and the processes that produce weather and climate.

5. A scientific _____ is a well-tested and widely accepted view that scientists agree best explains certain observable facts.

6. The dynamic water portion of Earth's physical environment is the _____.

7. The science that involves the application of all sciences in a comprehensive and interrelated study of the oceans in all their aspects and relationships is called _____.

8. A(n) _____ is a resource that can be replenished over a relatively short time span.

9. The "sphere" of Earth, called the _____, includes all life.

10. A resource, such as oil, that cannot be replenished is classified as a(n) _____.

11. _____ is the study of the universe.

12. The life-giving gaseous envelope surrounding Earth is the _____.

13. A scientific _____ is a generalization about the behavior of nature from which there has been no known deviation after numerous observations or experiments.

14. The _____ refers to the rigid, outer layer of Earth, which includes the crust and part of the upper mantle.

Comprehensive Review

1. What are the sciences that are traditionally included in Earth science?

2. Describe the difference between renewable and nonrenewable resources.

3. List and briefly describe Earth's four "spheres."

    1)

    2)

    3)

    4)

4. How would you explain to a friend what the science of Earth science involves?

5. What is the difference between a scientific theory and a scientific law?

6. Beginning with the Big Bang and ending with the formation of the solar system, summarize the sequence of major events that scientists believe have taken place in the universe.

7. What are the four steps that are often used in science to gain knowledge?

    1)

    2)

    3)

    4)

8. Label each of the four units of the solid
   Earth at the appropriate letter in Figure I.1.

9. Describe the following two broad areas
   of geology:

    a) Physical geology:

    b) Historical geology:

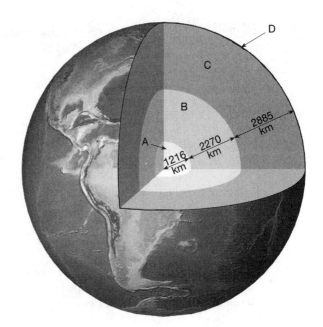

Figure I.1

10. Briefly describe how Earth acts as a system.

Practice Test

Multiple choice. Choose the best answer for the following multiple choice questions.

1. The _____ strongly influences the other three "spheres" because without life their makeups
   and natures would be much different.
    a) atmosphere
    b) hydrosphere
    c) solid Earth
    d) biosphere

2. Earth science attempts to relate our planet to the larger universe by investigating the science of
   _____.
      a) geology   c) meteorology   e) biology
      b) oceanography   d) astronomy

3. The science that studies the processes that produce weather and climate is _____.
      a) geology   c) meteorology   e) biology
      b) oceanography   d) astronomy

4. The global ocean covers _____ percent of Earth's surface.
      a) 71   c) 42   e) 83
      b) 38   d) 65

5. A scientific _____ is a preliminary untested explanation which tries to explain how or why
   things happen in the manner observed.
      a) estimate   c) fact   e) idea
      b) law   d) hypothesis

6. The Earth "sphere," called the _____, includes the fresh water found in streams, lakes,
   glaciers, as well as that existing underground.
      a) atmosphere   c) biosphere
      b) hydrosphere   d) solid Earth

7. The science that includes the study of the composition and movements of seawater, as well as
   coastal processes, sea-floor topography, and marine life is _____.
      a) geology   c) meteorology   e) biology
      b) oceanography   d) astronomy

8. Which one of the following is NOT an example of a nonrenewable resource?
      a) solar energy   c) copper   e) aluminum
      b) natural gas   d) coal

9. An understanding of Earth is essential for _____.
      a) location and recovery of basic resources
      b) minimizing the effects of natural hazards
      c) dealing with the human impact on the environment
      d) all the above

10. This science is divided into two broad divisions—physical and historical.
       a) geology   c) meteorology   e) biology
       b) oceanography   d) astronomy

11. The annual per capita consumption of metallic and nonmetallic mineral resources for the United
    States is nearly _____ tons.
       a) 2   c) 8   e) 15
       b) 6   d) 11

12. The conditions that surround and influence an organism are referred to as the _____.
       a) resource   c) technosphere
       b) social sphere   d) environment

13. Which one of the following is NOT an example of a renewable resource?
       a) water energy   c) iron   e) chickens
       b) lumber   d) cotton

14. In scientific inquiry, when competing hypotheses have been eliminated, a hypothesis may be elevated to the status of a scientific _____.
    a) estimate   c) idea   e) truth
    b) theory   d) understanding

15. The science whose name literally means "study of Earth" is _____.
    a) geology   c) meteorology   e) biology
    b) oceanography   d) astronomy

16. A scientific _____ is a generalization about the behavior of nature from which there has been no known deviation after numerous observations or experiments.
    a) explanation   c) idea   e) law
    b) hypothesis   d) understanding

17. Scientific advancements can be made _____.
    a) by applying systematic steps
    b) by accepting purely theoretical ideas which stand up to extensive investigation
    c) from unexpected happenings during an experiment
    d) all the above

18. Earth is approximately _____ billion years old.
    a) 2.5   c) 4.6   e) 9.8
    b) 3.0   d) 6.2

19. The most striking feature of the ocean floor is the _____.
    a) seamount system   d) mid-ocean ridge system
    b) reef system   e) continental shelf
    c) Hudson canyon

*True/false. For the following true/false questions, if a statement is not completely true, mark it false. For each false statement, change the **italicized** word to correct the statement.*

1. ___ The life-giving gaseous envelope surrounding Earth is called the *atmosphere*.

2. ___ The *biosphere* is a dynamic mass of water that is continually on the move from the oceans to the atmosphere, precipitating back to the land, and running back to the ocean.

3. ___ Although some *renewable* resources such as aluminum can be used over and over again, others, such as oil, cannot be recycled.

4. ___ The aim of *physical* geology is to understand the origin of Earth and the development of the planet since its formation.

5. ___ Science is based on the assumption that the natural world behaves in a(n) *unpredictable* manner.

6. ___ The most obvious difference between the continents and the ocean basins is their relative *levels*.

7. ___ *Biosphere* refers to everything that surrounds and influences an organism.

8. ___ To determine what is happening in the natural world, scientists collect *facts* through observation and measurement.

9. ___ The light and very thin outer skin of Earth is called the *mantle*.

10. ___ By the year 2000 nearly *seven* billion people will inhabit the planet.

11. ___ All life on Earth is included in the *biosphere*.

12. ___ A scientific *theory* is a well-tested and widely accepted view that scientists agree best explains certain observable facts.

Word choice. Complete each of the following statements by selecting the most appropriate response.

1. Earth is a [static/dynamic] body with many interacting parts and a long and complex history.

2. The universe began about [10/20] billion years ago.

3. A gently sloping platform of continental material called the continental [plain/shelf] extends seaward from the shore.

4. In the United States, about [6/20] percent of the world's population uses [10/30] percent of the annual production of mineral and energy resources.

5. Earth's crust is [thinnest/thickest] beneath the oceans.

6. The ocean floor contains extremely deep grooves called [guyots/trenches].

Written questions

1. List the four steps that scientists often use to conduct experiments and gain scientific knowledge.

2. List and briefly describe Earth's four "spheres."

3. Define Earth science. List the four areas commonly included in Earth science.

MINERALS:

Building Blocks of Rocks

Minerals: Building Blocks of Rocks begins with an explanation of the difference between a mineral and a rock, followed by a formal definition of a mineral. Elements, atoms, compounds, ions, and atomic bonding are explained. Also investigated are isotopes and radioactivity. Following descriptions of the properties used in mineral identification, the silicate and nonsilicate mineral groups are examined. The chapter concludes with a discussion of mineral resources, reserves, and ores.

Learning Objectives

After reading, studying, and discussing this chapter, you should be able to:

- Explain the difference between a mineral and a rock.
- Describe the basic structure of an atom and explain how atoms combine.
- List the most important elements that compose Earth's continental crust.
- Explain isotopes and radioactivity.
- Describe the physical properties of minerals and how they can be used for mineral identification.
- List the basic compositions and structures of the silicate minerals.
- List the economic use of some nonsilicate minerals.
- Distinguish between mineral resources, reserves, and ores.

Chapter Summary

- A *mineral* is a naturally occurring inorganic solid that possesses a definite chemical structure, which gives it a unique set of *physical properties*. Most rocks are aggregates composed of two or more minerals.

- The building blocks of minerals are *elements*. An *atom* is the smallest particle of matter that still retains the characteristics of an element. Each atom has a nucleus which contains protons and neutrons. Orbiting the nucleus of an atom are electrons. The number of protons in an atom's nucleus determines its atomic number and the name of the element. Atoms bond together to form a *compound* by either gaining, losing, or sharing electrons with another atom.

- *Isotopes* are variants of the same element, but with a different *mass number* (the total number of neutrons plus protons found in an atom's nucleus). Some isotopes are unstable and disintegrate naturally through a process called *radioactivity*.

- The properties of minerals include crystal form, luster, color, streak, hardness, cleavage, fracture, and specific gravity. In addition, a number of special physical and chemical properties (taste, smell, elasticity, malleability, feel, magnetism, double refraction, and chemical reaction to hydro-chloric acid) are useful in identifying certain minerals. Each mineral has a unique set of properties which can be used for identification.

- The eight most abundant elements found in Earth's continental crust (oxygen, silicon, aluminum, iron, calcium, sodium, potassium, and magnesium) also compose the majority of minerals.

- The most common mineral group is the *silicates*. All silicate minerals have the *silicon-oxygen tetrahedron* as their fundamental building block. In some silicate minerals the tetrahedra are joined in chains; in others, the tetrahedra are arranged into sheets, or three-dimensional networks. Each silicate mineral has a structure and a chemical composition that indicates the conditions under which it was formed.

- The nonsilicate mineral groups include the *oxides* (e.g., magnetite, mined for iron), *sulfides* (e.g., sphalerite, mined for zinc), *sulfates* (e.g., gypsum, used in plaster and frequently found in sedimentary rocks), *native elements* (e.g., graphite, a dry lubricant), *halides* (e.g., halite, common salt and frequently found in sedimentary rocks), and *carbonates* (e.g., calcite, used in portland cement and a major constituent in two well-known rocks: limestone and marble).

- The term *ore* is used to denote useful metallic minerals, like hematite (mined for iron) and galena (mined for lead), that can be mined for a profit, as well as some nonmetallic minerals, such as fluorite and sulfur, that contain useful substances.

Chapter Outline

I. Minerals versus rocks
  A. Mineral definition
    1. Naturally occurring
    2. Inorganic
    3. Solid
    4. Have a definite chemical structure
  B. Nearly 4000 known minerals
  C. Rocks are aggregates (mixtures) of minerals

II. Composition and structure of minerals
  A. Elements
    1. Basic building blocks of minerals
    2. Over 100 are known
  B. Atoms
    1. Smallest particles of matter
    2. Have all the characteristics of an element

III. How atoms are constructed
  A. Nucleus, which contains
    1. Protons—positive electrical charges
    2. Neutrons—neutral electrical charges
  B. Energy levels, or shells
    1. Surround nucleus
    2. Contain electrons—negative electrical charges
  C. Atomic number is the number of protons in an atom's nucleus
  D. Bonding of atoms
    1. Forms a compound with two or more elements

    2. Ions are atoms that gain or lose electrons
  E. Isotopes
    1. Have varying number of neutrons
    2. Have different mass numbers—the sum of the neutrons plus protons
    3. Many isotopes are radioactive and emit energy and particles

IV. Minerals
  A. Properties of minerals
    1. Crystal form
    2. Luster
    3. Color
    4. Streak
    5. Hardness
    6. Cleavage
    7. Fracture
    8. Specific gravity
    9. Other
      a. Taste
      b. Smell
      c. Elasticity
      d. Malleability
      e. Feel
      f. Magnetism
      g. Double refraction
      h. Reaction to hydrochloric acid
  B. A few dozen minerals are called the rock-forming minerals
    1. The eight elements that compose most rock-forming minerals are

oxygen (O), silicon (Si), aluminum (Al), iron (Fe), calcium (Ca), sodium (Na), potassium (K), and magnesium (Mg)
  2. Most abundant atoms in Earth's crust are
    a. Oxygen (46.6 percent by weight)
    b. Silicon (27.7 percent by weight)
C. Mineral groups
  1. Silicate minerals
    a. Most common mineral group
    b. Contain silicon-oxygen tetrahedron
    c. Groups based upon tetrahedron arrangement
      1. Olivine—independent tetrahedron
      2. Pyroxene group—tetrahedron are arranged in chains
      3. Amphibole group—tetrahedron are arranged in double chains
      4. Micas
        a. Tetrahedron are arranged in sheets
        b. Two types of mica
          1. Biotite (dark) and
          2. Muscovite (light)
      5. Feldspars
        a. Three-dimensional network of tetrahedron

    b. Two types of feldspars
      1. Orthoclase and
      2. Plagioclase
    6. Quartz—three-dimensional network of tetrahedron
  d. Feldspars are the most plentiful mineral group
  e. Crystallize from molten material
  2. Nonsilicate minerals
    a. Major groups
      1. Oxides
      2. Sulfides
      3. Sulfates
      4. Halides
      5. Carbonates
      6. "Native" elements
    b. Carbonates
      1. A major rock-forming group
      2. Found in the rocks limestone and marble
    c. Halite and gypsum are found in sedimentary rocks
    d. Many have economic value
D. Mineral resources
  1. Reserves are already identified deposits
  2. Ores are useful metallic minerals that can be mined at a profit
  3. Economic factors may change and influence a resource

Key Terms

Page numbers shown in ( ) refer to the textbook page where the term first appears.

atom (p. 20)	hardness (p. 22)	ore (p. 28)
atomic number (p. 20)	ion (p. 20)	proton (p. 20)
cleavage (p. 23)	isotope (p. 21)	radioactivity (p. 21)
color (p. 22)	luster (p. 21)	reserve (p. 27)
compound (p. 20)	mass number (p. 21)	rock (p. 19)
crystal form (p. 21)	mineral (p. 19)	silicate (p. 24)
electron (p. 20)	mineral resource (p. 27)	silicon-oxygen tetrahedron (p. 25)
element (p. 20)	Mohs hardness scale (p. 22)	specific gravity (p. 23)
energy level (p. 20)	neutron (p. 20)	streak (p. 22)
fracture (p. 23)	nucleus (p. 20)	

Vocabulary Review

Choosing from the list of key terms, furnish the most appropriate response for the following statements.

1. A(n) _____ is an electrically neutral subatomic particle found in the nucleus of an atom.

2. The smallest part of an element that still retains the element's properties is a(n) _____.

3. A(n) _____ is a useful metallic mineral that can be mined at a profit.

4. A naturally occurring, inorganic solid that possesses a definite chemical structure, which gives it a unique set of physical properties is a(n) _____.

5. A(n) _____ is an atom that has an electric charge because of a gain or loss of electrons.

6. The tendency of a mineral to break along planes of weak bonding is the property called _____.

7. The subatomic particle that contributes mass and a positive electrical charge to an atom is a(n) _____.

8. A(n) _____ is composed of two or more elements bonded together in definite proportions.

9. Silicon and oxygen combine to form the framework of the most common mineral group, the _____ minerals.

10. An already identified deposit from which minerals can be extracted profitably is called a(n) _____.

11. A(n) _____ is a subatomic particle with a negative electrical charge.

12. A(n) _____ can be defined simply as an aggregate of minerals.

13. Each atom has a central region called the _____.

14. The number of protons in an atom's nucleus determines its _____ and the name of the element.

15. A(n) _____ is a large collection of electrically neutral atoms, all having the same atomic number.

16. The sum of the neutrons and protons in the nucleus is the atom's _____.

17. Electrons are located at a given distance from an atom's nucleus in a region called the _____.

18. A variation of the same element but with a different mass number is called a(n) _____.

19. The external expression of the internal orderly arrangement of atoms is called the mineral's _____.

20. The process called _____ involves the natural decay of unstable isotopes.

21. The property of a mineral which involves the appearance or quality of light reflected from its surface is termed _____.

22. The fundamental building block of all silicate minerals is the _____.

23. A(n) _____ is a beneficial mineral that can be recovered for use.

24. The property called _____ is the color of a mineral in the powdered form.

25. The resistance of a mineral to abrasion or scratching, called _____, is one of the most useful diagnostic properties.

26. Minerals that do not exhibit cleavage when broken are said to _____.

27. Comparing the weight of a mineral to the weight of an equal volume of water determines the mineral's _____.

28. To assign a value to a mineral's resistance to abrasion or scratching, geologists use a standard scale called _____.

29. Due to impurities, a mineral's _____ is often an unreliable diagnostic property.

Comprehensive Review

1. List the four characteristics that any Earth material must exhibit in order to be considered a mineral.

   1)

   2)

   3)

   4)

2. List the three main subatomic particles found in an atom and describe how they differ from one another. Also indicate where each is located in an atom by selecting the proper letter in Figure 1.1.

   1)   Letter: ____

   2)   Letter: ____

   3)   Letter: ____

3. Briefly explain the difference between a mineral and a rock.

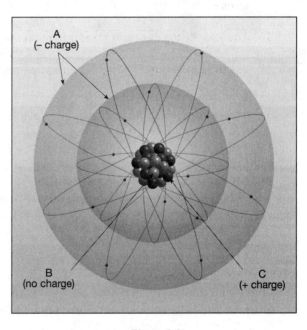

Figure 1.1

4. If an atom has 6 protons, 6 electrons, and 8 neutrons, what is the atom's

    a) Atomic number? _____

    b) Mass number? _____

5. If an atom has 17 electrons and its mass number is 35, calculate the following:

    a) Number of protons _____

    b) Atomic number _____

    c) Number of neutrons _____

6. Briefly explain the formation of the chemical bond involving a sodium atom and chlorine atom to produce the compound, sodium chloride.

7. Examine the diagram of an atom shown in Figure 1.2A. The atom's nucleus contains 8 protons and 5 neutrons. Is the atom shown in diagram 1.2A an ion? Explain your answer.

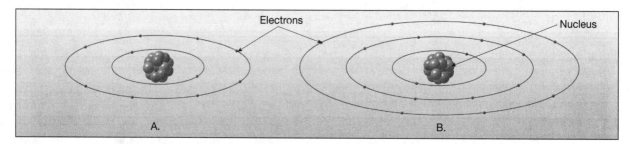

Figure 1.2

    a) What is the mass number of the atom shown in Figure 1.2A? _____

    b) What is the atomic number of the atom shown in Figure 1.2A? _____

8. Examine the diagram of an atom shown in Figure 1.2B. The atom's nucleus contains 17 protons. Is the atom shown in diagram 1.2B an ion? Explain your answer.

9. Define a mineral.

10. In what way does an isotope vary from the common form of the same element?

11. Briefly describe each of the following properties of minerals.

   a) Luster:

   b) Crystal form:

   c) Streak:

   d) Hardness:

   e) Cleavage:

   f) Fracture:

   g) Specific gravity:

12. The minerals represented by the diagrams in Figure 1.3 exhibit cleavage. Describe the cleavage of each mineral in the space below its diagram.

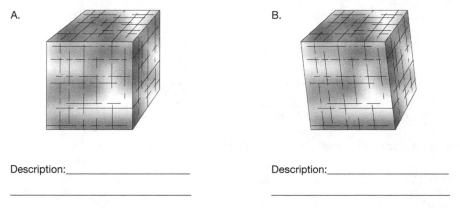

A.   B.

Description:_____   Description:_____

_____   _____

Figure 1.3

13. What are the two most common elements that compose Earth's continental crust? Also list the approximate percentage (by weight) of the continental crust that each element composes, along with its chemical symbol.

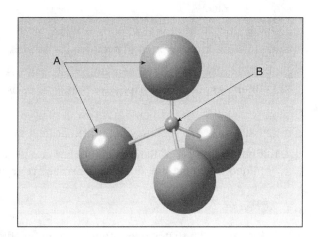

14. Figure 1.4 represents the silicon-oxygen tetrahedron. Letter A identifies _____ atoms, while letter B is a _____ atom.

Figure 1.4

15. Name the silicate mineral group(s) that exhibits each of the following silicate structures.

    a) Three-dimensional networks:

    b) Sheets:

    c) Double chains:

    d) Single chains:

    e) Single tetrahedron:

16. Name three common rock-forming silicate minerals and the most common rock-forming nonsilicate mineral.

    a) Three common rock-forming silicate minerals:

    b) Most common rock-forming nonsilicate mineral:

17. Define a rock.

18. How do most silicate minerals form?

19. List a mineral that is an ore of each of the following elements.

    a) Mercury:

    b) Lead:

    c) Zinc:

    d) Iron:

20. Briefly explain what happens to an isotope as it decays.

21. Could a sample of Earth material that contains 10 percent aluminum by weight be profitably mined to extract the aluminum? Explain your answer.

22. Figure 1.5 illustrates two different common silicate mineral structures. Write the name of the silicate mineral group that each structure represents by the corresponding figure.

A.   B.

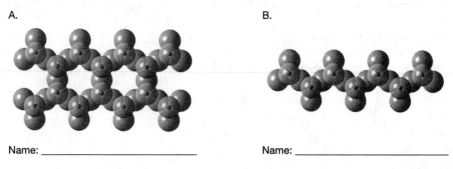

Name: _____   Name: _____

Figure 1.5

23. What are the primary elements that connect the large silicate structures to one another in the silicate minerals?

Practice Test ▬▬▬▬▬▬▬▬▬▬▬▬▬▬▬▬▬▬▬▬▬▬▬

Multiple choice. Choose the best answer for the following multiple choice questions.

1. Which one of the following is NOT a silicate mineral group?
   - a) micas
   - b) feldspars
   - c) carbonates
   - d) pyroxenes
   - e) quartz

2. The basic building block of the silicate minerals _____.
   - a) has the shape of a cube
   - b) contains 1 silicon atom and 4 oxygen atoms
   - c) always occurs independently
   - d) contains 2 iron atoms for each silicon atom
   - e) contains 1 oxygen atom and 4 silicon atoms

3. Which one of the following is NOT one of the eight most abundant elements in Earth's continental crust?
   - a) oxygen
   - b) iron
   - c) aluminum
   - d) calcium
   - e) hydrogen

4. Atoms of the same element possess the same number of _____.
   - a) isotopes
   - b) ions
   - c) protons
   - d) neutrons
   - e) compounds

5. Atoms containing the same numbers of protons and different numbers of neutrons are _____.
   - a) isotopes
   - b) ions
   - c) protons
   - d) neutrons
   - e) compounds

6. Which subatomic particle is most involved in chemical bonding?
   - a) electrons
   - b) ions
   - c) protons
   - d) neutrons
   - e) nucleus

7. The number of _____ in an atom's nucleus determines which element it is.
   - a) electrons
   - b) ions
   - c) protons
   - d) neutrons
   - e) isotopes

8. Oxygen comprises about _____ percent by weight of Earth's continental crust.
   a) 12   c) 47   e) 85
   b) 27   d) 63

9. The smallest particle of an element that still retains all the element's properties is a(n)
   _____.
   a) compound   c) atom   e) isotope
   b) rock   d) neutron

10. Atoms that gain or lose electrons become _____.
    a) electrons   c) protons   e) compounds
    b) ions   d) neutrons

11. When two or more elements bond together in definite proportions they form a(n) _____.
    a) rock   c) atom   e) nucleus
    b) ion   d) compound

12. The sum of the neutrons and protons in an atom's nucleus is the atom's _____.
    a) specific gravity   c) atomic number   e) isotope number
    b) atomic mass   d) energy-levels

13. Radioactivity occurs when _____.
    a) unstable nuclei disintegrate   d) electrons change energy-levels
    b) ions form   e) atoms bond
    c) a mineral is cleaved

14. The property which is a measure of the resistance of a mineral to abrasion or scratching
    is _____.
    a) crystal form   c) cleavage   e) hardness
    b) streak   d) fracture

15. The property which involves the color of a mineral in its powdered form is _____.
    a) crystal form   c) cleavage   e) hardness
    b) streak   d) fracture

16. The tendency of a mineral to break along planes of weak bonding is called _____.
    a) luster   c) cleavage   e) bonding
    b) crystal form   d) fracture

17. Minerals that break into smooth curved surfaces, like those seen in broken glass, have a unique
    type of fracture, called _____ fracture.
    a) conchoidal   c) rounded   e) uneven
    b) irregular   d) vitreous

18. The basic building block of the silicate minerals is the silicon-oxygen _____.
    a) cube   c) octagon   e) sphere
    b) tetrahedron   d) rhombohedron

19. Which of the following is NOT a primary element that joins silicate structures?
    a) calcium   c) potassium   e) oxygen
    b) iron   d) magnesium

20. Most silicate minerals form from _____.
    a) other minerals   c) erosion   e) water
    b) radioactive decay   d) molten rock

21. Which silicate mineral group has a double chain silicate structure?
    a) pyroxenes   c) micas   e) quartz
    b) amphiboles   d) feldspars

22. Which major rock-forming mineral is the chief constituent of the rocks limestone and marble?
    a) gypsum   c) mica   e) calcite
    b) halite   d) feldspar

23. The scale used by geologists to measure the hardness of a mineral is called _____ scale.
    a) Mohs   c) Richters   e) Playfairs
    b) Smiths   d) Bowens

24. Which nonsilicate mineral group contains the ores of iron?
    a) oxides   c) sulfates   e) carbonates
    b) sulfides   d) halides

25. Which one of the following statements concerning minerals is NOT true?
    a) naturally occurring   d) definite chemical structure
    b) solid   e) formed of atoms
    c) organic

*True/false. For the following true/false questions, if a statement is not completely true, mark it false. For each false statement, change the **italicized** word to correct the statement.*

1. ___ Ions form when atoms gain or lose *neutrons*.

2. ___ *Protons* have an opposite electrical charge from electrons.

3. ___ Nearly *eighty* minerals have been named.

4. ___ Only *eight* elements compose the bulk of the rock-forming minerals.

5. ___ The elements silicon and *carbon* comprise nearly three-fourths of Earth's continental crust.

6. ___ The most common mineral group is the *silicate* group.

7. ___ The structure of the silicon-oxygen tetrahedron consists of *four* oxygen atoms surrounding a smaller silicon atom.

8. ___ Each silicate mineral group has a particular silicate *structure*.

9. ___ Silicate minerals tend to cleave *through* the silicon-oxygen structures.

10. ___ Each *silicate* mineral has a structure and chemical composition that indicate the conditions under which it formed.

11. ___ Nonsilicate minerals make up about *one-tenth* of the continental crust.

12. ___ The mineral calcite is a *sulfide* mineral.

13. ___ *Quartz* is the mineral from which plaster and other similar building materials are composed.

14. ___ *Reserves* are already-identified mineral deposits from which minerals can be extracted profitably.

15. ___ *Luster* is the external expression of a mineral's internal orderly arrangement of atoms.

16. ___ Cleavage is defined by the number of planes exhibited and the *angles* at which they meet.

17. ___ Vitreous, pearly, and silky are types of *metallic* lusters.

18. ___ Isotopes of the same element have *different* mass numbers.

19. ___ A(n) *ion* is composed of two or more elements bonded together.

20. ___ The number of *protons* in an atom's nucleus determines its atomic number.

21. ___ Uncombined atoms have the same number of *neutrons* as protons.

22. ___ Individual *electrons* are located at given distances from an atom's nucleus in regions called energy levels.

23. ___ A(n) *compound* is the smallest particle of matter that has all the characteristics of an element.

24. ___ A mineral is a naturally occurring *inorganic* solid that possesses a definite chemical structure.

25. ___ Quartz is *softer* than calcite.

Word choice. Complete each of the following statements by selecting the most appropriate response.

1. The number of protons in an atom's nucleus determines its [atomic number/mass number]; while the [atomic number/mass number] is the sum of the neutrons and protons in the nucleus.

2. The [nucleus/atom] is the smallest particle of matter that has all characteristics of an element.

3. Isotopes of the same element have different [atomic/mass] numbers, the same [atomic/mass] number, and very [similar, different] chemical behavior.

4. Adjectives such as metallic, glassy, and earthy are used to describe the mineral property called [luster/color].

5. Minerals that do not exhibit cleavage when broken are said to [crystallize/fracture].

6. All silicate minerals contain the elements oxygen and [silicon/aluminum] joined together in the form of a [triangle/tetrahedron].

7. The crystal structure of a mineral has a direct influence on the mineral's [luster/cleavage].

8. The most prominent mineral in the carbonate group of minerals is [calcite/quartz].

9. The mineral name for table salt is [hematite/halite]; and the major constituent of the rock limestone is [calcite/gypsum].

10. An ore of iron is the mineral [pyrite/hematite]; while the primary ore of lead is [galena, sphalerite].

Written questions

1. What are the three main particles of an atom? How do they differ from one another?

2. Why don't most mineral samples demonstrate visibly their crystal form?

3. Describe the basic building block of all silicate minerals.

4. Define a mineral.

5. Explain the difference between a mineral resource and a mineral reserve.

ROCKS:

Materials of the Lithosphere

2

Rocks: Materials of the Lithosphere opens with a discussion of the rock cycle as part of the Earth system. The origins and processes involved in forming the three major rock groups—igneous rock, sedimentary rock, and metamorphic rock—are also reviewed. A discussion of the crystallization of magma precedes an examination of the classification, textures, and compositions of igneous rocks. Following an investigation of the origin of sediment, the classification of sedimentary rocks, as well as some of their common features, is discussed. The chapter also examines the agents of metamorphism, the textural and mineralogical changes that take place during metamorphism, and some common metamorphic rocks. In conclusion, resources from rocks and minerals are explained.

Learning Objectives

After reading, studying, and discussing this chapter, you should be able to:

- Diagram and discuss the rock cycle.
- List the geologic processes involved in the formation of each rock group.
- Briefly explain crystallization of magma.
- List the criteria used to classify igneous rocks.
- List the names, textures, and environments of formation for the most common igneous rocks.
- Discuss the origin of materials that accumulate as sediment.
- List the criteria used to classify sedimentary rocks.
- Explain the difference between detrital and chemical sedimentary rocks.
- List the names, textures, and environments of formation for the most common sedimentary rocks.
- List the common features of sedimentary rocks.
- Describe the agents of metamorphism.
- List the criteria used to classify metamorphic rocks.
- List the names, textures, and environments of formation for the most common metamorphic rocks.
- Discuss metallic and nonmetallic mineral resources.

Chapter Summary

• *Igneous rock* forms from *magma* that cools and solidifies in a process called *crystallization*. *Sedimentary rock* forms from the *lithification* of *sediment*. *Metamorphic rock* forms from rock that has been subjected to great pressure and heat in a process called *metamorphism*.

• The rate of cooling of magma greatly influences the size of mineral crystals in igneous rock—the faster the rate of cooling, the smaller the crystals. The four basic igneous rock textures are 1) *fine-grained*, 2) *coarse-grained*, 3) *porphyritic*, and 4) *glassy*.

• The mineral makeup of an igneous rock is ultimately determined by the chemical composition of the magma from which it crystallized. N.L. Bowen showed that as magma cools, minerals crystallize in an orderly fashion. *Crystal settling* can change the composition of magma and cause more than one rock type to form from a common parent magma.

• Igneous rocks are classified by their *texture* and *mineral composition*.

• The process of *weathering* produces sediment—unconsolidated particles of rock created by the weathering and erosion of rock, by chemical precipitation from solution in water, or from the secretions of organisms. Sediment is transported and deposited by water, wind, or glaciers. *Compaction* and *cementation* transform sediment into sedimentary rock.

• *Detrital sediments* are materials that originate and are transported as solid particles derived from weathering. *Chemical sediments* are soluble materials produced largely by chemical weathering that are precipitated by either inorganic or organic processes. *Detrital sedimentary rocks*, which are classified by particle size, contain a variety of mineral and rock fragments, with clay minerals and quartz the chief constituents. *Chemical sedimentary rocks* often contain the products of biological processes such as shells or mineral crystals that form as water evaporates and minerals precipitate. *Lithification* refers to the processes by which sediments are transformed into solid sedimentary rocks.

• Common detrital sedimentary rocks include *shale* (the most common sedimentary rock), *sandstone*, and *conglomerate*. The most abundant chemical sedimentary rock is *limestone*, composed chiefly of the mineral *calcite*. *Rock gypsum* and *rock salt* are chemical rocks that form as water evaporates and triggers the deposition of chemical precipitates.

• Some of the features of sedimentary rocks that are often used in the interpretation of Earth history and past environments include *strata*, or *beds* (the single most characteristic feature), *bedding planes*, and *fossils*.

• Two types of metamorphism are 1) *regional metamorphism* and 2) *contact metamorphism*. The agents of metamorphism include *heat, pressure*, and *chemically active fluids*. Heat is perhaps the most important because it provides the energy to drive the reactions that result in the *recrystallization* of minerals. Metamorphic processes cause many changes in rocks, including *increased density*, growth of *larger mineral crystals, reorientation of the mineral grains* into a layered or banded appearance known as *foliation*, and the formation of *new minerals*.

• Some common metamorphic rocks with a *foliated texture* include *slate, schist*, and *gneiss*. Metamorphic rocks with a *nonfoliated texture* include *marble* and *quartzite*.

• Some of the most important accumulations of *metallic mineral resources* are produced by igneous and metamorphic processes. *Vein deposits* (deposits in fractures or bedding planes) and *disseminated deposits* (deposited distributed throughout the entire rock mass) are produced from *hydrothermal solutions*—hot, metal-rich fluids associated with cooling magma bodies.

• *Nonmetallic mineral resources* are mined for the nonmetallic elements they contain or for the physical and chemical properties they possess. The two groups of nonmetallic mineral resources are 1) *building materials* (e.g., limestone and gypsum) and 2) *industrial minerals* (e.g., fluorite and corundum).

Chapter Outline ▬▬▬▬▬▬▬▬▬▬▬▬▬▬▬▬▬▬▬▬

I. Rock cycle
  A. Shows the relations among the three rock types
  B. Proposed by James Hutton in the late 1700s
  C. The cycle
    1. Magma
      a. Crystallization
    2. Igneous rock
      a. Weathering
      b. Transportation
      c. Deposition
    3. Sediment
      a. Lithification
    4. Sedimentary rock
      a. Metamorphism
    5. Metamorphic rock
      a. Melting
    6. Magma
  D. Full cycle does not always take place due to "shortcuts" or interruptions
    1. e.g., Sedimentary rock melts
    2. e.g., Igneous rock is metamorphosed
    3. e.g., Sedimentary rock weathers
    4. e.g., Metamorphic rock weathers
II. Igneous rocks
  A. Form as magma cools and crystallizes
    1. Rocks formed inside Earth are called plutonic or intrusive rocks
    2. Rocks formed on the surface
      a. Formed from lava (a material similar to magma, but without gas)
      b. Called volcanic or extrusive rocks
  B. Crystallization of magma
    1. Ions are arranged into orderly patterns
    2. Crystal size of the rock is determined by the rate of cooling
      a. Slow rate forms large crystals
      b. Fast rate forms microscopic crystals
      c. Very fast rate forms glass
  C. Classification is based on the rock's texture and mineral composition
    1. Texture
      a. Size and arrangement of crystals
      b. Types
        1. Fine-grained—fast rate of cooling
        2. Coarse-grained—slow rate of cooling
        3. Porphyritic (two crystal sizes)—two rates of cooling
        4. Glassy—very fast rate of cooling
    2. Mineral composition
      a. Explained by Bowen's reaction series which shows the order of mineral crystallization
      b. Influenced by crystal settling in the magma
  D. Naming igneous rocks
    1. Basaltic rocks
      a. Derived from the first minerals to crystallize
      b. Rich in iron and magnesium
      c. Low in silica
      d. Common rock is basalt
    2. Granitic rocks
      a. From the last minerals to crystallize
      b. Mainly feldspar and quartz
      c. High silica content
      d. Common rock is granite
III. Sedimentary rocks
  A. Form from sediment (weathered products)
  B. Form about 75 percent of the rock outcrops on the continents
  C. Used to construct much of Earth's history
    1. Clues to past environments
    2. Provide information about sediment transport
    3. Rocks often contain fossils
  D. Economic importance
    1. Coal
    2. Petroleum and natural gas
    3. Sources of iron and aluminum
  E. Classification
    1. Two groups based on the source of the material
      a. Detrital rocks
        1. Material is solid particles
        2. Classified by particle size
        3. Common rocks are
          a. Shale (most abundant)
          b. Sandstone
          c. Conglomerate
          d. Siltstone
      b. Chemical rocks

1. Derived from material that was once in solution and precipitates to form sediment
   a. Directly precipitated or
   b. Through life processes (biochemical origin)
2. Common rocks are
   a. Limestone—the most abundant chemical rock
   b. Travertine
   c. Microcrystalline quartz
      1. Chert
      2. Flint
      3. Jasper
      4. Agate
   d. Evaporites
      1. Rock salt
      2. Gypsum
   e. Coal
      1. Lignite
      2. Bituminous
F. Produced through lithification
   1. Loose sediments are transformed into solid rock
   2. Lithification processes
      a. Compaction
      b. Cementation by the materials
         1. Calcite
         2. Silica
         3. Iron oxide
G. Features
   1. Strata, or beds (most characteristic)
   2. Bedding planes separate strata
   3. Fossils
      a. Traces or remains of prehistoric life
      b. Are the most important inclusions
      c. Help determine past environments
      d. Used as time indicators
      e. Used for matching rocks from different places

IV. Metamorphic rocks
A. "Changed form" rocks
B. Can form from
   1. Igneous rocks
   2. Sedimentary rocks
   3. Other metamorphic rocks
C. Degrees of metamorphism
   1. Show in the rock's texture and mineralogy
   2. Types
      a. Low-grade (e.g., shale becomes slate)
      b. High-grade (causes the original features to be obliterated)

D. Metamorphic settings
   1. Regional metamorphism
      a. Over extensive areas
      b. Produces the greatest volume of metamorphic rock
   2. Contact metamorphism
      a. Near a mass of magma
      b. "Bakes" the surrounding rock
E. Metamorphic agents
   1. Heat
   2. Pressure
      a. From burial
      b. From stress
   3. Chemically active fluids
      a. Water (most common fluid)
      b. Ion exchange among minerals
F. Textures
   1. Foliated
      a. Minerals are in a parallel alignment
      b. Minerals are perpendicular to the force
   2. Nonfoliated
      a. Contain equidimensional crystals
      b. Resembles a coarse igneous rock
G. Classification
   1. Based on texture
   2. Two groups
      a. Foliated rocks
         1. Slate
            a. Fine-grained
            b. Splits easily
         2. Schist
            a. Strongly foliated
            b. "Platy"
            c. Types based on composition (e.g., mica schist)
         3. Gneiss
            a. Strong segregation of silicate minerals
            b. "Banded" texture
      b. Nonfoliated rocks
         1. Marble
            a. Parent rock—limestone
            b. Calcite crystals
            c. Used as a building stone
            d. Variety of colors
         2. Quartzite
            a. Parent rock—quartz sandstone
            b. Quartz grains are fused

V. Resources from rocks and minerals
A. Metallic mineral resources
   1. e.g., Gold, silver, copper
   2. Produced by

a. Igneous processes
b. Metamorphic processes
3. Hydrothermal (hot-water) solutions
   a. Hot
   b. Contain metal-rich fluids
   c. Associated with cooling magma bodies
   d. Types
      1. Vein deposits occur in fractures or bedding planes
      2. Disseminated deposits are distributed throughout the rock

B. Nonmetallic mineral resources
1. Make use of the materials
   a. Nonmetallic elements
   b. Physical or chemical properties
2. Two broad groups
   a. Building materials (e.g., limestone, gypsum)
   b. Industrial minerals (e.g., fluorite, corundum, sylvite

Key Terms

Page numbers shown in ( ) refer to the textbook page where the term first appears.

chemical sedimentary rock (p. 42)
coarse-grained texture (p. 37)
contact metamorphism (p. 48)
crystallization (p. 35)
detrital sedimentary rock (p. 42)
disseminated deposit (p. 54)
evaporite deposit (p. 45)
extrusive (volcanic) (p. 35)
fine-grained texture (p. 37)

foliated texture (p. 49)
glassy texture (p. 37)
hydrothermal solution (p. 52)
igneous rock (p. 35)
intrusive (plutonic) (p. 35)
lava (p. 35)
lithification (p. 35)
magma (p. 35)
metamorphic rock (p. 35)

nonfoliated texture (p. 49)
porphyritic texture (p. 37)
regional metamorphism (p. 48)
rock cycle (p. 34)
sediment (p. 35)
sedimentary rock (p. 35)
strata (beds) (p. 46)
texture (p. 37)
vein deposit (p. 52)

Vocabulary Review

Choosing from the list of key terms, furnish the most appropriate response for the following statements.

1. The _____ illustrates the origin of the three basic rock types and the role of geologic processes in transforming one rock type into another.

2. A metal-rich accumulation of mineral matter that occurs along a fracture or bedding plane is called a(n) _____.

3. _____ refers to the processes by which sediments are transformed into solid sedimentary rock.

4. Sedimentary rocks, when subjected to great pressures and heat, will turn into the rock type _____, providing they do not melt.

5. _____ is molten material found inside Earth.

6. The process whereby magma cools and solidifies, resulting in the formation and growth of a crystalline solid, is called _____.

7. Metamorphic rocks that have formed in response to large-scale mountain building processes are said to have undergone _____.

8. The rock type that forms when sediment is lithified is _____.

9. Magma that reaches Earth's surface is called _____.

10. The term _____ describes the overall appearance of an igneous rock, based on the size and arrangement of its interlocking crystals.

11. The texture of a metamorphic rock that gives it a layered appearance is called _____.

12. Igneous rocks that result when lava solidifies are classified as _____ rocks.

13. Unconsolidated material consisting of particles created by weathering and transported by water, glaciers, wind, or waves is called _____.

14. The rock type that forms from the crystallization of magma is _____.

15. Metamorphic rocks composed of only one mineral that forms equidimensional crystals often have a type of texture called a(n) _____.

16. Igneous rocks that form rapidly at the surface often have a type of texture, called _____, where the individual crystals are too small to be seen with the unaided eye.

17. A(n) _____ is a hot, metal rich fluid that is associated with a cooling magma body.

18. Detrital sedimentary rocks form from rock fragments, however, a(n) _____ forms when dissolved substances are precipitated back into solids.

19. An igneous rock that has large crystals embedded in a matrix of smaller crystals is said to have a type of texture called a(n) _____.

20. The single most characteristic feature of sedimentary rocks are layers called _____.

21. A(n) _____ is a sedimentary deposit that forms as a shallow arm of the sea evaporates.

22. When rock is in contact with, or near, a mass of magma, a type of metamorphism, called _____, takes place.

23. When igneous rock cools rapidly and ions do not have time to unite into an orderly crystalline structure, a type of texture called a _____ results.

24. A sedimentary rock that forms from sediments that originate as solid particles from weathered rocks is called a(n) _____.

25. When large masses of magma solidify far below the surface, they form igneous rocks which exhibit a type of texture where the crystals are roughly equal in size and large enough to be identified with the unaided eye, called a(n) _____.

26. A(n) _____ forms when hydrothermal activity distributes minute quantities of ore throughout a large rock mass.

Comprehensive Review

1. Using Figure 2.1, label the Earth material or chemical/physical process that occurs at each lettered position in the rock cycle.

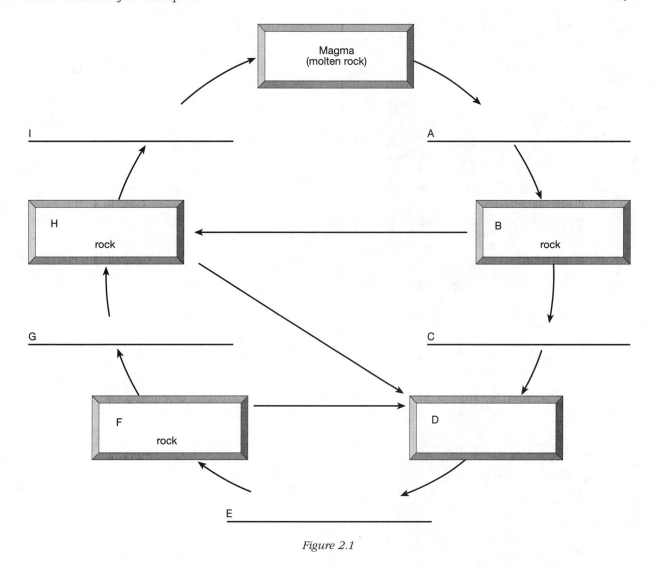

Figure 2.1

2. The rock cycle was initially proposed by _____ in the late _____.

3. Referring to the rock cycle, which processes are restricted to Earth's surface?

4. What are the various erosional agents that can pick up, transport, and deposit the products of weathering?

5. Briefly describe the formation of each of the three rock types.

a) Igneous rocks:

b) Sedimentary rocks:

c) Metamorphic rocks:

6. What factor most influences the size of mineral crystals in igneous rocks? How does·it influence the size of crystals?

7. What is the difference between magma and lava?

8. List and briefly describe the two criteria used to classify igneous rocks.

1)

2)

9. Describe the conditions that will result in the following igneous rock textures. Then, select the igneous rock photograph in Figure 2.2 that illustrates each of the textures.

a) Fine-grained texture:

Photograph letter: _____

b) Glassy texture:

Photograph letter: _____

c) Coarse-grained texture:

Photograph letter: _____

d) Porphyritic texture:

Photograph letter: _____

10. What discovery did N.L. Bowen make about the formation of minerals in a cooling magma?

11. According to Bowen's reaction series, which mineral is the first mineral to form from magma? If the reaction series is completed, which mineral will be the last to crystallize?

A.

B.

C.

D.

Figure 2.2

12. Is it possible for two igneous rocks to have the same mineral constituents, but different names? Explain your answer.

13. Using Bowen's reaction series as a guide, list the minerals that will most likely occur in an igneous rock within each of the following rock groups.

   Basaltic group:

   Granitic group:

14. Describe how the solid and liquid components of magma can separate.

15. What are two processes that cause sediment to be lithified into solid sedimentary rock?

16. Describe the two major groups of sedimentary rocks. Give an example of a rock found in each group.

     a) Detrital sedimentary rocks:

        Example:

     b) Chemical sedimentary rocks:

        Example:

17. What criterion is used to subdivide the detrital sedimentary rocks?

18. What does the presence of angular fragments in a detrital sedimentary rock suggest about the transportation of the particles?

19. What basis is used to subdivide the chemical sedimentary rocks?

20. What is the most abundant chemical sedimentary rock? What are the two origins for this rock?

21. Beginning with a swamp environment, list the successive stages in the formation of coal.

22. Selecting from the sedimentary rock photographs in Figure 2.3, answer the following questions.

     a) The rock in Figure 2.3A is made of _____ [chemical, detrital] material.

     b) The rock in Figure 2.3B is _____ [conglomerate, breccia].

     c) The rock in Figure 2.3 ___ [A, B, C] would most likely be associated with a quiet water environment.

23. What are fossils? As important tools for interpreting the geologic past, what are some uses of fossils?

A.

B.

C.

Figure 2.3

24. Briefly describe the two settings in which metamorphism most often occurs.

    a) Regional metamorphism:

    b) Contact metamorphism:

25. What are the three agents of metamorphism?

26. Describe the two textures of metamorphic rocks. Give an example of a rock that exhibits each texture.

    a) Foliated texture:

        Example:

    b) Nonfoliated texture:

        Example:

27. What is the texture of the metamorphic rock in Figure 2.4?

28. How did the gold deposits of the Homestake mine in South Dakota as well as the lead, zinc, and silver ores near Coeur d'Alene originate?

Figure 2.4

Practice Test

Multiple choice. Choose the best answer for the following multiple choice questions.

1. Which one of the following is NOT an agent of metamorphism?
   a) heat
   b) lithification
   c) chemically active fluids
   d) pressure

2. The great majority of rocks exposed at Earth's surface are of which type?
   a) igneous rocks
   b) sedimentary rocks
   c) metamorphic rocks

3. Molten material found inside Earth is called _____.
   a) lava
   b) rock fluid
   c) plasma
   d) mineraloid
   e) magma

4. The greatest volume of metamorphic rock is produced during _____.
   a) contact metamorphism
   b) lithification
   c) crystallization
   d) regional metamorphism
   e) weathering

5. The rock cycle was initially proposed in the late 1700s by _____.
   a) James Hutton
   b) John Playfair
   c) N.L. Bowen
   d) William Smith
   e) James Smith

6. Which one of the following is NOT a common cement found in sedimentary rocks?
   a) silica
   b) gypsum
   c) calcite
   d) iron oxide

7. The process of igneous rock formation is called _____.
   a) weathering
   b) decomposition
   c) lithification
   d) crystallization
   e) foliation

8. Igneous rocks that contain the last minerals to crystallize from magma and consist mainly of feldspars and quartz are said to have a(n) _____ composition.
   - a) granitic
   - b) gneissic
   - c) basaltic
   - d) metamorphic
   - e) lithic

9. Fossils found in sedimentary rocks can be used to _____.
   - a) interpret past environments
   - b) indicate certain periods of time
   - c) match rocks of the same age that are found at different places
   - d) all of the above

10. When large masses of magma solidify far below the surface, they form igneous rocks that exhibit a(n) _____ texture.
   - a) fine-grained
   - b) glassy
   - c) coarse-grained
   - d) porphyritic
   - e) fragmental

11. Which one of the following is NOT a detrital sedimentary rock?
   - a) shale
   - b) limestone
   - c) sandstone
   - d) conglomerate
   - e) siltstone

12. The surface process that slowly disintegrates and decomposes rock is called _____.
   - a) erosion
   - b) weathering
   - c) exfoliation
   - d) metamorphism
   - e) crystallization

13. Which one of the following is NOT composed of microcrystalline quartz?
   - a) travertine
   - b) flint
   - c) jasper
   - d) chert
   - e) agate

14. The single most characteristic feature of sedimentary rocks are _____.
   - a) crystals
   - b) mud cracks
   - c) ripple marks
   - d) strata

15. Which one of the following is NOT a primary element found in magma?
   - a) silicon
   - b) aluminum
   - c) iron
   - d) carbon
   - e) oxygen

16. Metamorphic rocks can form from _____.
   - a) igneous rocks
   - b) sedimentary rocks
   - c) other metamorphic rocks
   - d) all of the above

17. Vein deposits are usually produced by _____.
   - a) cementation and compaction
   - b) hydrothermal solutions
   - c) foliation
   - d) decomposition
   - e) weathering

18. Which one of the following metamorphic rocks has a nonfoliated texture?
   - a) marble
   - b) slate
   - c) mica schist
   - d) gneiss

19. The term meaning "conversion into rock" is _____.
    a) metamorphism   c) ionization   e) weathering
    b) crystallization   d) lithification

20. Accumulation of silts and clays, and eventually the rocks siltstone and shale, is generally best associated with which one of the following environments?
    a) swiftly flowing river   c) glacier   e) windy desert
    b) quiet swamp water   d) beach

21. Which one of the following is the metamorphic form of coal?
    a) lignite   c) anthracite
    b) bituminous   d) peat

22. Most metallic ores are produced by which two processes?
    a) cementation and compaction
    b) igneous and metamorphic
    c) igneous and sedimentary
    d) sedimentary and metamorphic
    e) decomposition and disintegration

23. Why can two igneous rocks have the same minerals but different names?
    a) they may have different colors
    b) names of igneous rocks are arbitrary
    c) the rocks may be of different sizes
    d) they may have different textures
    e) they may have been found in different places

24. Detrital sedimentary rocks are subdivided according to _____.
    a) particle size   d) the place where they were found
    b) color   e) their age
    c) hardness

25. Igneous rocks formed from magma that crystallized at depth are called _____ or intrusive rocks.
    a) plutonic   c) volcanic   e) foliated
    b) metamorphic   d) porphyritic

*True/false. For the following true/false questions, if a statement is not completely true, mark it false. For each false statement, change the **italicized** word to correct the statement.*

1. ___ The nonfoliated metamorphic equivalent of limestone is *marble*.

2. ___ The most common extrusive igneous rock is *granite*.

3. ___ Many metallic ore deposits are formed by the precipitation of minerals from *hydrothermal* solutions.

4. ___ The *rock cycle* shows the relations among the three rock types, and is essentially an outline of physical geology.

5. ___ About seventy-five percent of all rock outcrops on the continents are *sedimentary*.

6. ___ *Granite* is a classic example of an igneous rock that exhibits a coarse-grained texture.

7. ___ *Igneous* rocks are the rock type most likely to contain fossils.

8. ___ Igneous rock, when subjected to heat and pressure far below Earth's surface, will change to *sedimentary* rock.

9. ___ Limestone, a common building material, is a *nonmetallic* mineral resource.

10. ___ Because magma's density is *greater* than the surrounding rocks, it works its way to the surface over time spans from thousands to millions of years.

11. ___ Mineral alignment in a metamorphic rock usually gives the rock a *foliated* texture.

12. ___ Most sediment ultimately comes to rest in the *ocean*.

13. ___ The process called *weathering*, whereby magma cools, solidifies, and forms igneous rocks, may take place either beneath Earth's surface, or on the surface following a volcanic eruption.

14. ___ Separating *strata* are bedding planes, flat surfaces along which rocks tend to separate or break.

15. ___ *Lithification* literally means to "change form."

16. ___ Molten material found on Earth's surface is called *magma*.

17. ___ Rock salt and rock gypsum form when *evaporation* causes minerals to precipitate from water.

18. ___ The texture of an igneous rock is based in the size and *arrangement* of its interlocking crystals.

19. ___ The two major groups of *metamorphic* rocks are detrital and chemical.

20. ___ The *lithification* of sediment produces sedimentary rock.

21. ___ The agents of *metamorphism* include heat, pressure, and chemically active fluids.

22. ___ According to Bowen's reaction series, quartz is often the *last* mineral to crystallize from a melt.

23. ___ The mineral *sylvite* is the primary ingredient of plaster and wallboard.

24. ___ *Sedimentary* rocks are major sources of energy resources, iron, aluminum, and manganese.

25. ___ Slow cooling of magma results in the formation of *small* mineral crystals.

Word choice. Complete each of the following statements by selecting the most appropriate response.

1. The very first rocks to form on Earth's surface were [igneous/sedimentary/metamorphic].

2. Because a magma body is [less/more] dense than the surrounding rocks, it works its way [toward/away from] the surface over time spans of thousands to millions of years.

3. The molten minerals in magma are primarily members of the [carbonate/silicate/sulfide] group.

4. Detrital sedimentary rocks are primarily classified by particle [size/shape/composition].

5. Coal forms in swamp water that is oxygen-[enriched/deficient], therefore complete decay of the plant material is [not/always] possible.

6. During low-grade metamorphism, the sedimentary rock shale becomes the metamorphic rock called [marble/slate].

7. A rock that will effervesce with dilute hydrochloric acid most likely contains the mineral [quartz/calcite/feldspar].

8. The most important agent of metamorphism is [heat/pressure].

9. Sediment derived indirectly through the life processes of water-dwelling organisms is referred to as [detrital/chemical/biochemical] material.

10. Iron oxide, silica, and [calcite/olivine] are the most common cements found in sedimentary rocks.

11. A banded metamorphic rock that contains mostly elongated and granular minerals is called [schist/gneiss].

12. Hydrothermal solutions [can/cannot] migrate great distances through the surrounding rock before they are deposited.

Written questions

1. Referring to the rock cycle, explain why any one rock can be the raw material for another.

2. How is Bowen's reaction series related to the classification of igneous rocks?

3. What are the most common minerals in detrital sedimentary rocks? Why are these minerals so abundant?

4. In what ways do metamorphic rocks differ from the igneous and sedimentary rocks from which they formed?

Weathering, Soil, and Mass Wasting

3

Weathering, Soil, and Mass Wasting begins with a brief examination of the external processes of weathering, mass wasting, and erosion. The two forms of weathering, mechanical and chemical, are investigated in detail—including the types, conditions, rates, and net effect of each.

The soils section of the chapter begins with a description of the general composition, texture, and structure of soil. After examining the factors that influence soil formation, development, and classification, soil erosion, as well as some ore deposits produced by weathering, are presented.

Mass wasting begins with a look at the role the process plays in landform development. Following a discussion of the controls and triggers of mass wasting, a general presentation of the various types of mass wasting concludes the chapter.

Learning Objectives

After reading, studying, and discussing this chapter, you should be able to:

- Describe the processes of weathering, erosion, and mass wasting.
- Explain the difference between mechanical and chemical weathering.
- Discuss soil composition, texture, structure, formation, and classification.
- Describe the controls of mass wasting.
- List and describe the various types of mass wasting.

Chapter Summary

- Earth's external processes include 1) *weathering*—the disintegration and decomposition of rock at or near the surface, 2) *mass wasting*—the transfer of rock material downslope under the influence of gravity, and 3) *erosion*—the incorporation and transportation of material by a mobile agent, usually water, wind, or ice.

- *Mechanical weathering* is the physical breaking up of rock into smaller pieces. *Chemical weathering* alters a rock's chemistry, changing it into different substances. Rocks can be broken into smaller fragments by *frost wedging, unloading, thermal expansion,* and *biological activity.* Water is by far the most important agent of chemical weathering. Oxygen in water can *oxidize* some materials, while carbon dioxide (CO_2) dissolved in water forms *carbonic acid.* The chemical weathering of silicate minerals frequently produces 1) soluble products containing sodium, calcium, potassium, and magnesium, 2) insoluble iron oxides, and 3) clay minerals. The type and rate of rock weathering is influenced by a rock's mineral makeup—calcite readily dissolves in mildly acidic solutions and silicate minerals that form first from magma are least resistant to chemical weathering. Climatic factors, particularly temperature and moisture, are crucial to the rate of rock weathering.

• *Soil* is a combination of mineral and organic matter, water, and air—that portion of the *regolith* (the layer of rock and mineral fragments produced by weathering) that supports the growth of plants. *Soil texture* refers to the proportions of different particle sizes (clay, silt, and sand) found in soil. The most important *factors that control soil formation are parent material, time, climate, plants, animals, and slope.* Soil-forming process operate from the surface downward and produce zones or layers in the soil that soil scientists call *horizons.* From the surface downward the horizons are designated as O, A, E, B, and C respectively. Although there are hundreds of soil types and subtypes worldwide, the three very generic types are 1) *pedalfer*—characterized by an accumulation of iron oxides and aluminum-rich clays in the B horizon, 2) *pedocal*—characterized by an accumulation of calcium carbonate, and 3) *laterite*— deep soils that develop in the hot, wet tropics that are poor for growing because they are highly leached and bricklike. Soil erosion by wind and water is the ultimate fate of practically all soils. *Rates of soil erosion vary* from one place to another and depend on the soil's characteristics as well as such factors as climate, slope, and type of vegetation.

• Weathering creates mineral deposits by consolidating metals into economical concentrations. The process, called *secondary enrichment*, is accomplished by either 1) removing undesirable materials and leaving the desired elements enriched in the upper zones of the soil or 2) removing and carrying the desirable elements to lower soil zones where they are redeposited and become more concentrated. *Bauxite*, the principal ore of aluminum, is one important ore created by secondary enrichment.

• In the evolution of most landforms, mass wasting is the step that follows weathering. The combined effect of mass wasting and erosion by running water produce stream valleys. *Gravity is the controlling force of mass wasting.* Other factors that play an important role in overcoming inertia and triggering downslope movements are saturation of the material with water, oversteepening of slopes, removal of anchoring vegetation, and ground vibrations from earthquakes. The various processes included under the name of mass wasting are defined by 1) the type of material involved (debris, mud, earth, or rock), 2) the kind of motion (fall, slide, or flow), and 3) the velocity of the movement (fast, slow). The various kinds of mass wasting include the fast forms called *slump, rockslide, mudflow,* and *earthflow,* as well as the slow movements referred to as *creep* and *solifluction.*

Chapter Outline

I.   Earth's external processes include
    A. Weathering—the disintegration and decomposition of material at or near the surface
    B. Mass wasting—the transfer of rock material downslope under the influence of gravity
    C. Erosion—the incorporation and transportation of material by a mobile agent, usually water, wind, or ice
II. Weathering
    A. Two kinds of weathering
        1. Mechanical weathering
            a. Breaking of rocks into smaller pieces
            b. Four processes
                1. Frost wedging
                2. Unloading
                3. Thermal expansion
                4. Biological activity

2. Chemical weathering
    a. Alters the internal structures of minerals by removing or adding elements
    b. Most important agent is water
        1. Oxygen dissolved in water oxidizes materials
        2. Carbon dioxide (CO_2) dissolved in water forms carbonic acid and alters the material
    c. Weathering of granite
        1. Weathering of K-feldspar produces
            a. Clay minerals
            b. Potassium bicarbonate
            c. Silica in solution
        2. Weathering of silicate minerals produces
            a. Soluble sodium, calcium, potassium

and magnesium
products
b. Insoluble iron oxides
c. Clay minerals
B. Rates of weathering
1. Advanced mechanical weathering aids chemical weathering by increasing the surface area
2. Other important factors are
a. Mineral makeup
1. Marble (calcite) readily dissolves in weakly acidic solutions
2. Silicate minerals weather in the same order as their order of crystallization
b. Climate
1. Temperature and moisture are the most crucial factors
2. Chemical weathering is most effective in areas of warm temperatures and abundant moisture

III. Soil
A. A combination of mineral matter, water, and air—that portion of the regolith (rock and mineral fragments) that supports the growth of plants
B. Soil texture and structure
1. Texture
a. Refers to the proportions of different particle sizes
1. Sand (large size)
2. Silt
3. Clay (small size)
b. Loam is best suited for plant life
2. Structure
a. Soil particles clump together to give a soil its structure
b. Four basic soil structures
1. Platy
2. Prismatic
3. Blocky
4. Spheroidal
C. Controls of soil formation
1. Parent material
a. Residual soil—parent material is the bedrock
b. Transported soil—parent material has been carried from elsewhere and deposited
2. Time
a. Important in all geologic processes
b. Amount of time to evolve varies for different soils
3. Climate

4. Plants and animals
a. Organisms influence the soil's physical and chemical properties
b. Furnish organic matter to soil
5. Slope
a. Steep slope—often poor soils
b. Optimum is a flat-to-undulating upland surface
D. Soil Profile
1. Soil forming processes operate from the surface downward
2. Horizons—zones or layers of soil
a. Horizons in temperate regions
1. *O*—organic matter
2. *A*—organic and mineral matter
3. *E*—little organic matter
4. *B*—zone of accumulation
5. *C*—partially altered parent material
b. *O* and *A* together called topsoil
c. *O, A, E,* and *B* together called solum, or "true soil"
E. Soil types
1. Hundreds of soil types worldwide
2. Three very generic types
a. Pedalfer
1. Accumulation of iron oxides and Al-rich clays in the B horizon
2. Best developed under forest vegetation
b. Pedocal
1. Accumulate calcium carbonate
2. Associated with drier grasslands
c. Laterite
1. Hot, wet, tropical climates
2. Intense chemical weathering
F. Soil Erosion
1. Recycling of Earth materials
2. Natural rates of erosion depend on
a. Soil characteristics
b. Climate
c. Slope
d. Type of vegetation

IV. Weathering creates ore deposits
A. Process called secondary enrichment
1. Concentrates metals into economical deposits
2. Two ways of enrichment
a. Removing undesired material from the decomposing rock, leaving the desired elements behind
b. Desired elements are carried to lower zones and deposited

B. Examples
  1. Bauxite, the principal ore of aluminum
  2. Many copper and silver deposits
V. Mass Wasting
  A. The downslope movement of rock, regolith, and soil under the direct influence of gravity
  B. Gravity is the controlling force
  C. Important triggering factors are
    1. Saturation of the material with water
      a. Destroys particle cohesion
      b. Water adds weight
    2. Oversteepening of slopes
      a. Stable slope angle is different for various materials
      b. Oversteepened slopes are unstable
    3. Removal of anchoring vegetation
    4. Ground vibrations from earthquakes
  D. Types of mass wasting
    1. Generally each type is defined by
      a. The material involved
        1. Debris
        2. Mud
        3. Earth
        4. Rock
      b. The movement of the material
        1. Fall (free-fall of pieces)
        2. Slide (material moves along a surface)
        3. Flow (material moves as a viscous fluid)
      c. The velocity of the movement
        1. Fast
        2. Slow
    2. Forms of mass wasting
      a. Slump
        1. Rapid
        2. Movement along a curved surface
        3. Along oversteepened slopes
      b. Rockslide
        1. Rapid
        2. Blocks of bedrock move down a slope
      c. Mudflow
        1. Rapid
        2. Flow of debris with water
        3. Often confined to channels
        4. Serious problem in dry areas with heavy rains
        5. Mudflows on the slopes of volcanoes are called lahars
      d. Earthflow
        1. Rapid
        2. On hillsides in humid regions
        3. Water saturates the soil
      e. Creep—the slow movement of soil and regolith downhill
      f. Solifluction
        1. Slow
        2. In areas underlain by permafrost

Key Terms

Page numbers shown in ( ) refer to the textbook page where the term first appears.

angle of repose (p. 78)
chemical weathering (p. 61)
creep (p. 83)
differential weathering (p. 64)
earthflow (p. 82)
eluviation (p. 71)
erosion (p. 58)
exfoliation dome (p. 60)
fall (p. 79)
flow (p. 79)
frost wedging (p. 60)
horizon (p. 71)
humus (p. 66)

lahar (p. 81)
laterite (p. 72)
leaching (p. 71)
mass wasting (p. 58)
mechanical weathering (p. 58)
mudflow (p. 80)
parent material (p. 69)
pedalfer (p. 72)
pedocal (p. 72)
permafrost (p. 83)
regolith (p. 65)
rockslide (p. 80)

secondary enrichment (p. 76)
sheeting (p. 60)
slide (p. 79)
slump (p. 79)
soil (p. 65)
soil profile (p. 71)
soil texture (p. 68)
solifluction (p. 83)
solum (p. 71)
spheroidal weathering (p. 63)
talus slope (p. 60)
weathering (p. 58)

Vocabulary Review

Choosing from the list of key terms, furnish the most appropriate response for the following statements.

1. The process called _____ is the incorporation and transportation of material by a mobile agent, usually water, wind, or ice.

2. The process where water works its way into every crack or void in a rock and, upon freezing, expands and enlarges the opening is called _____.

3. The layer of rock and mineral fragments produced by weathering that nearly everywhere covers Earth's surface is termed _____.

4. _____ refers to the depletion of soluble materials from the upper soil by the downward-percolating water.

5. Stone Mountain, Georgia, is an excellent example of a(n) _____ that has formed as large slabs of rock have separated and spalled off.

6. A(n) _____ is a soil characterized by the accumulation of iron oxides and aluminum-rich clays in the B horizon.

7. The downward slipping of a mass of rock or unconsolidated material moving as a unit along a curved surface is called _____.

8. The disintegration and decomposition of rock at or near Earth's surface is called _____.

9. Permanently frozen ground, called _____, occurs in Earth's harsh tundra and ice cap climates.

10. _____ occurs when weathering concentrates metals into economically valuable deposits.

11. _____ is that portion of the regolith that supports plant growth.

12. Organic matter in soil produced by the decomposition of plants and animals is called _____.

13. When a rock undergoes _____ it is broken into smaller and smaller pieces, each retaining the characteristics of the original material.

14. A(n) _____ is a highly leached soil type found in the tropics that is rich in oxides of iron and aluminum.

15. A(n) _____ occurs when blocks of bedrock break loose and slide down a slope.

16. The complex process that alters the internal structures of minerals by removing and/or adding elements is termed _____.

17. _____ refers to the washing out of fine soil components from the A horizon by downward-percolating water.

18. A(n) _____ is a mudflow that occurs on the slope of a volcano when unstable layers of ash and debris become saturated and flow.

19. The transfer of rock material downslope under the influence of gravity is referred to as _____.

20. _____ relates to the fact that rocks exposed at Earth's surface usually do not weather at the same rate.

21. The relative proportions of clay, silt, and sand in a soil determine the _____.

22. A(n) _____ is a soil characterized by the accumulation of calcium carbonate in the upper horizons.

23. The process called _____ occurs when large masses of igneous rock, particularly granite, begin to break loose like the layers of an onion.

24. A(n) _____ is a layer or zone in a soil profile.

25. _____ is the gradual downhill movement of soil and regolith.

26. The source of the weathered mineral matter from which soils develop is called the _____.

27. A large pile of debris that forms at the base of a slope is called a(n) _____.

28. Together the O, A, E, and B horizons of soil constitute the _____, or "true soil."

29. The stable slope of a material is called its _____.

30. A(n) _____ is a vertical section through a soil showing the succession of soil horizons.

Comprehensive Review

1. List and describe the three general processes that are continually removing materials from higher elevations and transporting them to lower elevations.

   1)

   2)

   3)

2. List and briefly describe the two types of weathering.

   1)

   2)

3. What are four natural processes that break rocks into smaller fragments?

4. Write a brief statement describing the two natural ways that water becomes mildly acidic.

5. What are the three factors that influence the rate of weathering?

   1)

   2)

   3)

6. Referring to the process of weathering, describe what is happening to the rocks in Figure 3.1.

7. What are the three products of the weathering of potassium feldspar?

8. What is the difference between regolith and soil?

9. What are the three particle size classes used for determining soil texture?

Figure 3.1

10. Referring to Figure 3.2, what are the proportions of clay, silt, and sand of the soil shown at point B?

    Clay: _____   Silt: _____   Sand: _____

11. Referring to Figure 3.2, if a soil has 20 percent clay, 60 percent silt, and 20 percent sand it is called a(n):

12. What are the four basic soil structures?

13. List the five controls of soil formation.

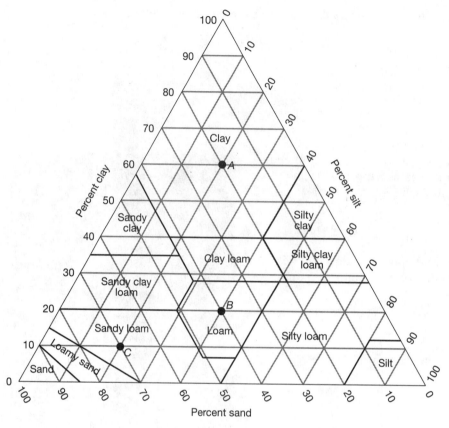

Figure 3.2

14. On Figure 3.3, label each of the five soil horizons with the appropriate identifying letter (e.g., O, A, etc.).

15. Use Figure 3.4 to answer the following questions.

    a) At which location would the thickest transported soils be located? ____

    b) Which location consists primarily of unconsolidated deposits? ____

    c) At which location would the thickest residual soil be located? ____

16. List the soil type that would most likely be found in each of the following climates.

    a) Dry, mid-latitude climate with grass and brush vegetation:

    b) Hot, wet, tropical climate:

    c) Humid, mid-latitude climate with forest vegetation:

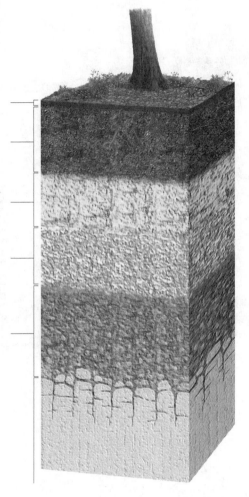

Figure 3.3

17. In addition to prolonged drought, what other factors contributed to the "Dust Bowl" of the 1930s?

18. Briefly describe how secondary enrichment produces ore deposits.

19. The combined effects of which two processes produce stream valleys?

20. Although gravity is the controlling force of mass wasting, what factors play an important role in overcoming inertia and triggering downslope movements of materials?

21. Identify each of the forms of mass wasting illustrated in Figure 3.5 by writing the name of the process by the diagram.

Figure 3.4

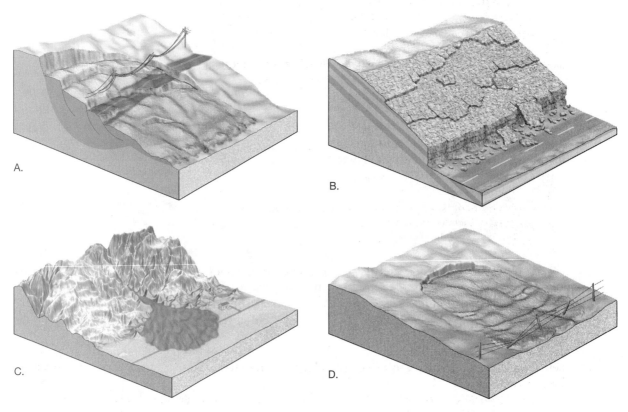

Figure 3.5

Practice Test

Multiple choice. Choose the best answer for the following multiple choice questions.

1. The most influential control of soil formation is _____.
   - a) parent material
   - c) slope orientation
   - e) plants and animals
   - b) climate
   - d) time

2. The process responsible for moving material downslope without the aid of running water, wind, or ice is _____.
   - a) erosion
   - c) weathering
   - e) eluviation
   - b) mass wasting
   - d) laterization

3. When weathering increases the concentration of an economically valuable resource, the process is referred to as _____.
   - a) eluviation
   - d) deposition
   - b) solifluction
   - e) secondary enrichment
   - c) primary enrichment

4. The disintegration and decomposition of rock at or near Earth's surface is called _____.
   - a) erosion
   - c) eluviation
   - e) solifluction
   - b) weathering
   - d) mass wasting

5. Compared to the distant past, current rates of soil erosion are _____.
   - a) about the same
   - c) impossible to determine
   - b) slower
   - d) faster

6. Decayed organic matter in soil is called _____.
   - a) humus
   - c) regolith
   - e) caliche
   - b) loam
   - d) laterite

7. Platy, prismatic, blocky, and spheroidal are each a type of _____.
   - a) humus
   - d) weathering
   - b) soil structure
   - e) soil texture
   - c) mineral composition

8. Chemical weathering will be most rapid in _____ climates.
   - a) cold, dry
   - c) cold, wet
   - b) warm, dry
   - d) warm, wet

9. Which soil type is most often associated with the wet tropics?
   - a) pedalfer
   - c) pedocal
   - b) laterite
   - d) mature

10. When a rock is subjected to _____ it breaks into smaller pieces having the same characteristics as the original rock.
    - a) mechanical weathering
    - d) oxidation
    - b) creep
    - e) eluviation
    - c) chemical weathering

11. Which one of the following sequences is the correct order that water will follow prior to forming a stream channel?
    - a) gullies, sheet erosion, rills
    - c) rills, sheet erosion, gullies
    - b) sheet erosion, rills, gullies
    - d) sheet erosion, gullies, rills

12. Clay minerals are produced from the chemical weathering of the _____ minerals.
    a) feldspar   c) carbonate   e) sulfide
    b) quartz   d) oxide

13. The combination of organic matter, rock and mineral fragments, water, and air that covers much of Earth's surface is referred to as _____.
    a) regolith   c) soil   e) humus
    b) laterite   d) caliche

14. Which one of the following minerals is most susceptible to chemical weathering?
    a) pyroxene   c) quartz
    b) biotite   d) potassium feldspar

15. Whenever the characteristics and internal structures of weathered minerals have been altered, they have undergone _____.
    a) creep   d) chemical weathering
    b) mechanical weathering   e) mass wasting
    c) eluviation

16. Which one of the following is least important in mass wasting?
    a) gravity   c) soil water   e) vegetation
    b) slope angle   d) slope orientation

17. The atmospheric gas that forms a mild acid when dissolved in water is _____.
    a) oxygen   c) nitrogen   e) silica
    b) argon   d) carbon dioxide

18. Soils resting on the bedrock from which they formed are called _____ soils.
    a) transported   c) major   e) weathered
    b) static   d) residual

19. Which of the following is NOT associated with mechanical weathering?
    a) frost wedging   c) oxidation   e) thermal expansion
    b) unloading   d) organic activity

20. The layer of weathered rock and mineral fragments that covers much of Earth's surface is called _____.
    a) humus   c) alluvium   e) laterite
    b) clay   d) regolith

21. The most important agent of chemical weathering is _____.
    a) air   c) soil   e) erosion
    b) water   d) gravity

22. When a mass of material moves downslope along a curved surface the process is called _____.
    a) mudflow   c) creep   e) solifluction
    b) rockslide   d) slump

23. The process of _____ produces a yellow to reddish-brown surface on iron-rich minerals.
    a) lithification   c) acidification   e) calcification
    b) oxidation   d) hydrolysis

24. The process called _____ is responsible for the formation of exfoliation domes.
    a) oxidation   c) creep   e) unloading
    b) frost wedging   d) slump

25. A soil's texture is determined by its _____.
    a) mineral composition   c) particle sizes   e) age
    b) type of humus   d) water content

26. In a well-developed soil profile the _____ horizon is the topmost layer.
    a) A   c) C   e) E
    b) B   d) O

27. The accumulation of rock fragments found at the base of many cliffs is referred to as the
    _____.
    a) talus slope   c) transported residue   e) base horizon
    b) lahar   d) earthflow

28. Which is the least effective form of mechanical weathering?
    a) thermal expansion   c) organic activity
    b) unloading   d) frost wedging

29. The removal of soluble materials from soil by percolating water is called _____.
    a) leaching   c) laterization   e) degradation
    b) erosion   d) eluviation

30. The most widespread form of mass wasting is _____.
    a) rockslide   c) earthflow   e) slump
    b) creep   d) mudflow

*True/false. For the following true/false questions, if a statement is not completely true, mark it false. For each false statement, change the **italicized** word to correct the statement.*

1. ___ Soils formed from bedrock are called *residual* soils.

2. ___ The B horizon of a soil is also called the *topsoil*.

3. ___ Removing vegetation will *increase* the rate of soil erosion.

4. ___ In Bowen's reaction series, the minerals that form *last* are the most susceptible to chemical weathering.

5. ___ The free descent of individual rock pieces of any size is called *fall*.

6. ___ A vertical section through all the layers of a soil is called the soil *horizon*.

7. ___ *Lahars* are mudflows on the sides of volcanoes.

8. ___ About *fifty* percent of the total volume of a good quality soil consists of pore spaces.

9. ___ Eluviation and leaching are most active in the *E* horizon.

10. ___ In general, the greater the slope the *thicker* the soil.

11. ___ The soil types that are most common in the eastern United States are *laterites.*

12. ___ *Climate* is the most influential control of soil formation.

13. ___ The soil layer composed of partially altered parent material is called the *C* horizon.

14. ___ A soil layer (or zone) is referred to as a(n) *strata.*

15. ___ Tiny, thread-like water channels are called *rills.*

16. ___ The downward movement of material as a viscous fluid is termed *slide.*

17. ___ The B horizon of a soil is also called the zone of *accumulation.*

18. ___ Of the three common soil types, *laterites* are the least suitable for agriculture.

19. ___ A sandy soil with an accumulation of aluminum-rich clays and iron oxides in the B horizon is called a(n) *pedalfer.*

20. ___ The greatest amount of soil erosion is caused by *wind.*

21. ___ The depletion of soluble materials from the upper soil is termed *leaching.*

22. ___ A soil with an accumulation of calcite is referred to as a(n) *pedalfer.*

Word choice. Complete each of the following statements by selecting the most appropriate response.

1. Quartz is [more/less] resistant to chemical weathering than feldspar.

2. Soils lacking clearly defined horizons are called [mature/hardpan/immature] soils.

3. Atmospheric nitrogen is fixed (altered) into soil nitrogen by [earthworms/microorganisms].

4. Frost wedging is most effective in the [tropical/middle/polar] latitudes.

5. The O and A horizons of a soil make up what is commonly called the [topsoil/subsoil].

6. Excessive clay in soil will result in [good/poor] drainage.

7. The primary ore of aluminum is [pyrite/bauxite].

8. When minerals containing iron are chemically weathered, the iron is usually [dissolved/oxidized/reduced].

9. Increasing the amount of water in the pore spaces of soil and regolith will usually [increase/decrease] the likelihood of downslope movement.

10. A compact and impermeable layer of accumulated clay in soil is called [caliche/hardpan].

11. When carbon dioxide (CO_2) dissolves in water, a [weak/strong] [acid/base] is produced.

12. Mass wasting will most likely occur when materials are at an angle that is [greater/less than] their angle of repose.

13. The same parent material [will/will not] always produce the same type of soil.

14. The lack of nutrients in laterite soils is due to [evaporation/leaching].

15. Most talus slopes form as the result of [falls/flows/slides].

16. Water [expands/contracts] when it freezes.

17. Caliche is composed of [calcium carbonate/silica/iron oxide] and is associated with [pedocal/pedalfer/laterite] soils.

18. During slump, movement of the material is along [flat/curved] surfaces.

19. Sheeting is believed to occur due to the [reduction/increase] in pressure when overlying rock is [removed/deposited].

20. The most common form of mass wasting is [slump/creep].

Written questions

1. Briefly describe the two types of weathering and how they are related.

2. What is soil? How does a soil profile develop?

3. What is the controlling force of mass wasting? What other factors are important?

Running Water and Groundwater

4

Running Water and Groundwater opens with an examination of the hydrologic cycle and the exchange of water between the oceans, atmosphere, and land. The discussion of running water includes an investigation of the factors that control streamflow and their influence on a stream's ability to erode and transport materials. Erosional and depositional features of narrow and wide valleys, as well as drainage patterns and stages of valley development, are also described.

After an examination of the importance, occurrence, and movement of groundwater, springs, geysers, wells, and artesian wells are investigated. Following a review of some of the environmental problems associated with groundwater, the chapter ends with a look at the formation and features of caves and karst topography.

Learning Objectives

After reading, studying, and discussing this chapter, you should be able to:

- Describe the movement of water through the hydrologic cycle.
- Describe the process of streamflow and list the factors that influence a stream's ability to erode and transport materials.
- List and describe the major features produced by stream erosion and deposition.
- Describe the two general types of stream valleys and their major features.
- Distinguish between the different types of drainage patterns.
- Describe how a stream valley and the surrounding landscape can change with time.
- Discuss the occurrence and movement of groundwater.
- Describe springs, geysers, wells, and artesian wells.
- List the major environmental problems associated with groundwater.
- Describe the major features of caves and karst topography.

Chapter Summary

- The *hydrologic cycle* describes the continuous interchange of water among the oceans, atmosphere, and continents. Powered by energy from the sun, it is a global system in which the atmosphere provides the link between the oceans and continents. The processes involved in the water cycle include *precipitation, evaporation, infiltration* (the movement of water into rocks or soil through cracks and pore spaces), *runoff* (water that flows over the land, rather than infiltrating into the ground), and *transpiration* (the release of water vapor to the atmosphere by plants). *Running water is the single most important agent sculpturing Earth's land surface.*

- The factors that determine a stream's *velocity* are *gradient* (slope of the stream channel), *shape, size* and *roughness* of the channel, and the stream's *discharge* (amount of water passing a given point per unit of time, frequently measured in cubic feet per second). Most often, the gradient and roughness of a stream decrease downstream, while width, depth, discharge, and velocity increase.

• The two general types of *base level* (the lowest point to which a stream may erode its channel) are 1) *ultimate base level* and 2) *temporary*, or *local base level*. Any change in base level will cause a stream to adjust and establish a new balance. Lowering base level will cause a stream to erode, while raising base level results in deposition of material in the channel.

• The work of a stream includes *erosion* (the incorporation of material), *transportation* (as *dissolved load*, *suspended load*, and *bed load*), and, whenever a stream's velocity decreases, *deposition*.

• Although many gradations exist, the two general types of stream valleys are 1) *narrow V-shaped valleys* and 2) *wide valleys with flat floors*. Because the dominant activity is downcutting toward base level, narrow valleys often contain *waterfalls* and *rapids*. When a stream has cut its channel closer to base level, its energy is directed from side to side, and erosion produces a flat valley floor, or *floodplain*. Streams that flow upon floodplains often move in sweeping bends called *meanders*. Widespread meandering may result in shorter channel segments, called *cutoffs*, and/or abandoned bends, called *oxbow lakes*.

• Common *drainage patterns* produced by streams include 1) *dendritic*, 2) *radial*, 3) *rectangular*, and 4) *trellis*.

• As a resource, *groundwater* represents the largest reservoir of freshwater that is readily available to humans. Geologically, the dissolving action of groundwater produces *caves* and *sinkholes*. Groundwater is also an equalizer of stream flow.

• Groundwater is that water which occupies the pore spaces in sediment and rock in a zone beneath the surface called the *zone of saturation*. The upper limit of this zone is the *water table*. The *zone of aeration* is above the water table where the soil, sediment, and rock are not saturated. Groundwater generally moves within the zone of saturation. The quantity of water that can be stored depends on the *porosity* (the volume of open spaces) of the material. However, the *permeability* (the ability to transmit a fluid through interconnected pore spaces) of a material is the primary factor controlling the movement of groundwater.

• *Springs* occur whenever the water table intersects the land surface and a natural flow of groundwater results. *Wells*, openings bored into the zone of saturation, withdraw groundwater and create roughly conical depressions in the water table known as *cones of depression*. *Artesian wells* occur when water rises above the level at which it was initially encountered.

• Some of the current environmental problems involving groundwater include 1) *overuse* by intense irrigation, 2) *land subsidence* caused by groundwater withdrawal, and 3) *contamination*.

• Most *caverns* form in limestone at or below the water table when acidic groundwater dissolves rock along lines of weakness, such as joints and bedding planes. *Karst topography* exhibits an irregular terrain punctuated with many depressions, called *sinkholes*.

Chapter Outline

I.   Hydrologic cycle
    A. Illustrates the circulation of Earth's
       water supply

B. Processes involved in the cycle
    1. Precipitation
    2. Evaporation

3. Infiltration
4. Runoff
5. Transpiration
C. Cycle is balanced
II. Running water
A. Streamflow
1. Factors that determine velocity
a. Gradient, or slope
b. Channel characteristics
1. Shape
2. Size
3. Roughness
c. Discharge
B. Upstream-downstream changes
1. Profile
a. Cross-sectional view of a stream
b. From head (source) to mouth
1. Profile is a smooth curve
2. Gradient decreases from the head to the mouth
2. Factors that increase downstream
a. Velocity
b. Discharge
c. Channel size
3. Factors that decrease downstream
a. Gradient, or slope
b. Channel roughness
C. Base level
1. Lowest point a stream can erode to
2. Two general types
a. Ultimate
b. Temporary, or local
3. Changing causes readjustment of the stream
D. Work of streams
1. Erosion
2. Transportation
a. Transported material is called the stream's load
1. Types of load
a. Dissolved load
b. Suspended load
c. Bed load
2. Load is related to a stream's
a. Competence
1. Maximum particle size
2. Determined by velocity
b. Capacity
1. Maximum load
2. Related to discharge
3. Deposition
a. Caused by a decrease in velocity
1. Competence is reduced

2. Sediment begins to drop out
b. Stream sediments
1. Called alluvium
2. Well-sorted deposits
c. Features produced by deposition
1. Deltas
a. Exist in oceans or lakes
b. Distributaries often form in the channel
2. Natural levees
a. Form parallel to the stream channel
b. Area behind the levees may contain
1. Back swamps
2. Yazoo tributaries
E. Stream valleys
1. Valley sides are shaped by
a. Weathering
b. Overland flow
c. Mass wasting
2. Characteristics of narrow valleys
a. V-shaped
b. Downcutting toward base level
c. Features often include
1. Rapids
2. Waterfalls
3. Characteristics of wide valleys
a. Stream is near base level
b. Downward erosion is less dominant
c. Stream energy is directed from side-to-side
d. Floodplain
e. Features often include
1. Meanders
2. Cutoffs
3. Oxbow lakes
F. Drainage basins and patterns
1. A divide separates drainage basins
2. Types of drainage patterns
a. Dendritic
b. Radial
c. Rectangular
d. Trellis
G. Stages of valley development
1. Youth
a. Rapids and waterfalls
b. V-shaped valleys
c. Vigorous downcutting
d. Steep gradient
2. Maturity
a. Downward erosion diminishes

b. Lateral erosion dominates
c. Floodplain beginning
d. Gradient lower than at youth
3. Old age
   a. Large floodplain
   b. Widespread shifting of the stream
   c. Characteristic features
      1. Natural levees
      2. Backswamps
      3. Yazoo tributaries
4. Rejuvenation
   a. "Made young again"
   b. Characteristic features
      1. Entrenched meanders
      2. Terraces

III. Water beneath the surface (groundwater)
A. Largest freshwater reservoir for humans
B. Geological roles
   1. As an erosional agent, dissolving by groundwater produces
      a. Sinkholes
      b. Caverns
   2. An equalizer of streamflow
C. Distribution and movement of groundwater
   1. Distribution of groundwater
      a. Zone of saturation
         1. All pore spaces in the material are filled with water
         2. Water within the pores is groundwater
      b. Water table—the upper limit of the zone of saturation
      c. Zone of aeration
         1. Area above the water table
         2. Pore spaces in the material are filled mainly with air
   2. Movement of groundwater
      a. Porosity
         1. Percentage of pore spaces
         2. Determines how much groundwater can be stored
      b. Permeability
         1. Ability to transmit water through connected pore spaces
         2. Aquiclude—an impermeable layer of material
         3. Aquifer—a permeable layer of material
D. Water features
   1. Springs
      a. Hot springs
         1. Water is 6-9°C warmer than the mean air temperature of the locality
         2. Heated by cooling of igneous rock
      b. Geysers
         1. Intermittent hot springs
         2. Water turns to steam and erupts
   2. Wells
      a. Pumping can cause a drawdown (lowering) of the watertable
      b. Pumping can form a cone of depression in the water table
   3. Artesian wells
      a. Water in the well rises higher than the initial groundwater level
      b. Types of artesian wells
         1. Nonflowing
         2. Flowing
E. Environmental problems associated with groundwater
   1. Treating it as a nonrenewable resource
   2. Land subsidence caused by its withdrawal
   3. Contamination
F. Features produced by groundwater
   1. Groundwater is often mildly acidic
      a. Contains weak carbonic acid
      b. Dissolves calcite in limestone
   2. Caverns
      a. Formed by dissolving rock beneath Earth's surface
      b. Formed in the zone of saturation
      c. Features found within caverns
         1. Form in the zone of aeration
         2. Composed of dripstone
            a. Calcite deposited as dripping water evaporates
            b. Common features
               1. Stalactites hanging from the ceiling
               2. Stalagmites developing on the cave floor
   3. Karst topography
      a. Formed by dissolving rock at, or near, Earth's surface
      b. Common features
         1. Sinkholes
            a. Surface depressions
            b. Formed by
               1. Dissolving bedrock
               2. Cavern collapse
         2. Caves and caverns
      c. Area lacks good surface drainage

Key Terms

Page numbers shown in ( ) refer to the textbook page where the term first appears.

alluvium (p. 95)	divide (p. 100)	radial pattern (p. 101)
aquiclude (p. 108)	drainage basin (p. 100)	rectangular pattern (p. 101)
aquifer (p. 108)	drawdown (p. 110)	rejuvenation (p. 102)
artesian well (p. 112)	entrenched meander (p. 102)	runoff (p. 88)
backswamp (p. 96)	floodplain (p. 98)	sinkhole (sink) (p. 117)
base level (p. 92)	geyser (p. 109)	sorting (p. 95)
bed load (p. 93)	gradient (p. 90)	spring (p. 109)
capacity (p. 94)	groundwater (p. 107)	stalactite (p. 116)
cavern (p. 116)	hot spring (p. 109)	stalagmite (p. 117)
competence (p. 94)	hydrologic cycle (p. 88)	suspended load (p. 93)
cone of depression (p. 111)	infiltration (p. 88)	transpiration (p. 88)
cutoff (p. 98)	karst topography (p. 117)	trellis pattern (p. 101)
delta (p. 95)	meander (p. 98)	water table (p. 107)
dendritic pattern (p. 101)	natural levee (p. 96)	well (p. 109)
discharge (p. 90)	oxbow lake (p. 98)	yazoo tributary (p. 96)
dissolved load (p. 93)	permeability (p. 108)	zone of aeration (p. 108)
distributary (p. 96)	porosity (p. 108)	zone of saturation (p. 107)

Vocabulary Review

Choosing from the list of key terms, furnish the most appropriate response for the following statements.

1. _____ is the movement of surface water into rocks or soil through cracks or pore spaces.

2. The land area that contributes water to a stream is referred to as the stream's _____.

3. The zone beneath Earth's surface where all open spaces in sediment and rock are completely filled with water is the _____.

4. The _____ of a material is the percentage of the total volume that consists of pore spaces.

5. _____ is the slope of a stream channel expressed as the vertical drop of a stream over a specified distance.

6. The general term for any well sorted stream-deposited sediment is _____.

7. The _____ of a stream is the maximum load it can carry.

8. Water beneath Earth's surface in the zone of saturation is known as _____.

9. A tributary that flows parallel to the main stream because a natural levee is present is called a _____.

10. The release of water vapor to the atmosphere by plants is referred to as _____.

11. A "looplike," sweeping bend in a stream is called a(n) _____.

12. The _____ is the upper limit of the zone of saturation.

13. The volume of water flowing past a certain point per unit of time is called a stream's _____.

14. The term _____ is applied to any well in which groundwater rises above the level where it was initially encountered.

15. The elevated landform that parallels some streams, called a(n) _____, acts to confine the stream's waters, except during flood stage.

16. The _____ of a material indicates its ability to transmit water through interconnected pore spaces.

17. The unending circulation of Earth's water supply from the oceans, to the atmosphere, to the land, and back to the oceans is called the _____.

18. A(n) _____ is the flat, low-lying portion of a stream valley that is subject to inundation during flooding.

19. The _____ of a stream measures the maximum size of particles it is capable of transporting.

20. The drainage basin of one stream is separated from the drainage basin of another by an imaginary line called a(n) _____.

21. An impermeable layer that hinders or prevents the movement of groundwater is termed a(n) _____.

22. _____ is the lowest point to which a stream can erode its channel.

23. Whenever the water table intersects the ground surface, a natural flow of groundwater, called a _____, results.

24. A rock strata or sediment that transmits groundwater freely is called a(n) _____.

25. The fine sediment transported within the body of flowing water is called a stream's _____.

26. Landscapes that, to a large extent, have been shaped by the dissolving power of groundwater are said to exhibit _____.

27. Water that flows over the land rather than infiltrating into the ground is referred to as _____.

Comprehensive Review

1. The combined effects of which two processes produce stream valleys?

2. Figure 4.1 illustrates the hydrologic cycle. Select the letter in the figure that represents each of the following processes.

   a) Runoff: ____   b) Evaporation: ____   c) Precipitation: ____   d) Infiltration: ____

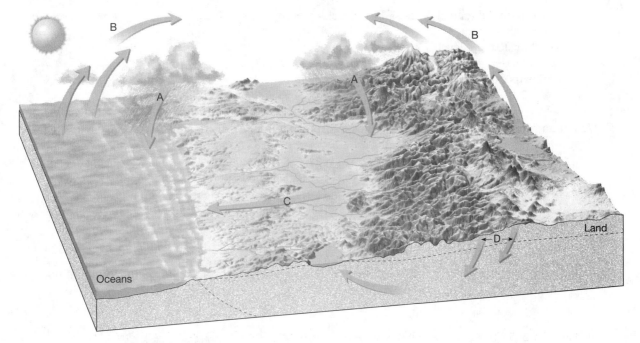

Figure 4.1

3. Briefly define the following measurements that are used to assess streamflow.

   a) Velocity:

   b) Gradient:

   c) Discharge:

4. What is the average gradient of a stream whose headwaters are 10 kilometers from the sea, at an altitude of 40 meters?

5. By circling the correct response, indicate how each of the following measurements changes from a stream's head to its mouth.

   a) Gradient [increases, decreases]   d) Discharge [increases, decreases]

   b) Velocity [increases, decrease]   e) Width and depth [increase, decrease]

   c) Channel roughness [increases, decreases]

6. Describe the two general types of base level.

    a) Ultimate base level:

    b) Temporary, or local, base level:

7. List and describe the three ways streams transport their load of sediment.

    1)

    2)

    3)

8. Why does the greatest erosion and transportation of sediment by a stream occur during a flood?

9. Figure 4.2 illustrates a wide stream valley. Select the letter in the figure that identifies each of the following features.

    a) Yazoo tributary: _____

    b) Back swamp: _____

    c) Natural levees: _____

    d) Oxbow lake: _____

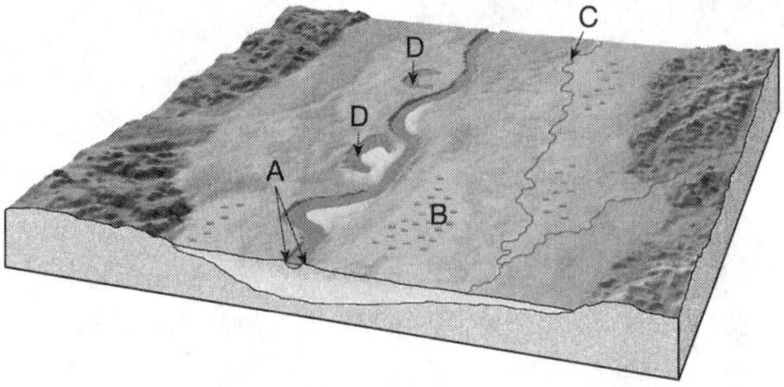

Figure 4.2

10. Briefly describe the appearance of the following types of valleys. Also list the major features you would expect to find associated with each type.

    1) Narrow valley:

        Features:

    2) Wide valley:

        Features:

11. Describe the type of drainage pattern that would develop on an isolated volcanic cone or domal uplift.

12. From the illustrations in Figure 4.3, select
    the diagram that best illustrates each of the
    following stages in the evolution of a
    valley.

    a) Youth: _____

    b) Maturity: _____

    c) Old age: _____

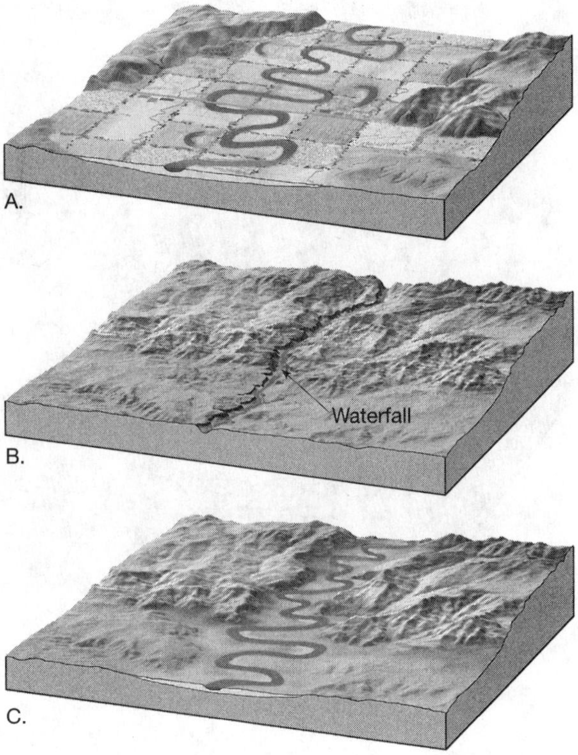

Figure 4.3

13. List some key words or features that describe or are associated with each of the three stages of
    valley development.

    a) Youth:

    b) Mature:

    c) Old age:

14. Using Figure 4.4, select the letter that illustrates each of the following.

    a) Zone of saturation: _____

    b) Aquiclude: _____

    c) Spring: _____

    d) Water table: _____

    e) Zone of aeration: _____

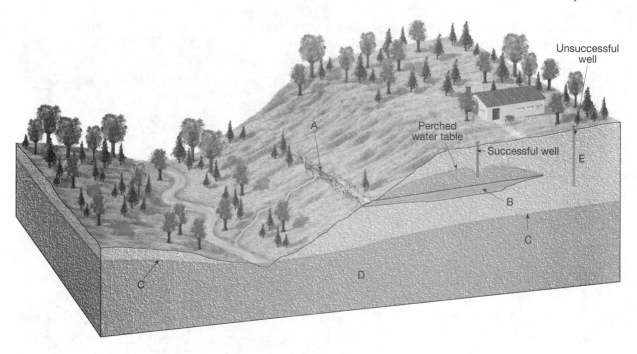

Figure 4.4

15. In general, how is the shape of the water table related to the shape of the surface?

16. What is the source of heat for most hot springs and geysers?

17. Using the terms porosity and permeability, describe an aquifer.

18. List two environmental problems that are associated with groundwater.

    1)

    2)

19. What are two ways that sinkholes form?

    1)

    2)

20. Identify the type of drainage pattern illustrated in each of the two diagrams in Figure 4.5 by writing the correct response by the diagram.

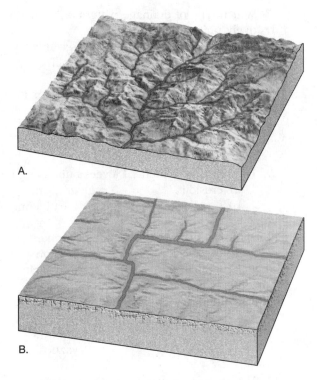

A.

B.

Figure 4.5

Practice Test

Multiple choice. Choose the best answer for the following multiple choice questions.

1. The largest reservoir of freshwater that is available to humans is _____.
   a) lakes and reservoirs   c) groundwater   e) atmospheric water vapor
   b) glaciers   d) river water

2. The single most important agent sculpturing Earth's land surface is _____.
   a) ice   c) running water
   b) wind   d) waves

3. Lowering a stream's base level will cause the stream to _____.
   a) deposit   c) change course   e) stop flowing
   b) meander   d) erode

4. Along a stream meander, the maximum velocity of water occurs _____.
   a) near the inner bank of the meander   c) near the outer bank of the meander
   b) in the center   d) near the stream's bed

5. A typical profile of a stream exhibits a constantly _____ gradient from the head to the mouth.
   a) increasing   c) uniform
   b) decreasing   d) arbitrary

6. The flat portion of a valley floor adjacent to a stream channel is called a _____.
   a) meander   c) yazoo   e) tributary
   b) floodplain   d) divide

7. Which type of drainage pattern will most likely develop where the underlying material is relatively uniform?
   a) dendritic   c) rectangular
   b) radial   d) trellis

8. Ice sheets and glaciers represent approximately _____ percent of the total volume of freshwater.
   a) 15   c) 55
   b) 35   d) 85

9. The function of artificial levees built along a river is to control _____.
   a) erosion   c) gradient   e) oxbow lakes
   b) flooding   d) meandering

10. Plants release water into the atmosphere through a process called _____.
    a) evaporation   c) erosion   e) precipitation
    b) infiltration   d) transpiration

11. Occasionally, deposition causes the main channel of a stream to divide into several smaller channels called _____.
    a) deltas   c) oxbows   e) distributaries
    b) meanders   d) yazoos

12. The average annual precipitation worldwide must equal the quantity of water _____.
    a) evaporated   c) that becomes runoff   e) locked in glaciers
    b) transpired   d) infiltrated

13. During periods when rain does not fall, rivers are sustained by _____.
    a) transpirated water   c) atmospheric moisture
    b) groundwater   d) runoff

14. Which one of the following would least likely be found in a wide valley?
    a) oxbow lake   c) rapids   e) meanders
    b) floodplain   d) cutoff

15. Along straight stretches, the highest velocities of water are near the _____ of the channel, just below the surface.
    a) banks   b) center

16. The combined effects of _____ and running water produce stream valleys.
    a) weathering   c) mass wasting   e) capacity
    b) tectonics   d) discharge

17. Which one of the following is a measure of a material's ability to transmit water through interconnected pore spaces?
    a) competence   c) capacity   e) permeability
    b) gradient   d) porosity

18. The capacity of a stream is directly related to its _____.
    a) velocity   c) gradient   e) discharge
    b) competence   d) meandering

19. Drawdown of a well creates a roughly conical lowering of the water table called the cone of
    _____.

    a) saturation   c) permeability   e) aeration

    b) lowering   d) depression

20. In a typical stream, where gradient is steep, discharge is _____.
    a) large   c) variable   e) reversed
    b) small   d) impossible to calculate

21. Most natural water contains the weak acid, _____ acid, which makes it possible for
    groundwater to dissolve limestone and form caverns.
    a) acetic   c) carbonic   e) nitric
    b) hydrochloric   d) sulfuric

22. The area above the water table where soil is not saturated is called the zone of _____.
    a) aeration   c) saturation   e) aquifers
    b) soil moisture   d) porosity

23. Most streams carry the largest part of their load _____.
    a) as bed load   c) in solution
    b) in suspension   d) near their head

24. The great majority of hot springs and geysers in the United States are found in the _____.
    a) North   c) East   e) Midwest
    b) South   d) West

25. Whenever the water table intersects the ground surface, a natural flow of water, called a
    _____, results.
    a) spring   c) aquiclude   e) geyser
    b) stalagmite   d) sinkhole

*True/false. For the following true/false questions, if a statement is not completely true, mark it false. For each false statement, change the **italicized** word to correct the statement.*

1. ___ Most of a stream's dissolved load is brought to it by *groundwater*.

2. ___ *Ultimate* base levels include lakes and resistant layers of rock.

3. ___ The unending circulation of Earth's water supply is called the *hydrologic* cycle.

4. ___ The reduced current velocity at the inside of a meander results in the deposition of coarse sediment, especially sand, called a *point bar*.

5. ___ Landscapes shaped by the dissolving power of groundwater are often said to exhibit *karst* topography.

6. ___ A flat valley floor, or floodplain, often occurs in *narrow* valleys.

7. ___ Since the water cycle is balanced, the average annual *precipitation* worldwide must equal the quantity of water evaporated.

8. ___ The gradient of a stream *increases* downstream.

9. ___ The drainage pattern that develops when the bedrock is crisscrossed by many right-angle joints and/or faults is the *radial* pattern.

10. ___ *Discharge* is one factor that determines the velocity of a stream.

11. ___ Much of the water that flows in rivers is not direct runoff from rain or snowmelt, but originates as *groundwater*.

12. ___ *Permeable* layers, such as clay, that hinder or prevent the movement of water beneath the surface are termed aquicludes.

13. ___ The term *artesian* is applied to any situation in which groundwater rises above the level where it was initially encountered in a well.

14. ___ A channel's shape, size, and *roughness* affect the amount of friction with the water in the channel.

15. ___ When rivers become rejuvenated, meanders often stay in the same place but become *deeper*.

16. ___ Most streams carry the largest part of their load in *solution*.

17. ___ The *competence* of a stream measures the maximum size of particles it is capable of transporting.

18. ___ A stream's velocity *increases* downstream.

19. ___ *Lowering* base level will cause a stream to gain energy and erode.

20. ___ Marshes called *backswamps* form because water on a floodplain cannot flow up the levee and into the river.

21. ___ The formation of caverns takes place in the zone of *aeration*.

22. ___ *Groundwater* represents the largest reservoir of freshwater that is readily available to humans.

23. ___ Some of the water that infiltrates the ground surface is absorbed by plants, which then release it into the atmosphere through a process called *infiltration*.

24. ___ Sorting of sediment occurs because each particle *size* has a critical settling velocity.

25. ___ Near the *mouth* of a river where discharge is great, gradient is small.

26. ___ The zone in the ground where all of the open spaces in sediment and rock are completely filled with water is called the zone of *saturation*.

27. ___ *Geysers* are intermittent hot springs that eject water with great force at various intervals.

28. ___ Two types of floodplains are erosional floodplains and *depositional* floodplains.

29. ___ Land *subsidence* is an environmental problem that can be caused by groundwater withdrawal.

30. ___ *Ice* is the single most important agent sculpturing Earth's land surface.

Word choice. Complete each of the following statements by selecting the most appropriate response.

1. The velocity of a stream in a straight channel is greatest at the [center/sides] of the channel.

2. A region with alternating bands of resistant and less resistant rock exposed at the surface will probably develop a [dendritic/radial/trellis] drainage pattern.

3. On a delta, the main channel of a river often divides into smaller channels called [tributaries/ distributaries/rills].

4. The upper limit of the zone of saturation is called the [water table/aquifer].

5. A stream's gradient usually [increases/decreases/remains unchanged] downstream.

6. A permeable rock layer that transmits water freely is referred to as an [aquifer/aquiclude].

7. Urbanization [decreases/increases] infiltration and [decreases/increases] runoff.

8. Karst topography is most likely to develop in regions with [basalt/limestone/sandstone] bedrock with a [dry/wet] climate.

9. Infiltration will be greatest on [steep/gentle] slopes following a [heavy/light] rain.

10. The amount of water passing a given point per unit of time is a stream's [velocity/discharge/ competence].

11. Building a dam will [raise/lower] the base level of a stream upstream from the dam.

12. The velocity of a stream is greatest at the [inside/outside] of a meander.

13. The acid that does most of the work of groundwater is [nitric/carbonic/sulfuric] acid.

14. A stream's discharge [increases/decreases] downstream.

15. The amount of groundwater a material can hold is controlled by the material's [permeability/ porosity].

Written questions

1. Briefly describe the movement of water through the hydrologic cycle.

2. List and briefly describe the three stages of valley development.

3. What is the difference between the porosity of a material and its permeability?

4. What are the conditions required for the development of karst topography?

Glaciers, Deserts, and Wind

5

Glaciers, Deserts, and Wind begins by examining the types, the movement, and the formation of glaciers. Also presented are the processes of glacial erosion and deposition, as well as the features associated with valley glaciers and ice sheets. The Pleistocene epoch and some indirect effects of Ice Age glaciers are also discussed. The section closes by examining some of the theories that attempt to explain the causes of glacial ages.

The study of deserts begins with a review of the roles of weathering and water in arid climates. A discussion of the evolution of much of the Basin and Range region of the United States provides insight into the processes that shape desert landscapes. The chapter ends with an investigation of wind erosion and deposition, including the various types of sand dunes.

Learning Objectives

After reading, studying, and discussing this chapter, you should be able to:

- Describe the types and locations of glaciers.
- Discuss glacial movement.
- List the types of glacial drift.
- Describe the features produced by glacial erosion and deposition.
- Define the Ice Age and the Pleistocene epoch.
- List the theories for the causes of glacial ages.
- Discuss the roles of weathering and water in arid climates.
- Describe the geologic evolution of the Basin and Range region.
- List the types of wind deposits and describe the features of wind deposition.

Chapter Summary

- A *glacier* is a thick mass of ice originating on the land as a result of the compaction and recrystallization of snow, and shows evidence of past or present flow. Today, *valley* or *alpine glaciers* are found in mountain areas where they usually follow valleys that were originally occupied by streams. *Ice sheets* exist on a much larger scale, covering most of Greenland and Antarctica.

- On the surface of a glacier, ice is brittle. However, below about 50 meters, pressure is great, causing ice to *flow* like a *plastic material*. A second important mechanism of glacial movement consists of the whole ice mass *slipping* along the ground.

- Glaciers erode land by *plucking* (lifting pieces of bedrock out of place) and *abrasion* (grinding and scraping of a rock surface). Erosional features produced by valley glaciers include *glacial troughs*, *hanging valleys*, *cirques*, *arêtes*, *horns*, and *fiords*.

71

• Any sediment of glacial origin is called *drift*. The two distinct types of glacial drift are 1) *till*, which is material deposited directly by the ice; and 2) *stratified drift*, which is sediment laid down by meltwater from a glacier.

• The most widespread features created by glacial deposition are layers or ridges of till, called *moraines*. Associated with valley glaciers are *lateral moraines*, formed along the sides of the valley, and *medial moraines*, formed between two valley glaciers that have joined. *End moraines*, which mark the former position of the front of a glacier, and *ground moraine*, an undulating layer of till deposited as the ice front retreats, are common to both valley glaciers and ice sheets.

• Perhaps the most convincing evidence for the occurrence of several glacial advances during the *Ice Age* is the widespread existence of *multiple layers of drift* and an uninterrupted record of climate cycles preserved in *sea-floor sediments*. In addition to massive erosional and depositional work, other effects of Ice Age glaciers included the *forced migration* of animals, *changes in stream and river courses, adjustment of the crust* by rebounding after the removal of the immense load of ice, and *climate changes* caused by the existence of the glaciers themselves. In the sea, the most far-reaching effect of the Ice Age was the *worldwide change in sea level* that accompanied each advance and retreat of the ice sheets.

• Any theory that attempts to explain the causes of glacial ages must answer the two basic questions: 1) What causes the onset of glacial conditions? and 2) What caused the alternating glacial and inter-glacial stages that have been documented for the Pleistocene epoch? Two of the many hypotheses for the cause of glacial ages involve 1) plate tectonics and 2) variations in Earth's orbit.

• The same geologic processes that operate in humid regions also operate in deserts, but under contrasting climatic conditions. In dry lands *rock weathering of any type is greatly reduced* because of the lack of moisture and the scarcity of organic acids from decaying plants. Practically all desert streams are dry most of the time and are said to be *ephemeral*. Nevertheless, *running water is responsible for most of the erosional work in a desert*. Although wind erosion is more significant in dry areas than elsewhere, the main role of wind in a desert is in the transportation and deposition of sediment.

• Many of the landscapes of the Basin and Range region of the western and southwestern United States are the result of streams eroding uplifted mountain blocks and depositing the sediment in interior basins. *Alluvial fans, playas,* and *playa lakes* are features often associated with these landscapes.

• In order for wind erosion to be effective, dryness and scant vegetation are essential. *Deflation*, the lifting and removal of loose material, often produces shallow depressions called *blowouts* and can also lower the surface by removing sand and silt, leaving behind a stony veneer, called *desert pavement*. *Abrasion*, the "sandblasting" effect of wind, is often given too much credit for producing desert features. However, abrasion does cut and polish rock near the surface.

• Wind deposits are of two distinct types: 1) extensive *blankets of silt*, called *loess*, that is carried by wind in *suspension*; and 2) *mounds and ridges of sand*, called *dunes*, which are formed from sediment that is carried as part of the wind's *bed load*. The *types of sand dunes* include 1) *barchan dunes*, solitary dunes shaped like crescents with their tips pointing downwind, 2) *transverse dunes,* which form a series of long ridges orientated at right angles to the prevailing wind, 3) *longitudinal dunes*, long ridges that are more or less parallel to the prevailing wind, 4) *parabolic dunes*, similar in shape to barchans except that their tips point into the wind, and 5) *star dunes*, isolated hills of sand that exhibit a complex form.

Chapter Outline

I. Glaciers
  A. A thick mass of ice that forms over land from the compaction and recrystallization of snow and shows evidence of past or present flow
  B. Types of glaciers
    1. Valley, or alpine glaciers—form in mountainous areas
    2. Ice sheets, or continental
      a. Large scale
      b. e.g., Over Greenland
    3. Other types
      a. Ice caps
      b. Piedmont glaciers
  C. Movement of glacial ice
    1. Types of glacial movements
      a. Flow
      b. Slipping along the ground
    2. Zone of fracture
      a. Uppermost 50 meters
      b. Crevasses form in brittle ice
    3. Zone of accumulation—the area where a glacier forms
    4. Zone of wastage—the area where there is a net loss due to melting
  D. Glaciers erode by
    1. Plucking—lifting of rock blocks
    2. Abrasion
      a. Rock flour (pulverized rock)
      b. Striations (grooves in the bedrock)
  E. Erosional features of valley glaciers
    1. Glacial trough
    2. Hanging valley
    3. Cirque
    4. Arête
    5. Horn
    6. Fiord
  F. Glacial deposits
    1. Glacial drift
      a. All sediments of glacial origin
      b. Types of glacial drift
        1. Till
          a. Material that is deposited directly by the ice
          b. Glacial erratics (boulders embedded in till)
        2. Stratified drift
          a. Deposited by meltwater
          b. Sediment is sorted

    2. Depositional features
      a. Moraines
        1. Layers or ridges of till
        2. Types
          a. Lateral
          b. Medial
          c. End
          d. Ground
      b. Outwash plain, or valley train
      c. Kettles
      d. Drumlins
      e. Eskers
      f. Kames
  G. Glaciers of the past
    1. Ice Age
      a. Began 2 to 3 million years ago
      b. Division of geological time is called the Pleistocene epoch
      c. Ice covered 30 percent of Earth's land area
    2. Indirect effects of Ice Age glaciers
      a. Migration of animals and plants
      b. Rebounding upward of the crust
      c. Worldwide change in sea level
      d. Climatic changes
  H. Causes of glaciation
    1. Successful theory must account for
      a. Cooling of Earth, as well as
      b. Short-term climatic changes
    2. Proposed possible causes
      a. Plate tectonics
        1. Continents were arranged differently
        2. Changes in oceanic circulation
      b. Variations in Earth's orbit
        1. Milankovitch hypothesis
          a. Shape (eccentricity) of Earth's orbit varies
          b. Angle of Earth's axis (obliquity) changes
          c. Axis wobbles (precession)
        2. Changes in climate over the past several hundred thousand years are closely associated with variations in Earth's orbit

II. Deserts
  A. Geologic processes in arid climates
    1. Weathering
      a. Not as effective as in humid regions

   b. Mechanical weathering forms unaltered rock and mineral fragments
   c. Some chemical weathering does occur
      1. Clay forms
      2. Thin soil forms
2. Role of water in arid climates
   a. Streams are dry most of the time
   b. Desert streams are said to be ephemeral
      1. Flow only during periods of rainfall
      2. Different names are used for desert streams
         a. e.g., Wash
         b. e.g., Arroyo
         c. e.g., Wadi, donga, or nullah
   c. Desert rainfall
      1. Rain often occurs as heavy showers
      2. Causes flash floods
   d. Poorly integrated drainage
   e. Most erosional work in a desert is done by running water
B. Evolution of a desert landscape
   1. Uplifted crustal blocks
   2. Interior drainage into basins produces
      a. Alluvial fans and bajadas
      b. Playas and playa lakes

C. Wind erosion
   1. By deflation
      a. Lifting of loose material
      b. Produces
         1. Blowouts
         2. Desert pavement
   2. By abrasion
D. Types of wind deposits
   1. Loess
      a. Deposits of windblown silt
      b. Extensive blanket deposits
      c. Primary sources are
         1. Deserts
         2. Glacial stratified drift
   2. Sand dunes
      a. Mounds and ridges of sand formed from the wind's bed load
      b. Characteristic features
         1. Slip face—the leeward slope of the dune
         2. Cross beds—sloping layers of sand in the dune
      c. Types of sand dunes
         1. Barchan dunes
         2. Transverse dunes
         3. Longitudinal dunes
         4. Parabolic dunes
         5. Star dunes

Key Terms

Page numbers shown in ( ) refer to the textbook page where the term first appears.

alluvial fan (p. 144)	glacial erratic (p. 130)	parabolic dune (p. 150)
alpine glacier (p. 122)	glacial striations (p. 127)	piedmont glacier (p. 122)
arête (p. 128)	glacial trough (p. 127)	playa lake (p. 144)
barchan dune (p. 148)	glacier (p. 122)	Pleistocene epoch (p. 135)
barchanoid dune (p. 149)	ground moraine (p. 132)	plucking (p. 127)
blowout (p. 145)	hanging valley (p. 127)	pluvial lake (p. 137)
cirque (p. 127)	horn (p. 128)	rock flour (p. 127)
crevasse (p. 123)	ice cap (p. 122)	slip face (p. 148)
cross beds (p. 148)	ice sheet (p. 122)	star dune (p. 150)
deflation (p. 145)	interior drainage (p. 143)	stratified drift (p. 129)
desert pavement (p. 145)	kame (p. 134)	till (p. 129)
drift (p. 129)	kettle (p. 133)	valley glacier (p. 122)
drumlin (p. 134)	lateral moraine (p.131)	valley train (p. 133)
end moraine (p. 131)	loess (p. 146)	zone of accumulation (p. 124)
ephemeral stream (p. 142)	longitudinal dune (p. 149)	zone of wastage (p. 124)
esker (p. 134)	medial moraine (p. 131)	
fiord (p. 128)	outwash plain (p. 133)	

Vocabulary Review

Choosing from the list of key terms, furnish the most appropriate response for the following statements.

1. The largest type of glacier is the _____, which often covers a large portion of a continent.

2. A(n) _____ is type of stream that carries water only in response to a specific episode of rainfall.

3. The all-embracing term for sediments of glacial origin, no matter how, where, or in what form they were deposited is _____.

4. A(n) _____ is a streamlined asymmetrical hill composed of till.

5. Most of the major glacial episodes occurred during a division of the geologic time scale known as the _____.

6. Embedded material in a glacier may gouge long scratches and grooves called _____ in the bedrock as the ice passes over.

7. Arid regions typically lack permanent streams and are often characterized by a type of drainage called _____.

8. A(n) _____ is a type of moraine that forms when two valley glaciers coalesce to form a single ice stream.

9. The area where there is net loss to a glacier due to melting is known as the _____.

10. A(n) _____ is a U-shaped valley produced by the erosion of a valley glacier.

11. A(n) _____ is a hollowed-out, bowl-shaped depression at the head of a valley produced by glacial erosion.

12. A sand dune shaped like a crescent with its tips pointing down wind is called a(n) _____.

13. The part of a glacier where snow accumulates and ice forms is called the _____.

14. As a stream emerges from a canyon and quickly loses velocity, it often deposits a cone of debris known as a(n) _____.

15. A long ridge of sand that forms more or less parallel to the prevailing wind direction is known as a(n) _____.

16. Deposits of windblown silt are called _____.

17. Materials deposited directly by glacial ice are known as _____.

18. A broad ramplike surface of stratified drift that is built adjacent to the downstream edge of an ice sheet is termed a(n) _____.

19. Deflation by wind often forms a shallow depression called a(n) _____.

20. The term applied to the sloping layers of sand that compose a dune is _____.

21. A(n) _____ is a thick mass of ice that forms over land from the accumulation, compaction, and recrystallization of snow.

22. A glacier that occupies the broad lowland at the base of a steep mountain is called a(n) _____.

Comprehensive Review

1. List and describe the two major types of glaciers.

   1)

   2)

2. Why are the uppermost 50 meters of a glacier appropriately referred to as the zone of fracture?

3. Briefly describe the following two ways that glaciers erode land.

   a) Plucking:

   b) Abrasion:

4. The area represented in Figure 5.1 was subjected to alpine glaciation. Select the appropriate letter in the figure that identifies each of the following features.

   a) Cirque: ____   c) Hanging valley: ____   e) Horn: ____

   b) Glacial trough: ____   d) Arête: ____

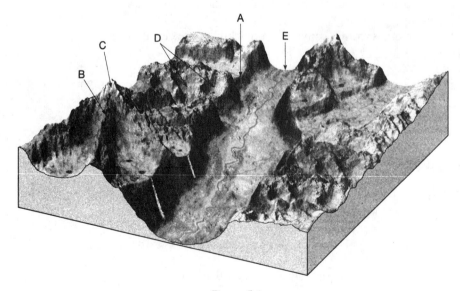

Figure 5.1

5. Distinguish between the two terms till and stratified drift.

6. What are two hypotheses relating to the possible causes of glacial periods?

7. In Figure 5.2, diagram _____ [A, B] illustrates the early stage in the evolution of the desert landscape. Select the appropriate letter in the figure that identifies each of the following features.

   a) Playa lake: _____

   b) Alluvial fan: _____

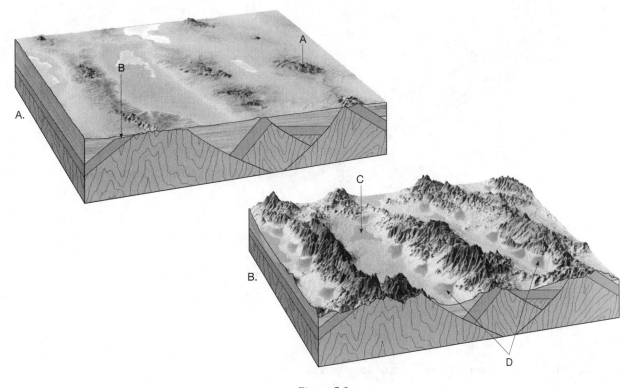

Figure 5.2

8. Describe each of the following types of moraines.

    a) End moraine:

    b) Lateral moraine:

    c) Ground moraine:

9. The area represented in Figure 5.3 is being subjected to glaciation by an ice sheet. Select the appropriate letter in the figure that identifies each of the following features.

    a) Drumlin: ____   c) Esker: ____

    b) Outwash plain: ____   d) End moraine: ____

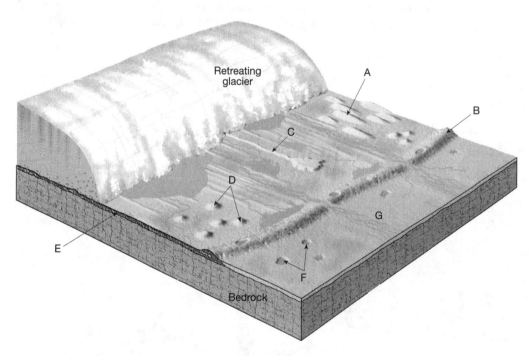

Figure 5.3

10. Briefly describe the process responsible for the migration of sand dunes.

11. What were some indirect effects of Ice Age glaciers?

12. What features would you look for in a mountainous region to determine if the area had been glaciated by alpine glaciers?

13. What evidence suggests that there were several glacial advances during the Ice Age?

14. In what two ways does the transport of sediment by wind differ from that by running water?

15. Referring to Figure 5.4, write the name of the type of dune illustrated by each diagram.

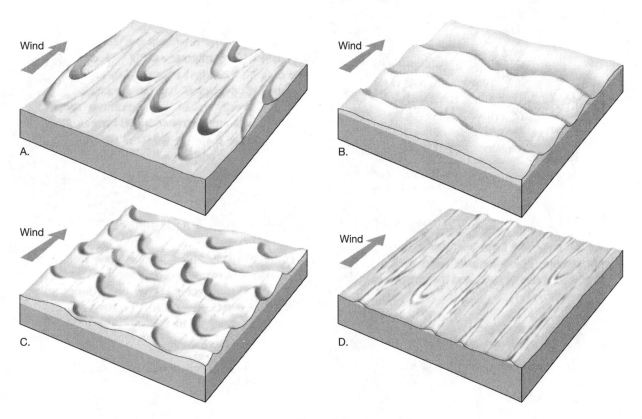

Figure 5.4

Practice Test

Multiple choice. Choose the best answer for the following multiple choice questions.

1. A broad, ramplike surface of stratified drift built adjacent to the downstream edge of most end moraines is called a(n) _____.
   - a) moraine
   - b) valley train
   - c) outwash plain
   - d) fiord
   - e) kettle

2. Icebergs are produced when large pieces of ice break off from the front of a glacier during a process termed _____.
   - a) ablation
   - b) deflation
   - c) calving
   - d) abrasion
   - e) plucking

3. Which type of dune will form at a right angle to the prevailing wind when there is abundant sand, little or no vegetation, and a constant wind direction?
   - a) barchan dune
   - b) transverse dune
   - c) longitudinal dune
   - d) parabolic dune
   - e) star dune

4. Most of the major glacial episodes during the Ice Age occurred during a division of the geologic time scale called the _____ epoch.
   - a) Pliocene
   - b) Eocene
   - c) Paleocene
   - d) Pleistocene
   - e) Miocene

5. The two major ways that glaciers erode land are abrasion and _____.
   - a) plucking
   - b) slipping
   - c) deflation
   - d) gouging
   - e) scouring

6. Which one of the following is NOT an effect that Pleistocene glaciers had upon the landscape?
   - a) mass extinctions
   - b) crustal depression and rebounding
   - c) animal and plant migration
   - d) adjustments of stream courses
   - e) worldwide change in sea level

7. A thick ice mass that forms over the land from the accumulation, compaction, and recrystallization of snow is a _____.
   - a) wadi
   - b) fiord
   - c) drumlin
   - d) glacier
   - e) loess

8. The _____ drainage of arid regions is characterized by intermittent streams that do not flow out of the desert to the ocean.
   - a) radial
   - b) interior
   - c) exterior
   - d) excellent
   - e) well-developed

9. The all-embracing term for sediments of glacial origin is _____.
   - a) outwash
   - b) drift
   - c) loess
   - d) silt
   - e) erratic

10. The end moraine that marks the farthest advance of a glacier is called the _____ moraine.
    - a) lateral
    - b) terminal
    - c) medial
    - d) ground
    - e) recessional

11. Cracks that form in the zone of fracture of a glacier are called _____.
    - a) fractures
    - b) crevasses
    - c) faults
    - d) gouges
    - e) kettles

12. Desert streams that carry water only in response to specific episodes of rainfall are said to be
    _____.
    a) episodic   c) occasional   e) youthful
    b) flowing   d) ephemeral

13. Sinuous ridges composed of sand and gravel deposited by streams flowing in tunnels beneath
    glacial ice are _____.
    a) eskers   c) drumlins   e) moraines
    b) cirques   d) horns

14. Materials that have been deposited directly by a glacier are called _____.
    a) outwash   c) till   e) stratified drift
    b) sediment   d) loess

15. Valleys that are left standing high above the main trough of a receding valley glacier are termed
    _____ valleys.
    a) ephemeral   c) glacial   e) hanging
    b) wadis   d) fiord

16. The thickest and most extensive loess deposits occur in western and northern _____.
    a) Canada   c) Australia   e) Egypt
    b) China   d) Libya

17. Glacial troughs that have become deep, steep-sided inlets of the sea are called _____.
    a) fiords   c) cirques   e) kames
    b) hanging valleys   d) washes

18. Moraines that form when two valley glaciers coalesce to form a single ice stream are termed
    _____ moraines.
    a) lateral   c) terminal   e) recessional
    b) medial   d) ground

19. Boulders found in glacial till or lying free on the surface are called glacial _____.
    a) remnants   c) drumlins   e) erratics
    b) striations   d) moraines

20. Which one of the following is NOT a feature associated with valley glaciers?
    a) arête   c) horn   e) cirque
    b) glacial trough   d) arroyo

21. Dry, flat lake beds located in the center of basins in arid areas are called _____.
    a) playas   c) kames   e) alluvial fans
    b) arroyos   d) deltas

22. The most noticeable result of deflation in some places are shallow depressions called
    _____.
    a) kettles   c) dunes   e) sinkholes
    b) blowouts   d) drumlins

23. Evidence indicates that, in addition to the Pleistocene epoch, there were at least _____
    earlier periods of glacial activity.
    a) two   c) four   e) six
    b) three   d) five

24. The leeward slope of a dune is called the _____ face.
    a) cross   c) wind   e) gradual
    b) slope   d) slip

25. Which one of the following states is included in the Basin and Range region?
    a) Nevada   c) Texas   e) Florida
    b) Missouri   d) New York

*True/false. For the following true/false questions, if a statement is not completely true, mark it false. For each false statement, change the **italicized** word to correct the statement.*

1. ___ During the Ice Age, ice sheets and alpine glaciers were far *more* extensive than they are today.

2. ___ Snow accumulation and glacial ice formation occur in an area known as the zone of *wastage*.

3. ___ Much of the weathered debris in deserts is the result of *mechanical* weathering processes.

4. ___ *Medial* moraines form along the sides of valley glaciers.

5. ___ The two major types of glaciers are *valley* glaciers and ice sheets.

6. ___ Under pressure equivalent to more than the weight of *ten* meters of ice, ice will behave as a plastic and flow.

7. ___ The Ice Age began between two and three *million* years ago.

8. ___ The position of the front of a glacier depends on the balance between *accumulation* and wastage.

9. ___ The combined areas of present-day continental ice sheets represents almost *thirty* percent of Earth's land area.

10. ___ Dry regions of the world encompass nearly *sixty* percent of Earth's land surface.

11. ___ The rates that glaciers advance vary considerably from one glacier to another but can be as great as several *kilometers* per day.

12. ___ Desert floods arrive suddenly and subside *quickly*.

13. ___ Glaciated valleys typically exhibit a *V-shaped* appearance.

14. ___ Streamlined asymmetrical hills composed of till occurring in clusters are called *eskers*.

15. ___ The movement of glacial ice is generally referred to as *flow*.

16. ___ *Glaciers* are capable of carrying huge blocks of material that no other erosional agent could budge.

17. ___ In *humid* regions, moisture binds particles together and vegetation anchors the soil so that wind erosion is negligible.

18. ___ The Matterhorn in the Swiss Alps is the most famous example of a *cirque*.

19. ___ Sediments laid down by glacial meltwater are called stratified *drift*.

20. ___ *Moraines* are bowl-shaped erosional features found at the heads of glacial valleys.

21. ___ In the United States, *loess* deposits are an indirect product of glaciation.

22. ___ Continued sand accumulation, coupled with periodic slides down the *slip* face, causes the slow migration of a sand dune in the direction of air movement.

23. ___ The tips of *parabolic* dunes point into the wind.

Word choice. Complete each of the following statements by selecting the most appropriate response.

1. Moraines are composed of [stratified drift/till/loess].

2. The pulverized rock produced by glacial abrasion is called rock [till/loess/flour].

3. Most of the weathered debris in deserts is the result of [mechanical/chemical] weathering.

4. Most deserts have [exterior/interior] drainage.

5. [Kettles/Fiords] are deep, steep-sided ocean inlets formed by glacial erosion.

6. An individual boulder, different from the local bedrock, that was deposited by a glacier is called a glacial [erratic/kame].

7. The breaking off of large pieces of ice at the front of a glacier is a process called [calving/crevassing].

8. Stratified drift is [sorted/unsorted] sediment.

9. Desert streams often have [many/few] tributaries.

10. Erosion by ice sheets tends to [accentuate/subdue] the topography of a region.

11. Desert pavement is created by [abrasion/deflation].

12. Stream erosion in most desert regions [is/is not] highly influenced by sea level.

13. Loess deposits [have/lack] visible layers.

Written questions

1. List and briefly describe at least four glacial depositional features.

2. List and briefly describe at least three erosional features you might expect to find in an area where valley glaciers exist, or have recently existed.

3. List three indirect effects of Ice Age glaciers.

4. Describe how sand dunes migrate.

Earthquakes and Earth's Interior

6

Earthquakes and Earth's Interior begins with a brief description of the effects of the major earthquakes that have taken place in California within the last decade. Following an explanation of how earthquakes occur, the types of seismic waves, their propagation, and how they appear on a typical seismic trace are presented. This is followed by a discussion of earthquake epicenters—how they are located and their worldwide distribution. Earthquake intensity and magnitude are also explained. The destruction caused by seismic vibrations and their associated perils introduces a discussion of earthquake prediction. The presentation of earthquakes closes with an explanation of how earthquakes are used to discover Earth's interior structure and a brief description of Earth's interior composition.

Learning Objectives

After reading, studying, and discussing this chapter, you should be able to:

- Describe the cause of earthquakes.
- List the types of seismic waves and describe their propagation.
- Describe how an earthquake epicenter is located.
- Describe the worldwide distribution of earthquake epicenters.
- Explain how the magnitude of an earthquake is determined.
- List the other destructive forces that can be triggered by an earthquake.
- Discuss the status of earthquake prediction.
- Describe Earth's interior structure and composition.

Chapter Summary

- *Earthquakes* are vibrations of Earth produced by the rapid release of energy from rocks that rupture because they have been subjected to stresses beyond their limit. This energy, which takes the form of waves, radiates in all directions from the earthquake's source, called the *focus*. The movements that produce most earthquakes occur along large fractures, called *faults*, that are associated with plate boundaries.

- Two main groups of *seismic waves* are generated during an earthquake: 1) *surface waves*, which travel along the outer layer of Earth; and 2) *body waves*, which travel through Earth's interior. Body waves are further divided into *primary*, or *P, waves*, which push (compress) and pull (dilate) rocks in the direction the wave is traveling, and *secondary*, or *S, waves*, which "shake" the particles in rock at right angles to their direction of travel. P waves can travel through solids, liquids, and gases. Fluids (gases and liquids) will not transmit S waves. In any solid material, P waves travel about 1.7 times faster than S waves.

- The location on Earth's surface directly above the focus of an earthquake is the *epicenter*. An epicenter is determined using the difference in velocities of P and S waves.

• *There is a close correlation between earthquake epicenters and plate boundaries.* The principal earthquake epicenter zones are along the outer margin of the Pacific Ocean, known as the*circum-Pacific belt*, and through the world's oceans along the *oceanic ridge system*.

• Earthquake *intensity* depends not only on the strength of the earthquake, but on other factors, such as distance from the epicenter, the nature of surface materials, and building design. The *Mercalli intensity scale* assesses the damage from a quake at a specific location. Using the *Richter scale*, the *magnitude* (a measure of the total amount of energy released) of an earthquake is determined by measuring the *amplitude* (maximum displacement) of the largest seismic wave recorded, with adjustments of the amplitude made for the weakening of seismic waves as they move from the focus, as well as for the sensitivity of the recording instrument. A logarithmic scale is used to express magnitude, in which a tenfold increase in recorded wave amplitude corresponds to an increase of one on the magnitude scale. Each unit of Richter magnitude equates to roughly a 30-fold energy increase.

• The most obvious factors that determine the amount of destruction accompanying an earthquake are the *magnitude* of the earthquake and the *proximity* of the quake to a populated area. *Structural damage* attributable to earthquake vibrations depends on several factors, including 1) *intensity*, 2) *duration* of the vibrations, 3) *nature of the material* upon which the structure rests, and 4) the *design* of the structure. Secondary effects of earthquakes include *tsunamis, landslides, ground subsidence*, and *fire*.

• Substantial research to predict earthquakes is underway in Japan, the United States, China, and Russia—countries where earthquake risk is high. No consistent method of short-range prediction has yet been devised. Long-range forecasts are based on the premise that earthquakes are repetitive or cyclical. Seismologists study the history of earthquakes for patterns, so their occurrences might be predicted.

• As indicated by the behavior of P and S waves as they travel through Earth, the four major zones of Earth's interior are the 1) *crust* (the very thin outer layer), 2) *mantle* (a rocky layer located below the crust with a thickness of 2885 kilometers), 3) *outer core* (a layer about 2270 kilometers thick, which exhibits the characteristics of a mobile liquid), and 4) *inner core* (a solid metallic sphere with a radius of about 1216 kilometers).

• The *continental crust* is primarily made of *granitic* rocks, while the *oceanic crust* is of *basaltic* composition. *Ultramafic rocks* such as *peridotite* are thought to make up the *mantle*. The *core* is composed mainly of *iron and nickel*.

Chapter Outline

I. Earthquakes
  A. General features
    1. Vibration of Earth produced by a sudden release of energy
    2. Associated with movements along faults
      a. Explained by the plate tectonics theory
      b. Mechanism for earthquakes was first explained by H. Reid
        1. Early 1900s
        2. Rocks "spring back"
          a. Phenomenon called elastic rebound

      b. Vibrations (earthquakes) occur as rock elastically returns to its original shape
      c. Fault creep
    3. Often preceded by foreshocks
    4. Often followed by aftershocks
  B. Earthquake waves
    1. Study of earthquake waves is called seismology
    2. Earthquake recording instrument
      a. Records movement of Earth
      b. Instrument is a seismograph
      c. Record is called a seismogram
    3. Types of earthquake waves

a. Surface waves
   1. Complex motion
   2. Slowest velocity of all waves
b. Body waves
   1. Primary (P) waves
      a. Push-pull (compressional) motion
      b. Travel through
         1. Solids
         2. Liquids
         3. Gases
      c. Greatest velocity of all earthquake waves
   2. Secondary (S) waves
      a. "Shake" motion
      b. Travel only through solids
      c. Slower velocity than P waves

C. Locating an earthquake
   1. Focus—the place within Earth where earthquake waves originate
   2. Epicenter
      a. Point on the surface, directly above the focus
      b. Located using the difference in the arrival times between P and S wave recordings, which are related to distance
      c. Three station recordings are needed to locate an epicenter
         1. Circle equal to the epicenter distance is drawn around each station
         2. Point where three circles intersect is the epicenter
   3. Earthquake zones are closely correlated with plate boundaries
      a. e.g., Circum-Pacific belt
      b. e.g., Oceanic ridge system

D. Earthquake intensity and magnitude
   1. Mercalli intensity scale
      a. Assesses damage at a specific location
      b. Depends on
         1. Strength of the earthquake
         2. Distance from the epicenter
         3. Nature of the surface material
         4. Building design
   2. Magnitude
      a. Concept introduced by Charles Richter in 1935
      b. Measured using the Richter scale
      c. Earthquake magnitude scale
         1. Amplitude of largest wave recorded

         2. Largest earthquakes near magnitude 8.6
         3. Magnitudes less than 2.0 are usually not felt
         4. Each unit of magnitude increase corresponds to
            a. A tenfold increase in wave amplitude
            b. About a 30-fold energy increase

E. Earthquake destruction
   1. Factors that determine destruction
      a. Magnitude of the earthquake
      b. Proximity to population
   2. Destruction from
      a. Ground shaking
      b. Liquefaction of the ground
         1. Saturated material turns fluid
         2. Underground objects may float to surface
      c. Tsunamis, or seismic sea waves
      d. Landslides and ground subsidence
      e. Fires

F. Prediction
   1. Short-range—no reliable method yet devised for predicting
   2. Long-range
      a. Premise is that earthquakes are repetitive
      b. Region is given a probability of a quake

II. Earth's interior
A. Most of our knowledge of the inteior comes from the study of P and S earthquake waves
   1. Travel times of P and S waves through Earth vary depending on the properties of the materials
   2. S waves travel only through solids
B. Earth's interior structure
   1. Crust
      a. Thin outer layer
      b. Varies in thickness
         1. 70 km in some mountainous regions
         2. Less than 5 km in oceanic regions
      c. Two parts
         1. Continental crust—lighter, granitic rocks
         2. Oceanic crust—basaltic composition
      d. Lithosphere
         1. Crust and uppermost mantle (about

100 km thick)
2. Cool, rigid, solid
e. Mohorovicic discontinuity (Moho)
   separates the crust from the mantle
2. Mantle
   a. Below crust
   b. 2885 km thick
   c. Composition similar to the
      igneous rock peridotite
   d Asthenosphere
      1. Upper mantle
      2. At a depth from 100 to 350
         kilometers
      3. Hot, weak rock
      4. Easily deformed
      5. Up to 10 percent is molten

6. Key to explaining
   plate movement
3. Outer core
   a. Below mantle
   b. 2270 km thick
   c. Mobile liquid
   d. Does not transmit S waves
   e. Mainly iron and nickel
      composition
   f. Related to Earth's magnetic field
4. Inner core
   a. 1216 km radius
   b. Solid
   c. Iron and nickel
      composition
   d. High density

Key Terms

Page numbers shown in ( ) refer to the textbook page where the term first appears.

aftershock (p. 159)
asthenosphere (p. 174)
body wave (p. 160)
crust (p. 173)
earthquake (p. 155)
elastic rebound (p. 157)
epicenter (p. 162)
fault (p. 157)
focus (p. 155)
foreshock (p. 159)

inner core (p. 173)
liquefaction (p. 166)
lithosphere (p. 174)
magnitude (p. 163)
mantle (p. 173)
Mercalli intensity (p. 162)
Mohorovicic discontinuity
   (Moho) (p. 173)
outer core (p. 173)
primary (P) wave (p. 160)

Richter scale (p. 163)
secondary (S) wave (p. 160)
seismic sea wave (tsunami)
   (p. 167)
seismogram (p. 159)
seismograph (p. 159)
seismology (p. 159)
shadow zone (p. 173)
surface wave (p. 160)

Vocabulary Review

Choosing from the list of key terms, furnish the most appropriate response for the following statements.

1. The vibration of Earth produced by a rapid release of energy is called a(n) _____.

2. A large ocean wave that often results from vertical displacement of the ocean floor during an earthquake is called a(n) _____.

3. A small earthquake, called a(n) _____, often precedes a major earthquake by days or, in some cases, by as much as several years.

4. The source of an earthquake is called the _____.

5. The "springing back" of rock after it has been deformed, much like a stretched rubber band does when released, is termed _____.

6. The type of earthquake wave that travels along the outer layer of Earth is the _____.

7. Earth's _____ is a layer of the interior which exhibits the characteristics of a mobile liquid composed of iron and nickel.

8. The _____ of an earthquake is the location on the surface directly above the focus.

9. The zone between about 100 kilometers and 350 kilometers within Earth that consists of hot, weak rock that is easily deformed is called the _____.

10. A fracture in rock along which displacement has occurred is termed a(n) _____.

11. Earth's very thin outer layer is called the _____.

12. The type of earthquake wave that "shakes" the particles at right angles to the direction it is traveling is the _____.

13. A recording, or trace, of an earthquake is called a(n) _____.

14. The boundary that separates the crust from the mantle is known as the _____.

15. Using Richter's scale, the _____ of an earthquake is determined by measuring the amplitude of the largest wave recorded on the seismogram.

16. Earth's _____ is a solid metallic sphere about 1216 kilometers in radius.

17. The type of earthquake wave that pushes and pulls rock in the direction it is traveling is the _____.

18. The cool, rigid layer of Earth, which includes the entire crust as well as the uppermost mantle, is the _____.

19. The _____ determines the magnitude of an earthquake by measuring the amplitude of the largest wave recorded on a seismogram.

20. The _____ is a belt from about 105 degrees to 140 degrees distance from an earthquake epicenter in which direct P waves are absent because of refraction by Earth's core.

21. The rocky layer of Earth located beneath the crust and having a thickness of 2885 kilometers is the _____.

22. An instrument that records earthquake waves is called a(n) _____.

Comprehensive Review

1. List and describe the three basic types of earthquake waves.

   1)

   2)

   3)

2. Using Figure 6.1, select the letter of the diagram that illustrates the characteristic motion of the following types of earthquake waves.

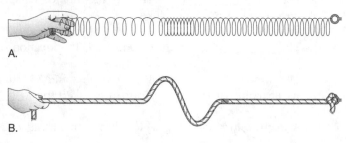

   a) P wave: ____

   b) S wave: ____

Figure 6.1

3. Figure 6.2 is a typical recording of an earthquake, called a seismogram. Using Figure 6.2, select the letter on the figure that identifies the recording of each of the following types of earthquake waves.

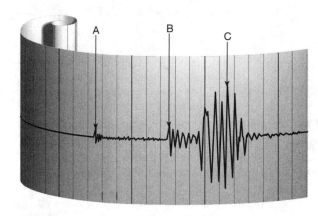

   a) Surface waves: ____

   b) S waves: ____

   c) P waves: ____

Figure 6.2

4. Describe the differences in velocity and mode of travel between P waves and S waves.

5. Briefly describe how the epicenter of an earthquake is located.

6. What are two factors that determine the degree of destruction that accompanies an earthquake?

   1)

   2)

7. List two major, continuous earthquake belts. With what general Earth feature are most earthquake epicenters closely correlated?

8. Using Figure 6.3, select the letter that identifies each of the following parts of Earth's interior.

    a) Mantle: ____

    b) Inner core: ____

    c) Continental crust: ____

    d) Oceanic crust: ____

    e) Asthenosphere: ____

    f) Outer core: ____

    g) Lithosphere: ____

9. Briefly describe the composition of the following zones of Earth's interior.

    a) Mantle:

    b) Outer core:

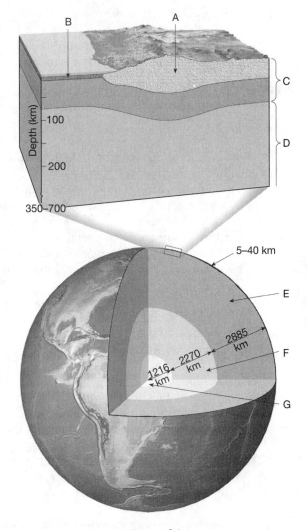

Figure 6.3

Practice Test

Multiple choice. Choose the best answer for the following multiple choice questions.

1. It is estimated that over _____ earthquakes that are strong enough to be felt occur worldwide annually.
    a) 500   c) 10000   e) 30000
    b) 1000   d) 20000

2. The location on the surface directly above the earthquake focus is called the _____.
    a) ephemeral   c) epicenter   e) epitaph
    b) epicycle   d) epinode

3. The cool, rigid layer of Earth that includes the entire crust as well as the uppermost mantle is called the _____.
    a) asthenosphere   c) oceanic crust   e) lithosphere
    b) lower crust   d) Moho

4. Underground storage tanks rising to the surface as the result of an earthquake is evidence of
   _____.

   a) tsunamis   c) subsidence   e) fault creep
   b) liquefaction   d) fracturing

5. Which earthquake body wave has the greatest velocity?
   a) P wave
   b) S wave

6. The belt from about 105 to 140 degrees away from an earthquake where no P waves are recorded
   is known as the _____.
   a) shadow zone   c) Moho zone   e) low velocity zone
   b) absent zone   d) reflective zone

7. The study of earthquakes is called _____.
   a) seismogram   c) seismology   e) seismogony
   b) seismicity   d) seismography

8. Which one of the following regions has the greatest amount of seismic activity?
   a) central Europe   d) the circum-Pacific belt
   b) southern Russia   e) the eastern United States
   c) the central Atlantic basin

9. The difference in _____ of P and S waves provides a method for determining the epicenter of an earthquake.
   a) magnitudes   c) sizes   e) foci
   b) velocities   d) modes of travel

10. The source of an earthquake is called the _____.
    a) fulcrum   c) epicenter   e) focus
    b) ephemeral   d) foreshock

11. Long-range earthquake forecasts are based on the premise that earthquakes are _____.
    a) random   c) fully understood   e) repetitive
    b) destructive   d) always occurring

12. The epicenter of an earthquake is located using the distances from a minimum of _____
    seismic stations.
    a) three   c) five   e) seven
    b) four   d) six

13. Which of the earthquake body waves cannot be transmitted through fluids?
    a) P waves
    b) S waves

14. Dense rocks like _____ are thought to make up the mantle and provide the lava for
    oceanic eruptions.
    a) limestone   c) peridotite   e) rhyolite
    b) granite   d) sandstone

15. In areas where unconsolidated materials are saturated with water, earthquakes can turn stable soil
    into a fluid during a phenomenon called _____.
    a) libation   c) leaching   e) localization
    b) lithification   d) liquefaction

16. The adjustments of materials that follow a major earthquake often generate smaller earthquakes called _____.

    a) tremors
    b) aftershocks
    c) foreshocks
    d) surface waves
    e) body waves

17. An earthquake with a magnitude of 6.5 releases _____ times more energy than one with a magnitude of 5.5.

    a) 10
    b) 20
    c) 30
    d) 40
    e) 50

18. Earthquake epicenters are most closely correlated with _____.

    a) continental interiors
    b) plate boundaries
    c) population centers
    d) continental shelves
    e) high latitudes

19. The amount of damage caused by an earthquake at a specific location is assessed using the _____.

    a) Mercalli scale
    b) Richter scale
    c) Mohs scale
    d) seismic scale
    e) Gutenberg scale

20. The greatest concentration of metals occurs in Earth's _____.

    a) crust
    b) asthenosphere
    c) mantle
    d) core
    e) lithosphere

*True/false. For the following true/false questions, if a statement is not completely true, mark it false. For each false statement, change the **italicized** word to correct the statement.*

1. ___ Earthquake waves that travel through Earth's interior are called *body* waves.

2. ___ The adjustments that follow a major earthquake often generate smaller earthquakes called *foreshocks*.

3. ___ Earthquakes with a Richter magnitude less than *eight* are usually not felt by humans.

4. ___ Most of our knowledge of Earth's interior comes from the study of *earthquakes*.

5. ___ Vibrations known as earthquakes occur as rock slips and *elastically* returns to its original shape.

6. ___ Earthquake *body* waves are divided into two types called primary (P) waves and secondary (S) waves.

7. ___ Most tsumanis result from *horizontal* displacement of the ocean floor during an earthquake.

8. ___ The study of earthquakes is called *seismography*.

9. ___ No reliable method of *short-range* earthquake prediction has yet been devised.

10. ___ Fluids (gases and liquids) *cannot* transmit P waves.

11. ___ P waves arrive at a recording station *after* S waves.

12. ___ The boundary that separates the crust from the underlying mantle is known as the *shadow* discontinuity.

13. ___ Most earthquakes occur along faults associated with *plate* boundaries.

14. ___ The lithosphere is situated *below* the asthenosphere.

15. ___ Earth's *inner* core is a solid metallic sphere.

16. ___ The *epicenter* of an earthquake is the location on the surface directly above the focus.

17. ___ To locate an epicenter, the distance from *three* or more different seismic stations must be known.

18. ___ The mantle is *solid* because both P and S waves travel through it.

19. ___ The farther an earthquake recording station is from an earthquake, the *greater* the difference in arrival times of the P and S waves.

20. ___ Earthquakes in the central and eastern United States occur *more* frequently than along plate-boundary areas.

21. ___ The continental crust is mostly made of *granitic* rocks.

22. ___ A refined Richter scale is used to describe earthquake *magnitude*.

Word choice. Complete each of the following statements by selecting the most appropriate response.

1. [S/P] waves cannot pass through Earth's outer core.

2. The location on Earth's surface above the source of an earthquake is called the [epicenter/focus].

3. Seismic waves that travel along Earth's outer layer are called [body/surface] waves.

4. The energy of an earthquake is released during a process called elastic [deformation/rebound].

5. The boundary between the crust and the mantle is called the [Moho/shadow zone].

6. Most of our knowledge about Earth's interior comes from the study of [rocks/seismic waves].

7. Large fractures in Earth's crust along which movement occurs are referred to as [epicenters/faults].

8. A scale used to measure earthquake magnitude is the [Mercalli/Richter] scale.

9. The [lithosphere/asthenosphere] consists of partially melted rock.

10. The instruments used to record earthquakes are called [seismographs/seismograms] and the records they produce are called [seismographs/seismograms].

11. The greatest damage done by a tsunami is [at sea/on land].

12. In general, the most earthquake resistant structures are [rigid/flexible].

13. [Inertia/Gravity] is the tendency of a stationary object to hold still, or a moving object to stay in motion.

14. Earth's [lithosphere/asthenosphere] is rigid.

15. A seismograph records the relative motion between a [stationary/moving] weight and a base that vibrates.

Written questions

1. Describe an earthquake and the circumstances that cause it to occur.

2. What are the major differences between P and S earthquake waves?

3. Describe the composition (mineral/rock makeup) of Earth's crust, mantle, and core.

Plate Tectonics

7

Plate Tectonics opens by examining the lines of evidence that Alfred Wegener used in the early 1900s to support his continental-drift hypothesis. This evidence included matching the outlines of the shorelines of continents that are now separated by vast ocean basins, fossils, rock types and structural similarities between continents, and paleoclimates. Also presented are the main objections to Wegener's ideas.

Following a brief overview, the theory of plate tectonics is discussed in detail. The movement of lithospheric plates and different types of plate boundaries are examined extensively. Paleomagnetism (polar wandering and magnetic reversals), the distribution of earthquakes, ages and distribution of ocean basin sediments, and hot spots are used to provide additional support for plate tectonics. The formation and eventual breakup of Pangaea are also investigated. The chapter closes with comments about the driving mechanism of plate tectonics.

Learning Objectives

After reading, studying, and discussing this chapter, you should be able to:

- List the evidence that was used to support the continental-drift hypothesis.
- Describe the theory of plate tectonics.
- Explain the differences between the continental-drift hypothesis and the theory of plate tectonics.
- List and describe the evidence used to support the plate tectonics theory.
- Explain the difference between divergent, convergent, and transform plate boundaries.
- Discuss the formation and breakup of Pangaea.
- Describe the models that have been proposed to explain the driving mechanism for plate motion.

Chapter Summary

- In the early 1900s *Alfred Wegener* set forth his *continental-drift* hypothesis. One of its major tenets was that a supercontinent called *Pangaea* began breaking apart into smaller continents about 200 million years ago. The smaller continental fragments then "drifted" to their present positions. To support the claim that the now-separate continents were once joined, Wegener and others used the *fit of South America and Africa, ancient climatic similarities, fossil evidence*, and *rock structures*.

- One of the main objections to the continental-drift hypothesis was its inability to provide an acceptable mechanism for the movement of continents.

- The theory of *plate tectonics*, a far more encompassing theory than continental drift, holds that Earth's rigid outer shell, called the *lithosphere*, consists of about twenty segments called *plates* that are in motion relative to each other. Most of Earth's *seismic activity, volcanism*, and *mountain building* occur along the dynamic margins of these plates.

• A major departure of the plate tectonics theory from the continental-drift hypothesis is that large plates contain both continental and ocean crust and the entire plate moves. By contrast, in continental drift, Wegener proposed that the sturdier continents "drifted" by breaking through the oceanic crust, much like ice breakers cut through ice.

• The three distinct types of plate boundaries are 1) *divergent boundaries*—where plates move apart, 2) *convergent boundaries*—where plates move together as in *oceanic-continental convergence*, *oceanic-oceanic convergence*, or *continental-continental* convergence, and 3) *transform boundaries* —where plates slide past each other.

• The theory of plate tectonics is supported by 1) *paleomagnetism*, the direction and intensity of Earth's magnetism in the geologic past, 2) the global distribution of *earthquakes* and their close association with plate boundaries, 3) the ages of *sediments* from the floors of the deep-ocean basins, and 4) the existence of island groups that formed over *hot spots* and provide a frame of reference for tracing the direction of plate motion.

• The gross details of the migrations of individual continents over the past 500 million years have been reconstructed. *Pangaea began breaking apart about 200 million years ago.* North America separated from Africa between 200 and 165 million years ago. Evidence indicates that prior to the formation of Pangaea the landmasses had probably gone through several episodes of fragmentation similar to what we see happening today.

• Several models for the driving mechanism of plates have been proposed. One model involves large *convection cells* within the mantle carrying the overlying plates. Another proposes that dense *oceanic material descends and pulls* the lithosphere along. No single driving mechanism can account for all of the major facets of plate motion.

Chapter Outline

I. Continental-drift
   A. Alfred Wegener
      1. First proposed hypothesis, 1915
      2. Published The Origin of Continents and Oceans
   B. Wegener's continental drift hypothesis
      1. Supercontinent called Pangaea began breaking apart about 200 million years ago
      2. Continents "drifted" to present positions
      3. Continents "broke" through the ocean crust
      4. Evidence used by Wegener
         a. Fit of South America and Africa
         b. Fossils
         c. Rock structures
         d. Ancient climates
      5. Main objection to Wegener's proposal was its inability to provide a mechanism for the movement of contents
II. Plate tectonics
   A. More encompassing than continental drift

   B. Associated with Earth's rigid outer shell
      1. Called the lithosphere
      2. Consists of about 20 slabs (plates)
         a. Plates are moving slowly
         b. Largest plate is the Pacific plate
         c. Plates are mostly beneath the ocean
   C. Asthenosphere
      1. Exists beneath the lithosphere
      2. Hotter and weaker than lithosphere
      3. Allows for motion of lithosphere
   D. Plate boundaries
      1. Associated with plate boundaries
         a. Seismic activity
         b. Volcanism
         c. Mountain building
      2. Types of plate boundaries
         a. Divergent (spreading) boundary
            1. Most exist along oceanic ridge crests
            2. Seafloor spreading occurs along the boundary
               a. Forms fractures (openings) on the ridge crests
               b. Fractures fill with molten material

3. When the boundary occurs on a continent, rifts or rift valleys form
b. Convergent boundary
 1. Lithosphere is subducted into the mantle
 2. Types of convergent boundaries
 a. Oceanic-continental boundary
  1. Forms a subduction zone with a deep-ocean trench
  2. Volcanic arcs form
   a. e.g., Andes
   b. e.g., Cascades
   c. e.g., Sierra Nevada system
 b. Oceanic-oceanic boundary
  1. Often forms volcanoes on the ocean floor
  2. Island arc forms as volcanoes emerge
   a. e.g., Aleutian islands
   b. e.g., Alaskan Peninsula
   c. e.g., Philippines
   d. e.g., Japan
 c. Continental-continental boundary
  1. Neither plate will subduct
  2. Can produce mountains
   a. e.g., Himalayas
   b. Other possibilities
    1. Alps
    2. Appalachians
    3. Urals
c. Transform boundary
 1. Plates slide past one another
 a. No new crust is created
 b. No crust is destroyed
 2. Transform faults
 a. Most are in oceanic crust
 b. Parallel the direction of plate movement
 c. Aids movement of crustal material
E. Evidence that supports plate tectonics
 1. Paleomagnetism
 a. Probably the most persuasive evidence
 b. Ancient magnetism preserved in rocks
 c. Paleomagnetic records show
  1. Polar wandering (evidence that continents moved)
  2. Earth's magnetic field reversals
   a. Recorded in the sea floor rocks

b. Confirms seafloor spreadng
2. Earthquake patterns
 a. Associated with plate boundaries
 b. Deep-focus earthquakes along trenches provide a method for tracking the plate's descent
3. Ocean drilling
 a. Deep Sea Drilling Project (ship: *Glomar Challenger*)
 b. Age of deepest sediments
  1. Youngest are near the ridges
  2. Older are at a distance from the ridge
 c. Ocean basins are geologically young
4. Hot spots
 a. Rising plumes of mantle material
 b. Volcanoes can form over them
  1. e.g., Hawaiian Island chain
  2. Chains of volcanoes mark plate movement
F. Breakup of Pangaea
 1. Migrations of continents over the past 500 million years has been determined
 2. Breakup begins about 200 million years ago
 a. North America and Africa began separating between 200 and 165 million years ago
 b. Africa and South America begin splitting apart about 135 million years ago
 3. Landmasses also had fragmented prior to the formation of Pangaea
 4. Fragments that formed Pangaea began collecting between 500 and 225 million years ago
G. Driving mechanism of plate tectonics
 1. No one model explains all plate motions
 2. Earth's heat is the driving force
 3. Several models have been proposed
 a. Convection currents in mantle
 b. Slab-pull and slab-push model
  1. Descending oceanic crust pulls the plate
  2. Elevated ridge system pushes the plate
 c. Hot plumes
  1. Extend from mantle-core boundary
  2. Spread laterally under lithosphere

Key Terms

Page numbers shown in ( ) refer to the textbook page where the term first appears.

continental drift (p. 181)
convergent boundary (p. 188)
deep-ocean trench (p. 192)
divergent boundary (p. 188)
hot spot (p. 201)
island arc (p. 193)

normal polarity (p. 197)
paleomagnetism (p. 197)
Pangaea (p. 181)
plate (p. 185)
plate tectonics (p. 185)
polar wandering (p. 197)

reverse polarity (p. 198)
rift (rift valley) (p. 190)
seafloor spreading (p. 188)
subduction zone (p. 192)
transform boundary (p. 188)
volcanic arc (p. 192)

Vocabulary Review

Choosing from the list of key terms, furnish the most appropriate response for the following statements.

1. In his 1915 book, *The Origin of Continents and Oceans,* Wegener set forth his radical hypothesis of _____.

2. Each of the rigid slabs of Earth's lithosphere is referred to as a(n) _____.

3. _____ is the mechanism responsible for producing the new sea floor between two diverging plates.

4. The type of plate boundary where plates move together, causing one of the slabs to be consumed into the mantle as it descends beneath an overriding plate, is referred to as a(n) _____.

5. Remnant magnetism in rock bodies is referred to as _____.

6. The theory of _____ holds that Earth's rigid outer shell consists of about twenty rigid slabs.

7. When rocks exhibit the same magnetism as the present magnetic field, they are said to possess _____.

8. The type of plate boundary where plates move apart, resulting in upwelling of material from the mantle to create new sea floor, is referred to as a(n) _____.

9. A rising plume of mantle material often causes a(n) _____, such as the one responsible for the intraplate volcanism that produced the Hawaiian Islands.

10. Wegener hypothesized that about 200 million years ago a supercontinent that he called _____ began breaking into smaller continents, which then drifted to their present positions.

11. The region where an oceanic plate descends into the asthenosphere because of convergence is called a(n) _____.

12. Material displaced downward along spreading centers on continents often creates a downfaulted valley called a(n) _____.

13. A(n) _____ occurs where plates grind past each other without creating or destroying lithosphere.

14. _____ is the idea that Earth's magnetic poles have migrated greatly through time.

15. Rocks that exhibit magnetism that is opposite of the present magnetic field are said to possess
_____.

16. A chain of small volcanic islands that forms when two oceanic slabs converge, one descending beneath the other, is called a(n) _____.

17. As an oceanic plate slides beneath an overriding plate, a deep, linear feature, called a(n) _____, often forms on the ocean floor adjacent to the zone of subduction.

Comprehensive Review

1. List four lines of evidence that were used to support the continental drift hypothesis.

    1)

    2)

    3)

    4)

2. What was one of the main objections to Wegener's continental-drift hypothesis?

3. Using Figure 7.1, select the letter of the diagram that portrays each of the following types of plate boundaries.

    a) Transform boundary: ____

    b) Convergent boundary: ____

    c) Divergent boundary: ____

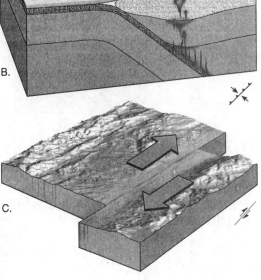

Figure 7.1

4. Briefly explain the theory of plate tectonics.

5. What are the three types of convergent plate boundaries?

    1)

    2)

    3)

6. Briefly explain how each of the following has been used to support the theory of plate tectonics.

    a) Paleomagnetism:

    b) Earthquake patterns:

    c) Ocean drilling:

7. Using Figure 7.2, select the appropriate letter that identifies each of the following features.

    a) Subducting oceanic lithosphere: ____   c) Volcanic arc: ____

    b) Trench: ____   d) Continental lithosphere: ____

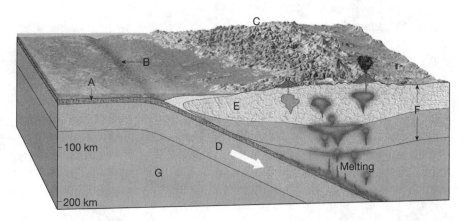

Figure 7.2

8. Briefly explain how the Appalachian Mountains are related to the formation of the supercontinent of Pangaea.

9. List and briefly explain the three hypotheses that have been proposed for the driving mechanism of plate motion.

   1)

   2)

   3)

10. Figure 7.3 illustrates the ancient supercontinent of Pangaea as it is thought to have existed about 300 million years ago. Using the figure, select the appropriate letter that identifies each of following current-day landmasses.

   a) Antarctica: _____   e) North America: _____

   b) Eurasia: _____   f) Africa: _____

   c) South America: _____   g) Australia: _____

   d) India: _____

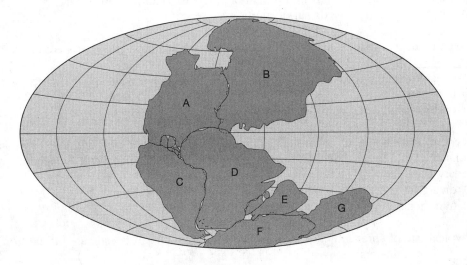

Figure 7.3

11. How are satellites being used to directly test the theory of plate tectonics?

12. What are the names of the two subcontinents that made up the supercontinent of Pangaea? Which present-day continents formed the southern-most of the two subcontinents?

Practice Test

Multiple choice. Choose the best answer for the following multiple choice questions.

1. The general term that refers to the deformation of Earth's crust and results in the formation of structural features such as mountains is _____.
   a) erosion   c) mass wasting   e) subduction
   b) tectonics   d) volcanism

2. During the last four million years, Earth's magnetic field has reversed _____.
   a) twice   c) hundreds of times
   b) several times   d) thousands of times

3. Most of Earth's seismic activity, volcanism, and mountain building occur along _____.
   a) lines of magnetism   c) parallels of latitude   e) hot spots
   b) plate boundaries   d) random trends

4. The apparent movement of Earth's magnetic poles over time is referred to as _____.
   a) magnetic drift   c) subduction   e) polar wandering
   b) perturbation   d) polar spreading

5. Alfred Wegener is best known for his hypothesis of _____.
   a) continental drift   c) atoll formation   e) seafloor spreading
   b) natural selection   d) subduction

6. Complex mountain systems such as the Alps, Appalachians, and Himalayas are the result of _____.
   a) oceanic-oceanic convergence   d) continental-continental convergence
   b) hot spots   e) island arcs
   c) oceanic-continental convergence

7. The type of plate boundary where plates move together, causing one of the slabs of lithosphere to be consumed into the mantle as it descends beneath an overriding plate, is called a _____ boundary.
   a) divergent   c) transform   e) convergent
   b) transitional   d) gradational

8. The typical rate of spreading along ridges in the Atlantic Ocean is estimated to be from 2 to _____ centimeters per year.
   a) 5   c) 15   e) 25
   b) 10   d) 20

9. The name given by Wegener to the supercontinent he believed existed prior to the current continents was _____.
   a) Pangaea   c) Euroamerica   e) Panamerica
   b) Atlantis   d) Pantheon

10. Magnetic minerals lose their magnetism when heated above a certain temperature, called the _____.
    a) magnetopoint   c) melting point   e) Bechtol point
    b) Curie point   d) break point

11. The best approximation of the true outer boundary of the continents is the seaward edge of the _____.

    a) deep-ocean trench   c) mid-ocean ridge   e) continental shelf
    b) abyssal plain   d) present-day shorelines

12. The Aleutian, Mariana, and Tonga islands are examples of _____.

    a) island arcs   c) transform boundaries e) mid-ocean ridges
    b) abyssal plains   d) hot spots

13. Beneath Earth's lithosphere, the hotter, weaker zone known as the _____ allows for motion of Earth's rigid outer shell.

    a) crust   c) outer core   e) oceanic crust
    b) asthenosphere   d) Moho

14. Mountains comparable in age and structure to the mountain belt that contains the Appalachians are found in _____.

    a) Africa   c) the British Isles   e) Asia
    b) Australia   d) South America

15. Which one of the following was NOT used in support of Wegener's continental drift hypothesis?

    a) fossil evidence   d) ancient climates
    b) paleomagnetism   e) rock structures
    c) fit of South America and Africa

16. The Red Sea is believed to be the site of a recently formed _____.

    a) divergent boundary   d) hot spot
    b) ocean trench   e) gradational boundary
    c) convergent boundary

17. The type of plate boundary where plates move apart, resulting in upwelling of material from the mantle to create new sea floor, is a _____ boundary.

    a) divergent   c) transform   e) convergent
    b) transitional   d) gradational

18. One of the main objections to Wegener's hypothesis was his inability to provide an acceptable _____ for continental drift.

    a) time   c) mechanism   e) place
    b) rate   d) direction

19. The spreading rate for the North Atlantic Ridge is _____ than the rates for the East Pacific Rise.

    a) greater
    b) less

20. Earth's rigid outer shell is called the _____.

    a) asthenosphere   c) outer core   e) lithosphere
    b) mantle   d) continental mass

21. Most deep-focus earthquakes occur in association with _____.

    a) hot spots   d) abyssal plains
    b) ocean trenches   e) transform boundaries
    c) oceanic ridge systems

22. The theory of plate tectonics holds that Earth's rigid outer shell consists of about _____ rigid slabs.

    a) five   c) fifteen   e) twenty-five

    b) ten   d) twenty

23. The age of the deepest sediment in an ocean basin _____ with increasing distance from the oceanic ridge.

    a) increases   c) remains the same

    b) decreases   d) varies

24. During oceanic-continental convergence, as the oceanic plate slides beneath the overriding plate, a(n) _____ is often produced adjacent to the zone of subduction.

    a) deep-ocean terrace   c) transform fault   e) divergent boundary

    b) deep-ocean ridge   d) deep-ocean trench

25. The chain of volcanic structures, extending from the Hawaiian Islands to Midway Island and then continuing northward toward the Aleutian trench has formed over a(n) _____ as the Pacific plate moved.

    a) subduction zone   c) island arc   e) divergent boundary

    b) hot spot   d) convergent boundary

26. Which one of the following is NOT a hypothesis that has been proposed for the mechanism of plate motion?

    a) hot-plumes hypothesis   c) convection-current hypothesis

    b) slab-push and slab-pull hypothesis   d) mantle-density hypothesis

27. The northern Appalachian Mountains formed during a collision between North America and

    _____.

    a) Africa   c) Europe   e) Greenland

    b) Australia   d) Asia

*True/false. For the following true/false questions, if a statement is not completely true, mark it false. For each false statement, change the **italicized** word to correct the statement.*

1. ___ The *lithosphere* consists of both crustal rocks and a portion of the upper mantle.

2. ___ *Transform* faults connect convergent and divergent plate boundaries in various combinations.

3. ___ The rate of plate movement is measured in *kilometers* per year.

4. ___ The Red Sea is believed to be the site of a recently formed *convergent* plate boundary.

5. ___ In North America, the *Cascade Range* and Sierra Nevada system are volcanic arcs associated with the subduction of oceanic lithosphere.

6. ___ Older portions of the sea floor are carried into Earth's *core* in regions where trenches occur in the deep-ocean floor.

7. ___ The island of Hawaii is *older* than Midway Island.

8. ___ The "Ring of Fire" is an area of earthquake and volcanic activity that encircles the *Pacific* ocean basin.

9. ___ *Transform* faults are roughly parallel to the direction of plate movement.

10. ___ To explain continental drift, Wegener proposed that the *continents* broke through the oceanic crust, much like ice breakers cut through ice.

11. ___ There is a close association between deep-focus earthquakes and ocean *ridges*.

12. ___ Beneath Earth's lithosphere is the hotter and weaker zone known as the *asthenosphere*.

13. ___ Seafloor spreading is the mechanism that has produced the floor of the *Atlantic* Ocean during the past 165 million years.

14. ___ The unequal distribution of heat inside Earth generates some type of thermal convection in the *crust* which ultimately drives plate motion.

15. ___ The supercontinent of *Pangaea* began breaking apart about 200 million years ago.

16. ___ When rocks exhibit the same magnetism as the present magnetic field they are said to possess *reverse* polarity.

17. ___ Deep-ocean trenches are located adjacent to *subduction* zones.

18. ___ The *oldest* oceanic crust is located at the oceanic ridge crests.

19. ___ The largest single rigid slab of Earth's outer shell is the *Pacific* plate.

20. ___ The oldest sediments found in the ocean basins are *less* than 160 million years old.

21. ___ Along a *transform* plate boundary, plates grind past each other without creating or destroying lithosphere.

22. ___ The Aleutian, Mariana, and Tonga islands are island arcs associated with *oceanic-oceanic* plate convergence.

23. ___ Lithospheric plates are *thickest* in the ocean basins.

24. ___ The idea that Earth's *magnetic* poles had migrated through time is known as polar wandering.

25. ___ Using the dates of the most recent *magnetic* reversals, the rate at which spreading occurs at the various ridges can be determined.

Word choice. Complete each of the following statements by selecting the most appropriate response.

1. Tectonic plates are large segments of Earth's [lithosphere/asthenosphere].

2. The best way to determine the true shape of a continent is to trace the outer boundary of its [continental shelf/shoreline/mountain ranges].

3. At convergent plate boundaries, oceanic lithosphere is being [created/consumed].

4. Most large tectonic plates containing continental crust [also/do not] contain oceanic crust.

5. Most divergent plate boundaries are associated with [continental/oceanic] ridges.

6. Tectonic plates are [flexible/rigid] slabs of Earth materials.

7. The supercontinent of Pangaea began breaking apart about [200/500] million years ago.

8. The coal fields of North America contain fossil evidence that these regions were once located in [tropical/polar] climates.

9. The region where one plate descends into the asthenosphere below another plate is called a [rifting/subduction] zone.

10. Most plates have [only/more than] one type of plate boundary.

11. At divergent plate boundaries, lithosphere is being [created/destroyed].

12. Wegener proposed that the portion of the supercontinent that is now South Africa was once centered over the [equator/South Pole].

13. The primary driving force for plate movement comes from the unequal distribution of [heat/gravity] within Earth.

14. Continental crust is [thicker/thinner] than oceanic crust.

15. As oceanic crust moves away from a spreading center, it becomes [warmer/cooler] and [less/more] dense.

16. Most deep-focus earthquakes are associated with [hot spots/subduction zones].

17. The breakup of Pangaea was initiated along two major [trenches/rifts].

18. The study of paleomagnetism has provided evidence about the [rate/depth] of seafloor spreading.

19. Hot spots are believed to be created by [converging plates/mantle plumes].

20. The age of the deepest ocean sediments [increases/decreases] with increasing distance from an oceanic ridge crest.

21. Once a rock forms, changing its position [can/will not] change the magnetic alignment of its minerals.

22. Transform faults are roughly [perpendicular/parallel] to the direction of plate movement.

23. Earthquakes associated with divergent and transform boundaries have [deep/shallow] foci.

24. When spreading centers develop within a continent, valleys called [trenches/rifts] form.

25. The magma produced in a subduction zone often produces [oceanic ridges/volcanic arcs].

26. Africa and South America separated [before/after] the North Atlantic ocean basin began forming.

27. The increasing density of a slab of oceanic crust as it cools and moves away from a spreading center explains the [slab-push/slab-pull] model for the driving force of plate tectonics.

28. The formation of the [Himalaya Mountains/Baja Peninsula] occurred less than 10 million years ago.

Written questions

1. What relation exists between the ages of the Hawaiian Islands, hot spots, and plate tectonics?

2. List and briefly describe the three major types of plate boundaries.

3. List the lines of evidence used to support the theory of plate tectonics.

Igneous Activity

8

Igneous Activity begins with a description of the catastrophic 1980 eruption of Mount St. Helens. A discussion of volcanism and the factors that determine the nature of volcanic eruptions (magma composition, temperature, and amount of dissolved gases) is followed by an examination of the materials that can be extruded during an eruption. The types of volcanic cones, their origins, shapes, and compositions, as well as the nature of fissure eruptions and volcanic landforms, are also presented.

An examination of intrusive igneous activity includes the classification and description of the major intrusive igneous bodies—dikes, sills, laccoliths, and batholiths. The chapter closes with a discussion of the relations between igneous activity, plate tectonics, and the origin and distribution of magma.

Learning Objectives

After reading, studying, and discussing this chapter, you should be able to:

- List the factors that determine the violence of volcanic eruptions.
- List the materials that are extruded from volcanoes.
- Describe the major features produced by volcanic activity.
- List and describe the major intrusive igneous features.
- Discuss the origin of magma.
- Describe the relation between igneous activity and plate tectonics.

Chapter Summary

- The primary factors that determine the nature of volcanic eruptions include the magma's *temperature*, its *composition*, and the *amount of dissolved gases* it contains. As lava cools, it begins to congeal, and as *viscosity* increases, its mobility decreases. *The viscosity of magma is directly related to its silica content. Granitic* lava, with its high silica content, is very viscous and forms short, thick flows. *Basaltic* lava, with a lower silica content, is more fluid and may travel a long distance before congealing. Dissolved gases provide the force which propels molten rock from the vent of a volcano.

- The materials associated with a volcanic eruption include *lava flows* (*pahoehoe* and *aa* flows for basaltic lavas), *gases* (primarily in the form of *water vapor*), and *pyroclastic material* (pulverized rock and lava fragments blown from the volcano's vent, which include *ash, pumice, lapilli, cinders, blocks,* and *bombs*).

- *Shield cones* are broad, slightly domed volcanoes built primarily of fluid, basaltic lava. *Cinder cones* have very steep slopes composed of pyroclastic material. *Composite cones,* or *stratovolcanoes,* are large, nearly symmetrical structures built of interbedded lavas and pyroclastic deposits. Composite cones represent the most violent type of volcanic activity. Often associated with violent eruptions are *nuée ardente*, a fiery cloud of hot gases infused with incandescent ash that races down steep volcanic slopes. Large composite cones often generate a type of mudflow known as a *lahar*.

• Other than volcanoes, regions of volcanic activity may contain *craters* (steep walled depressions at the summit of most volcanoes), *calderas* (craters that exceed one kilometer in diameter), *volcanic necks* (rocks that once occupied the vents of volcanoes, but are now exposed because of erosion), *lava plateaus* produced from *fissure eruptions* (volcanic material extruded from fractures in the crust), and *pyroclastic flows*.

• Igneous intrusive bodies are classified according to their *shape* and by their *orientation with respect to the host rock*, generally sedimentary rock. The two general shapes are *tabular* (sheetlike) and *massive*. Intrusive igneous bodies that cut across existing sedimentary beds are said to be *discordant*, whereas those that form parallel to existing sedimentary beds are *concordant*.

• *Dikes* are tabular, discordant igneous bodies produced when magma is injected into fractures that cut across rock layers. Tabular, concordant bodies, called *sills*, form when magma is injected along the bedding surfaces of sedimentary rocks. In many respects sills closely resemble buried lava flows. *Laccoliths* are similar to sills but form from less-fluid magma that collects as a lens-shaped mass that arches the overlying strata upward. *Batholiths*, the largest intrusive igneous bodies with surface exposures of more than 100 square kilometers (40 square miles), frequently compose the cores of mountains.

• *Temperature, pressure*, and *partial melting influence the formation of magma*. By raising its temperature, solid rock will melt and generate magma. One source of heat is that which is released by the decay of radioactive elements found in Earth's mantle and crust. Secondly, a drop in confining pressure can lower the melting temperature of rock sufficiently to trigger melting. Thirdly, igneous rock, which contains several different minerals, melts over a range of temperatures, with the lower temperature minerals melting first. This process, known as *partial melting*, produces most, if not all, magma. Partial melting often results in the production of a melt with a higher silica content than the parent rock.

• Active areas of *volcanism* are found *along the oceanic ridges* (spreading center volcanism), *adjacent to ocean trenches* (subduction zone volcanism), as well as *the interiors of plates* (intraplate volcanism) themselves. Most active volcanoes are associated with plate boundaries.

Chapter Outline

I. Volcanic eruptions
  A. Factors that determine the violence of an eruption
    1. Composition of the magma
    2. Temperature of the magma
    3. Dissolved gases in the magma
  B. Viscosity of magma
    1. Viscosity is a measure of a material's resistance to flow
    2. Factors affecting viscosity
      a. Temperature (hotter magmas are less viscous)
      b. Composition (silica content)
        1. High silica—high viscosity (e.g., granitic lava)
        2. Low silica—more fluid (e.g., basaltic lava)
      c. Dissolved gases
        1. Mainly water vapor and carbon dioxide
        2. Gases expand near the surface
        3. Provide the force to extrude lava
        4. Violence of an eruption is related to how easily gases escape from magma
          a. Easy escape from fluid magma
          b. Viscous magma produces a more violent eruption

II. Materials associated with volcanic eruptions
　　A. Lava flows
　　　　1. Basaltic lavas are more fluid
　　　　2. Types of lava
　　　　　　a. Pahoehoe lava (resembles braids in ropes)
　　　　　　b. Aa lava (rough, jagged blocks)
　　B. Gases
　　　　1. One to 5 percent of magma by weight
　　　　2. Mainly water vapor and carbon-dioxide
　　C. Pyroclastics
　　　　1. "Fire fragments"
　　　　2. Types of pyroclastic material
　　　　　　a. Ash—fine, glassy fragments
　　　　　　b. Pumice—from "frothy" lava
　　　　　　c. Lapilli—"walnut" size
　　　　　　d. Cinders—"pea-sized" with voids
　　　　　　e. Particles larger than lapilli
　　　　　　　　1. Blocks—hardened lava
　　　　　　　　2. Bombs—ejected as hot lava
III. Volcanoes
　　A. General features
　　　　1. Opening at summit
　　　　　　a. Crater (steep-walled depression at the summit)
　　　　　　b. Caldera (a summit depression greater than 1 km diameter)
　　　　2. Vent (the conduit that connects the crater to the magma chamber)
　　B. Types of volcanoes
　　　　1. Shield volcano
　　　　　　a. Broad, slightly domed
　　　　　　b. Primarily made of basaltic (fluid) lava
　　　　　　c. Generally large
　　　　　　d. Generally produce a large volume of lava
　　　　　　e. e.g., Mauna Loa in Hawaii
　　　　2. Cinder cone
　　　　　　a. Built from ejected lava fragments
　　　　　　b. Steep slope angle
　　　　　　c. Rather small size
　　　　　　d. Frequently occur in groups
　　　　3. Composite cone (or stratovolcano)
　　　　　　a. Most are adjacent to the Pacific Ocean (e.g., Fujiyama, Mt. Shasta)
　　　　　　b. Large size
　　　　　　c. Interbedded lavas and pyroclastics
　　　　　　d. Most violent type of activity (e.g., Vesuvius)
　　　　　　e. Often produce nuée ardente
　　　　　　　　1. Fiery cloud
　　　　　　　　2. Hot gases infused with ash
　　　　　　　　3. Flows down sides of a volcano
　　　　　　　　4. Speeds up to 200 km per hour
　　　　　　f. May produce a lahar, a type of mudflow
IV. Volcanic landforms
　　A. Crater or caldera
　　　　1. Steep walled depression at summit
　　　　2. Caldera—a crater that exceeds one kilometer in diameter
　　B. Volcanic neck
　　　　1. Resistant vent left standing after erosion
　　　　2. e.g., Ship Rock, New Mexico
　　C. Fissure eruption and lava plateau
　　　　1. Volcanic material extruded from fractures
　　　　2. e.g., Columbia Plateau
　　D. Pyroclastic flow
　　　　1. From silica-rich magma
　　　　2. Consists of ash and pumice fragments
　　　　3. Material is propelled from the vent at a high speed
　　　　4. e.g., Yellowstone plateau
V. Intrusive igneous activity
　　A. Magma emplaced at depth
　　B. Underground igneous body is called a pluton
　　C. Plutons are classified according to
　　　　1. Shape
　　　　　　a. Tabular (sheetlike)
　　　　　　b. Massive
　　　　2. Orientation with respect to the host (surrounding) rock
　　　　　　a. Discordant—cuts across sedimentary beds
　　　　　　b. Concordant—parallel to sedimentary beds
　　D. Types of igneous intrusive features
　　　　1. Dike, a tabular, discordant pluton
　　　　2. Sill, a tabular, concordant pluton (e.g., Palisades Sill, NY)
　　　　3. Laccolith
　　　　　　a. Forms in same way as sill
　　　　　　b. Lens shaped mass
　　　　　　c. Arches overlying strata upward
　　　　4. Batholith
　　　　　　a. Largest intrusive body
　　　　　　b. Surface exposure 100+ square kilometers (smaller bodies are termed stocks)
　　　　　　c. Frequently form the cores of mountains

VI. Igneous activity and plate tectonics
   A. Origin of magma
      1. Temperature and magma generation
         a. Magma must originate from solid rock
         b. Temperature melts solid rock
         c. Temperature increases with depth (one possible heat source is radioactive decay)
      2. Role of pressure
         a. Increase in pressure causes an increase in melting temperature
         b. Drop in confining pressure
            1. Lowers the melting temperature
            2. Occurs when rock ascends
      3. Partial melting
         a. Igneous rocks are mixtures of minerals
         b. Melting over a range of temperatures
         c. Forms a melt with a higher silica content
   B. Distribution of igneous activity
      1. Igneous activity along plate margins
         a. Oceanic ridge spreading center
            1. Lithosphere pulls apart
            2. Less pressure on underlying rocks
            3. Partial melting occurs
            4. Large quantities of basaltic magma are produced
         b. Convergent plate margin
            1. Subduction zone (trench)
            2. Descending plate partially melts
            3. Magma slowly rises upward
            4. Rising magma can form
               a. Island arc in an ocean
               b. Andesitic-granitic volcanoes on a continent
            5. Associated with the Pacific Basin
               a. Called "Ring of Fire"
               b. Explosive—high gas volcanoes
      2. Intraplate volcanism
         a. Activity within a rigid plate
         b. Basaltic magma source
            1. Partial melting of mantle rock
            2. Plumes of hot mantle material
               a. Form hot spots on the surface
               b. A plume may be located below Hawaii
            c. Granitic magma source when the continental crust is remelted over a mantle plume

---

Key Terms

Page numbers shown in ( ) refer to the textbook page where the term first appears.

aa flow (p. 217)	fissure eruption (p. 227)	pyroclastic flow (p. 227)
batholith (p. 229)	flood basalt (p. 227)	pyroclastics (p. 217)
caldera (p. 218)	hot spot (p. 233)	shield volcano (p. 219)
cinder cone (p. 221)	laccolith (p. 229)	sill (p. 228)
composite cone (stratovolcano) (p. 22)	lahar (p. 224)	vent (p. 218)
crater (p. 218)	nuée ardente (p. 223)	viscosity (p. 215)
dike (p. 228)	pahoehoe flow (p. 217)	volcanic neck (p. 226)
	partial melting (p. 232)	volcano (p. 218)

Vocabulary Review

Choosing from the list of key terms, furnish the most appropriate response for the following statements.

1. Particles of pulverized rock, lava, and glass fragments blown from the vent of a volcano are referred to as _____.

2. A(n) _____ is an unusually large volcanic summit depression that exceeds one kilometer in diameter.

3. A lava flow that has a surface of rough, jagged blocks is called a(n) _____.

4. Ship Rock, New Mexico, a feature produced by erosion, is an example of a(n) _____.

5. A(n) _____ is a sheetlike body that is produced when magma is injected into a fracture that cuts across rock layers.

6. The crater of a volcano is connected to a magma chamber via a pipelike conduit called a(n) _____.

7. A(n) _____ is a rather small volcano with steep slopes built from ejected lava fragments.

8. By far the largest intrusive igneous body is a(n) _____.

9. A volcano that takes the shape of a broad, slightly domed structure is called a(n) _____.

10. A flow of fluid basaltic lava that issues from cracks or fissures and commonly covers an extensive area to a thickness of hundreds of meters is called a(n) _____.

11. A lava flow with a smooth-to-ropy appearance that is produced from fluid basaltic lava is called a(n) _____.

12. A hot plume, which may extend to Earth's core-mantle boundary, produces a(n) _____, a volcanic region a few hundred kilometers across.

13. A(n) _____ is a large, nearly symmetrical volcano built of interbedded strata of lavas and pyroclastic deposits.

14. A(n) _____ is a mountainous accumulation of material formed by successive eruptions from a central vent.

15. A(n) _____ is an igneous intrusive feature that forms from a lens-shaped mass of magma that arches the overlying strata upward.

16. An eruption in which volcanic material is extruded from a long, narrow crack in the crust is known as a(n) _____.

17. A(n) _____ is a tabular pluton formed when magma is injected along sedimentary bedding surfaces.

18. Hot gases infused with incandescent ash ejected from a volcano produce a fiery cloud, called a(n) _____, that flows down steep slopes at high speed.

19. Most igneous rocks melt over a temperature range of a few hundred degrees, a process known as _____, which produces most, if not all, magma.

20. A material's _____ is a measure of its resistance to flow.

21. A type of mudflow associated with violent eruptions is known as a(n) _____.

Comprehensive Review

1. List three factors that determine whether a volcano extrudes magma violently or "gently."

    1)

    2)

    3)

2. Using Figure 8.1, select the letter of the diagram that illustrates each of the following types of volcanoes. Name a volcano that is an example of each type. Also describe the eruptive pattern most commonly associated with each type.

    a) Shield volcano: ____ (Example:)

    b) Cinder cone: ____ (Example:)

    c) Composite cone (stratovolcano): ____ (Example:)

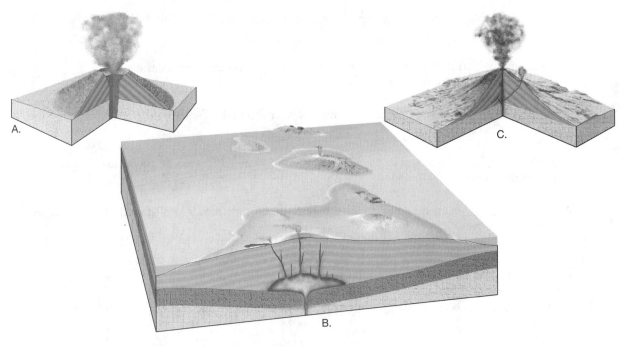

Figure 8.1

3. What are the major gases released during a volcanic eruption?

4. Describe the meaning of the following terms as they are related to igneous plutons.

    a) Discordant:

    b) Concordant:

5. What factors affect the viscosity of magma?

6. Using Figure 8.2, select the letter that illustrates each of the following igneous intrusive features.

    a) Sill: ____   c) Laccolith: ____

    b) Batholith: ____   d) Dike: ____

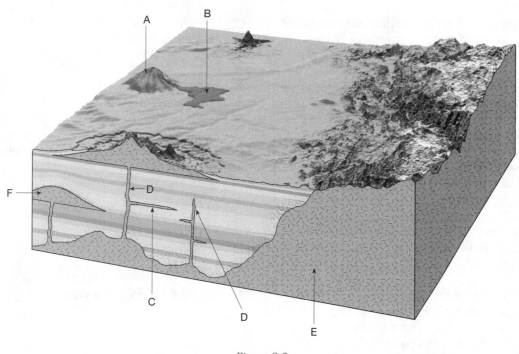

Figure 8.2

7. Briefly comment on the role of each of the following in the origin of magma.

    a) Temperature:

    b) Pressure:

    c) Partial melting:

8. Compare basaltic magma to granitic magma in terms of

    a) Silica content:

    b) Viscosity:

    c) Melting temperature:

9. List the three major zones of volcanic activity and relate each to global tectonics.

1)

2)

3)

Practice Test

Multiple choice. Choose the best answer for the following multiple choice questions.

1. Underground igneous rock bodies are called _____.
   a) aquifers   c) playas   e) placers
   b) plutons   d) pluvials

2. The greatest volume of volcanic material is produced by _____.
   a) cinder cones   c) laccoliths   e) explosive eruptions
   b) fissure eruptions d) shield cones

3. Highly viscous magmas tend to impede the upward migration of expanding gases, which often results in _____ eruptions.
   a) relatively quiet
   b) explosive

4. The most violent type of volcanic activity is associated with _____.
   a) cinder cones   c) composite cones   e) shield cones
   b) sills   d) intermediate cones

5. Which one of the following is NOT a factor that determines the violence of a volcanic eruption?
   a) temperature of the magma
   b) size of the volcanic cone
   c) the magma's composition
   d) amount of dissolved gases in the magma

6. The most abundant gas produced during Hawaiian eruptions is _____.
   a) nitrogen   c) oxygen   e) water vapor
   b) carbon dioxide   d) chlorine

7. The force that extrudes magma from a volcanic vent is provided by _____.
   a) dissolved gases   c) the magma's heat   e) discordant plutons
   b) gravity   d) the volcano's slope

8. The area of igneous activity commonly called the *Ring of Fire* surrounds the _____.
   a) Indian Ocean   c) Atlantic Ocean   e) Pacific Ocean
   b) Coral Sea   d) Sea of Japan

9. A magma's viscosity is directly related to its _____.
   a) depth   c) volcanic cone   e) color
   b) age   d) silica content

10. Pulverized rock, lava, and glass fragments produced from the vent of a volcano are known as _____.
    a) nuée ardentes   c) pyroclastics   e) pahoehoes
    b) sills   d) craters

11. Which type of volcano consists of interbedded strata of lavas and pyroclastic material?
    a) cinder cone   c) intermediate cone   e) pyro-cone
    b) composite cone   d) shield cone

12. Fluid basaltic lavas of the Hawaiian type commonly form _____.
    a) aa flows   c) pyroclastic flows   e) lapilli
    b) pahoehoe flows   d) nuée ardente

13. Basaltic lava tends to be _____ fluid than granitic lava.
    a) more
    b) less

14. When silica-rich magma is extruded, ash and pumice fragments may be propelled from the vent at high speeds and produce _____.
    a) flood basalts   c) batholiths   e) pyroclastic flows
    b) pahoehoe flows   d) a shield volcano

15. Unusually large volcanic summit depressions that exceed one kilometer in diameter are known as _____.
    a) calderas   c) craters   e) laccoliths
    b) vents   d) sills

16. Large particles of hardened lava ejected from a volcano are termed _____.
    a) bombs   c) welded tuff   e) blocks
    b) lapilli   d) cinders

17. Eruptions of fluid basaltic lavas, such as those that occur in Hawaii, tend to be _____.
    a) relatively quiet   c) extremely violent
    b) unpredictable   d) explosive

18. Hot gases infused with incandescent ash ejected from a volcano often produce a fiery cloud called a(n) _____.
    a) nuée ardente   c) lapilli   e) volcanic bomb
    b) laccolith   d) pyroclastic

19. The type of volcano produced almost entirely of pyroclastic material is the _____.
    a) shield volcano   c) composite cone   e) batholith
    b) cinder cone   d) pyro-cone

20. A(n) _____ is a tabular, concordant pluton.
    a) dike   c) stock   e) batholith
    b) laccolith   d) sill

21. Which type of volcanoes are generally small and occur in groups?
    a) composite cones   c) shield cones   e) intermediate cones
    b) laccoliths   d) cinder cones

22. In a near-surface environment, silica-rich rocks of granitic composition melt at a _____ temperature than basaltic rocks.
    a) higher
    b) lower

23. The largest intrusive igneous bodies are _____.
    a) dikes   c) batholiths   e) sills
    b) stocks   d) laccoliths

24. Intraplate volcanism may be associated with the formation of _____ over rising plumes of hot mantle material.
    a) hot spots   c) subduction zones   e) volcanic necks
    b) dikes   d) ocean ridges

25. In general, an increase in the confining pressure _____ a rock's melting temperature.
    a) increases   c) stabilizes
    b) decreases

*True/false. For the following true/false questions, if a statement is not completely true, mark it false. For each false statement, change the **italicized** word to correct the statement.*

1. ___ Most of Earth's active volcanoes are near *divergent* plate margins.

2. ___ The more viscous a material, the *greater* its resistance to flow.

3. ___ Plutons similar to but smaller than batholiths are termed *stocks*.

4. ___ Water vapor is the *most* abundant gas in magma.

5. ___ The smallest volcanoes are *composite* cones.

6. ___ The greatest volume of volcanic rock is produced along the oceanic *ridge* system.

7. ___ Magmas that produce basaltic rocks contain *more* silica than those that form granitic rocks.

8. ___ Reducing confining pressure *lowers* a rock's melting temperature.

9. ___ Located at the summit of most volcanoes is a steep-walled depression called a *sill*.

10. ___ An important consequence of partial melting is the production of a melt with a *higher* silica content than the parent rock.

11. ___ *Dikes* are tabular, discordant plutons.

12. ___ With increasing depth in Earth's interior, there is a gradual *decrease* in temperature.

13. ___ A magma's viscosity is directly related to its *iron* content.

14. ___ Most of the volcanoes of the Cascade Range in the northwestern United States are *shield* cones.

15. ___ When *subduction* volcanism occurs in the ocean, a chain of volcanoes called an island arc is produced.

16. ___ The large expanse of granitic rock exposed in the interior of North America is called the *Canadian* Shield.

17. ___ The viscosity of magma, plus the quantity of dissolved *gases* and the ease with which they can escape, determines the nature of volcanic eruptions.

18. ___ One of the best known *batholiths* in North America is along the Hudson River near New York.

19. ___ Secondary volcanic vents that emit only gases are called *fumaroles*.

20. ___ The Columbia River Plateau in the northwestern United States formed from very fluid *basaltic* lava that erupted from numerous fissures.

Word choice. Complete each of the following statements by selecting the most appropriate response.

1. [Increasing /Decreasing] the temperature of magma will increase its viscosity.

2. Most calderas form when the [side/summit] of a volcano collapses into a partially emptied [magma chamber/vent].

3. Most of the pungent odor associated with a volcano is from [carbon/nitrogen/sulfur] compounds.

4. The most explosive volcanoes are produced by [high/low] viscosity magma containing a [large/small] quantity of dissolved gases.

5. Partial melting of a rock usually results in a magma with a [higher/lower] silica content than the parent rock.

6. The greatest volume of volcanic material is produced along the [ocean trenches/oceanic ridge system].

7. The magma of spreading center volcanism is produced primarily by the partial melting of [continental crust/upper mantle rock/oceanic crust] and is mostly of [basaltic/granitic] composition.

8. A volcanoes effect on climate, if any, would be caused by ash and [lava/gases] being ejected into the [upper/lower] atmosphere.

9. In subduction-zone volcanism, rocks melt due to a(n) [increase/decrease] in [temperature/pressure].

10. [Increasing/Decreasing] the silica content of magma will increase its viscosity.

11. The smallest volcanic cones are typically [cinder/composite/shield] cones.

12. In spreading center volcanism, rocks melt due to a(n) [increase/decrease] in [temperature/pressure].

13. The most fluid magmas have a [basaltic/granitic] composition.

14. The most explosive volcanism is associated with [spreading center/subduction zone] volcanism due to the relatively high [crystal/basalt/water] content of the associated magma.

15. Rocks can melt without changing temperature when their confining pressure is [increased/reduced].

Written questions

1. What is the difference between magma and lava?

2. List the three main types of volcanoes and describe the shape of each.

3. List and describe three igneous intrusive features.

Mountain Building

9

Mountain Building begins with a brief examination of the processes of crustal uplifting, including the concepts of isostasy and isostatic adjustment and rock deformation. The various types of folds (anticlines, synclines, domes, and basins) and faults (both dip-slip and strike-slip) are investigated. Following an overview of the structural characteristics of the four main mountain types, the chapter concludes with an extensive presentation of mountain building and its association with plate boundaries.

Learning Objectives

After reading, studying, and discussing this chapter, you should be able to:

- Describe the process of isostasy and the role of isostatic adjustment during crustal uplifting.
- Explain the difference between elastic and plastic rock deformation.
- List the major types of folds and faults and describe how they form.
- Describe the four main mountain types and give examples of each.
- Describe the relation between mountain building and plate tectonics.

Chapter Summary

- The name for the processes that collectively produce a mountain system is *orogenesis*. Earth's less dense crust is believed to float on top of the denser rocks in the mantle, much like wooden blocks floating in water. This concept of a floating crust in gravitational balance is called *isostasy*. As erosion lowers the peaks of mountains, *isostatic adjustment* gradually raises the mountains in response.

- Rocks that are subjected to stresses greater than their own strength begin to deform, usually by *folding* or *fracturing*. During *elastic deformation* a rock behaves like a rubber band and will return to nearly its original size and shape when the stress is removed. Once the *elastic limit* is surpassed, rocks either deform plastically or fracture. *Plastic deformation* results in permanent changes in the size and shape of a rock through folding and flowing.

- The most basic geologic structures associated with rock deformation are *folds* (flat-lying sedimentary and volcanic rocks bent into a series of wavelike undulations) and *faults*. The two most common types of folds are *anticlines*, formed by the upfolding, or arching, of rock layers, and *synclines*, which are downfolds, or troughs. Most folds are the result of horizontal *compressional stresses*. Folds may be *symmetrical*, *asymmetrical*, or, if one limb has been tilted beyond the vertical, *overturned*. *Domes* (upwarped structures) and *basins* (downwarped structures) are circular or somewhat elongated folds formed from vertical displacements of rocks.

- Faults are fractures in the crust along which appreciable displacement has occurred. Faults in which the movement is primarily vertical are called *dip-slip faults*. Dip-slip faults include both *normal* and *reverse faults.* Low angle reverse faults are also called *thrust faults.* In *strike-slip faults,* horizontal movement causes displacement along the trend, or *strike,* of the fault. Normal faults indicate *tensional stresses* that pull the crust apart. Along the spreading centers of plates, divergence can cause a central block called a *graben,* bounded by normal faults, to drop as the plates separate. Reverse and thrust faulting indicate that *compressional forces* are at work. A *joint* is a fracture along which no appreciable displacement has occurred.

• Mountains can be classified according to their structural characteristics. The four main mountain types are 1) *fault block mountains* (e.g., Basin and Range Province and Sierra Nevada of California), associated with tensional stresses and at least on one side by normal faults, 2) *folded mountains* (e.g., Alps, Urals, Himalayas, Appalachians), the most complex which form most of the major mountain belts, 3) *upwarped mountains* (e.g., Black Hills, Adirondack Mountains), caused by broad arching, and 4) *volcanic mountains.* Some regions, such as plateaus deeply dissected by erosion, exhibit mountainous topography without appreciable crustal deformation.

• *Major mountain systems form along convergent plate boundaries. Andean-type mountain building* along continental margins involves the convergence of an oceanic plate and a plate whose leading edge contains continental crust. At some point in the formation of Andean-type mountains a subduction zone forms along with a volcanic arc. Sediment from the land as well as that scraped from the subducting plate becomes plastered against the landward side of the trench, forming what is called an *accretionary wedge.* The best example of an Andean-type mountain belt is found in the western United States, including the Sierra Nevada and the Coast Ranges in California. *Continental collisions,* in which both plates may be carrying continental crust, have resulted in the formation of the Himalaya Mountains and Tibetan Highlands. Recent investigations indicate that *accretion,* a third mechanism of orogenesis, takes place where *smaller crustal fragments collide and merge with continental margins* along some plate boundaries. Many of the mountainous regions rimming the Pacific have formed in this manner. The accreted crustal blocks are referred to as *terranes.*

Chapter Outline

I. Crustal uplift
  A. Evidence for crustal uplift
    1. Marine fossils at high elevations in mountains
    2. Wave-cut platforms high above sea level
  B. Reasons for crustal uplift
    1. Not so easy to determine
    2. Floating crust in gravitational balance
      a. Called isostasy
      b. When weight is removed from the crust, crustal uplifting occurs
        1. Called isostatic adjustment
        2. Considerable vertical crustal movement can occur
II. Rock deformation
  A. Rocks subjected to stresses greater than their own strength will deform by either
    1. Folding, or
    2. Fracturing
  B. Types of rock deformation

    1. Elastic deformation—where the rock returns to nearly its original size and shape when the stress is removed
    2. Plastic deformation
      a. Once elastic limit is surpassed, rocks either deform plastically or fracture
      b. Size and shape of rock are permanently altered through folding or flowing
  C. Folds
    1. Rocks bent into a series of waves
    2. Most folds result from compressional stresses which shorten and thicken the crust
    3. Two common types of folds
      a. Anticline—upfolded, or arched, rock layers
      b. Syncline—downfolded, or trough, rock layers

4. Anticlines and synclines can be
  a. Symmetrical
  b. Asymmetrical
  c. Overturned
5. Where folds die out they are said to be plunging
6. Other types of folds
  a. Dome
    1. Circular, or slightly elongated
    2. Upwarped displacement of rocks
    3. Oldest rocks in core
  b. Basin
    1. Circular, or slightly elongated
    2. Downwarped displacement of rocks
    3. Youngest rocks in core

D. Faults
1. Faults are fractures (breaks) in rocks along which appreciable displacement has occurred
2. Types of faults
  a. Dip-slip fault
    1. Movement along the inclination (dip) of fault plane
    2. Parts of a dip-slip fault
      a. Hanging wall—the rock above the fault surface
      b. Foot-wall—the rock below the fault surface
    3. Types of dip-slip faults
      a. Normal fault
        1. Tensional force—pulls rock apart
        2. Hanging wall moves down
      b. Reverse fault
        1. Compressional force—squeezes rock
        2. Hanging wall moves up
      c. Thrust fault
        1. Strong compressional force
        2. Hanging wall moves up
        3. Low-angle reverse fault
        4. More horizontal
  b. Strike-slip fault
    1. Movement along trend, or strike
    2. Transform fault
      a. Large strike-slip fault
      b. Often associated with plate boundaries
3. Pulling of crust apart (tension) can form

  a. Graben
    1. Down-dropped block
    2. Bounded by normal faults
    3. Can produce an elongated valley (e.g., Great Rift Valley, Africa)
  b. Horst
    1. Flanking a graben
    2. Relatively uplifted block

E. Joints
1. Fractures along which no appreciable displacement has occurred—has no movement
2. Most are associated with stresses during mountain building
3. Other types
  a. Columnar joints caused by igneous rock shrinking and producing pillar like columns
  b. Sheeting caused by curved joints in large igneous bodies like batholiths

III. Mountain types
A. Mountains are classified according to their most dominant characteristics
B. Four main categories
1. Fault-block mountains
  a. Produced by tensional stresses that stretch the crust
  b. Mountains bounded at least on one side by normal faults of high-to-moderate angle
  c. e.g., Basin and Range Province, Teton Range, and Sierra Nevada
2. Folded mountains
  a. Largest—most complex mountains
  b. e.g., Alps, Urals, Himalayas, Appalachians, and northern Rockies
3. Upwarped mountains
  a. Broad arching of crust
  b. e.g., Black Hills, Adirondacks
4. Volcanic mountains

IV. Mountain building
A. Refers to processes that uplift mountains
B. Major mountain systems are related to plate tectonics where most mountains form along convergent plates
C. Mountains are associated with
1. Convergent boundaries
  a. Oceanic-continental crust convergence
    1. Andean-type mountains
    2. Subduction zone forms

3. Deformation of continental
   margin
4. Volcanic arc forms
5. Accretionary wedge
6. Examples
   a. Sierra Nevada Range
   b. California's Coast Ranges
b. Where continental crusts converge
   1. e.g., India and Eurasian plate
      collision
      a. Himalaya Mountains
      b. Tibetan Highlands

2. Other examples
   a. Alps
   b. Appalachians
2. Continental accretion
   a. Third mechanism of mountain
      building
   b. Small crustal fragments collide
      with and accrete to continental
      margins
   c. Accreted crustal blocks are called
      terranes
   d. Occurred along the Pacific Coast

Key Terms

Page numbers shown in ( ) refer to the textbook page where the term first appears.

accretionary wedge (p. 255)
anticline (p. 242)
basin (p. 243)
dip-slip fault (p. 244)
dome (p. 243)
fault (p. 244)
fault-block mountains
   (p. 248)
fold (p. 242)

folded mountains (complex
   mountains) (p. 249)
graben (p. 244)
horst (p. 244)
isostasy (p. 239)
isostatic adjustment
   (p. 240)
joint (p. 245)
mountain building (p. 252)

normal fault (p. 244)
oblique-slip fault (p. 244)
orogenesis (p. 238)
reverse fault (p. 244)
strike-slip fault (p. 244)
syncline (p. 242)
terrane (p. 257)
thrust fault (p. 244)
upwarped mountains (p. 252)

Vocabulary Review

Choosing from the list of key terms, furnish the most appropriate response for the following statements.

1. A fault in which the dominant displacement is along the trend of the fault is called a(n)
   _____.

2. A(n) _____ is an elongate, uplifted block of crust bounded by faults.

3. _____ is the concept that Earth's crust is "floating" in gravitational balance upon the material of the mantle.

4. A wavelike layer of rock that was originally horizontal and subsequently deformed is a(n)
   _____.

5. A(n) _____ is a circular or somewhat elongated downfolded structure.

6. _____ is the name for the processes that collectively produce a mountain system.

7. A fracture in rock along which displacement has occurred is termed a(n) _____.

8. A reverse fault having a very low angle is also called a(n) _____.

9. A(n) _____ is a valley formed by the downward displacement of a fault-bounded block.

10. The largest and most complex mountain systems are _____.

11. A(n) _____ is the type of fold most commonly formed by the upfolding, or arching, of rock layers.

12. A downfold, or trough, of rock layers forms a type of fold called a(n) _____.

13. A(n) _____ is a type of fault with both vertical and horizontal movement.

14. Mountains formed by displacement of rock along a fault are referred to as _____.

15. A fault where the primary movement is vertical is called a(n) _____.

16. Mountains that are caused by a broad arching of the crust are called _____.

17. A dip-slip fault is classified as a(n) _____ when the hanging wall moves down relative to the footwall.

18. When weight is removed from the crust and uplifting occurs the response is called a(n) _____.

19. A(n) _____ is a fracture in rock along which no appreciable displacement has occurred.

20. An accumulation of sedimentary and metamorphic rocks, including occasional scraps of ocean crust, that forms in association with a subduction zone is called a(n) _____.

21. A crustal block whose geologic history is distinct from the histories of adjoining crustal blocks is termed a(n) _____.

22. A circular or somewhat elongated upfolded structure is called a(n) _____.

23. A(n) _____ is the type of fault that is produced when the hanging wall moves upward relative to the foot-wall.

Comprehensive Review

1. What evidence supports the fact that sedimentary rocks found at high elevations in mountains were once below sea level?

2. Why is the crust beneath the oceans at a lower elevation than the continental crust?

3. Describe the concept of isostasy.

4. Write a brief paragraph explaining the difference between elastic and plastic rock deformation.

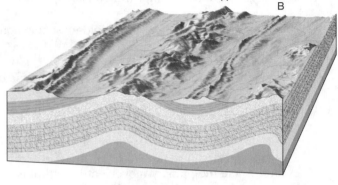

Figure 9.1

5. Briefly describe each of the following types of folds and select the letter of the diagram in Figure 9.1 that illustrates it.

    a) Anticline: _____

    b) Syncline: _____

6. Write the name of the geologic structure illustrated by each diagram in Figure 9.2 below the appropriate illustration.

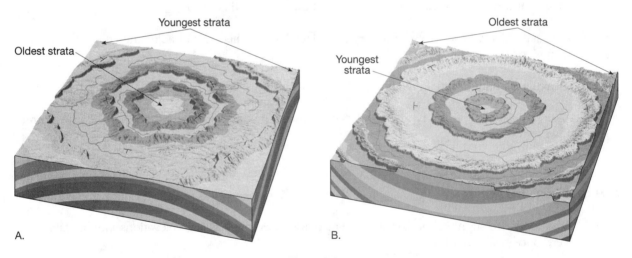

Figure 9.2

7. Briefly describe each of the following types of faults and select the letter of the diagram in Figure 9.3 that illustrates it.

    a) Reverse fault: _____

    b) Strike-slip fault: _____

    c) Normal fault: _____

    d) Thrust fault: _____

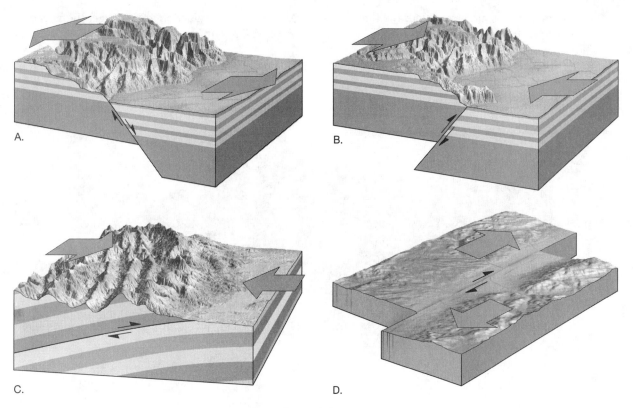

Figure 9.3

8. Using Figure 9.4, select the letter that illustrates each of the following.

    a) Graben: ____

    b) Horst: ____

9. Which type of fault are those illustrated in Figure 9.4?

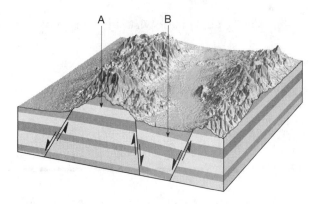

Figure 9.4

10. What are the two types of faults that are most often produced by compressional forces?

11. Briefly explain the events that produce mountains at convergent boundaries where

    a) Oceanic and continental crusts converge:

    b) Continental crusts converge:

12. Figure 9.5 illustrates mountain building along an Andean-type subduction zone. Select the appropriate letter in the figure that identifies each of the following features.

   a) Ocean trench: ____

   d) Accretionary wedge: ____

   b) Asthenosphere: ____

   e) Subducting oceanic lithosphere: ____

   c) Volcanic arc: ____

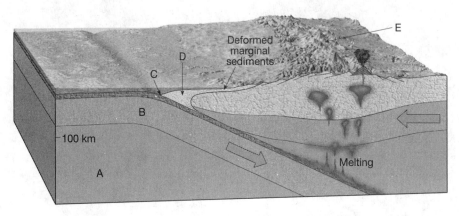

Figure 9.5

13. What are two mountain systems that have formed as the result of continental collision?

14. What is an oil trap? What are the two basic conditions that all oil traps have in common?

15. Why is the San Andreas fault more appropriately referred to as the San Andreas fault system?

Practice Test

Multiple choice. Choose the best answer for the following multiple choice questions.

1. Which one of the following is NOT a form of rock deformation?
   a) elastic deformation   c) plastic deformation
   b) fracturing   d) erosion

2. The thickest part of the crust occurs in _____.
   a) the asthenosphere   d) eastern Canada
   b) the ocean basin   e) old eroded mountains
   c) young mountain ranges

3. In the Gulf coastal plain region of the United States important accumulations of oil occur in association with geologic structures called _____.
   a) normal faults   c) hot spots   e) salt domes
   b) parallel joints   d) ocean trenches

4. The two most common types of folds are anticlines and _____.
   - a) domes
   - b) synclines
   - c) basins
   - d) superclines
   - e) batholiths

5. Compared to the elevation of a thin piece of continental crust, the highest elevation of a thick piece in isostatic balance will be _____.
   - a) higher
   - b) the same
   - c) lower
   - d) overturned
   - e) older

6. Orogenesis refers to those processes that collectively produce _____.
   - a) a mountain system
   - b) earthquakes
   - c) oceanic plates
   - d) trenches
   - e) subduction zones

7. A fracture with horizontal displacement parallel to its surface trend is called a(n) _____ fault.
   - a) dip-slip
   - b) joint
   - c) strike-slip
   - d) oblique-slip
   - e) overturned

8. The rock immediately above a fault surface is commonly called the _____.
   - a) anticline
   - b) foot-wall
   - c) syncline
   - d) hanging wall
   - e) dip-slip

9. The removal of material by erosion will cause the crust to _____.
   - a) subduct
   - b) rise
   - c) fold
   - d) subside
   - e) thicken

10. Folding is usually the result of _____.
    - a) tensional forces
    - b) shear forces
    - c) faulting
    - d) jointing
    - e) compressional forces

11. As heat and pressure increase, plastic deformation _____.
    - a) stops occurring
    - b) becomes less likely
    - c) is replaced by elastic deformation
    - d) can cause erosion
    - e) becomes more likely

12. Which one of the following pairs is properly matched?
    - a) normal fault-tensional forces
    - b) thrust fault-tensional forces
    - c) reverse fault-tensional forces
    - d) normal fault-compressional forces

13. The total accumulated displacement from earthquakes and creep along the San Andreas fault system is approximately _____ kilometers.
    - a) 120
    - b) 255
    - c) 380
    - d) 415
    - e) 560

14. Where two oceanic plates converge, _____ subduction zones often occur.
    - a) Andean-type
    - b) Himalayan-type
    - c) American-type
    - d) Aleutian-type
    - e) Arctic-type

15. Which one of the following features is formed by crustal upwarping?
    - a) syncline
    - b) basin
    - c) anticline
    - d) graben
    - e) accretionary wedge

16. The most important difference between faults and joints is that joints _____.
    - a) occur along folds
    - b) are often parallel
    - c) have no displacement
    - d) are usually vertical
    - e) are very rare

17. The period of deformation known as the Laramide Orogeny that produced a portion of the Rocky Mountains peaked about _____ million years ago.
    a) 10   c) 40   e) 80
    b) 20   d) 60

18. It is assumed that many of the terranes found in the North American Cordillera were once crustal fragments scattered throughout the eastern _____ ocean basin.
    a) Pacific   b) Atlantic   c) Indian

19. The San Andreas fault system is a well-known example of a(n) _____ fault.
    a) normal   c) thrust   e) overturned
    b) transform   d) reverse

20. Faults where the movement is primarily vertical are called _____ faults.
    a) transform   c) dip-slip   e) strike-slip
    b) oblique   d) random

21. The collision and joining of crustal fragments to a continent is called continental _____.
    a) subduction   c) accretion   e) aggradation
    b) isostasy   d) suturing

22. Where oceanic crust is being thrust beneath a continental mass, _____ subduction zones often occur.
    a) Andean-type   c) American-type   e) Arctic-type
    b) Himalaya-type   d) Aleutian-type

23. Dip-slip faults are classified as _____ faults when the hanging wall moves up relative to the footwall.
    a) normal   c) reverse   e) strike-slip
    b) transform   d) tensional

24. The _____ Mountains are primarily a volcanic arc produced by a subducting plate.
    a) Himalaya   c) Adirondacks   e) Appalachian
    b) Andes   d) Ural

25. Which one of the following mountain ranges has formed where continental crusts have converged?
    a) Sierra Nevada   c) Himalaya Mountains
    b) Andes Mountains   d) Coast Ranges

*True/false. For the following true/false questions, if a statement is not completely true, mark it false. For each false statement, change the **italicized** word to correct the statement.*

1. ___ As erosion lowers the peaks of mountains, isostatic adjustment gradually *lowers* the mountains in response.

2. ___ At high temperatures and pressures, most rocks deform *plastically* once their elastic limit is surpassed.

3. ___ The first encompassing explanation of orogenesis came as part of the *plate-tectonics* theory.

4. ___ *Joints* are fractures in rock along which appreciable displacement has occurred.

5. ___ The rock in a fault that is higher than the fault surface is referred to as the *hanging* wall.

6. ___ Strike-slip faults that are associated with plate boundaries are called *transform* faults.

7. ___ Under surface conditions, rocks that exceed their elastic limit behave like a brittle solid and *fracture.*

8. ___ Most mountain building occurs in *tensional* environments.

9. ___ The crustal thickness for some mountain chains is greater than *twice* the average thickness of the continental crust.

10. ___ Where the axis of an anticline descends to the ground, the fold is said to be *plunging.*

11. ___ In large basins that contain sedimentary rock sloping at low angles, the *oldest* rocks are found near the center of the structure.

12. ___ When stress is applied, rocks first respond by deforming *plastically.*

13. ___ Faults in which the movement is primarily *horizontal* are called dip-slip faults.

14. ___ In a plunging *anticline*, the outcrop pattern "points" in the direction of the plunge.

15. ___ Wavelike undulations of sedimentary and volcanic rocks are called *grabens.*

16. ___ *Aleutian-type* subduction zones occur where two oceanic plates converge.

17. ___ In a normal fault, the hanging wall moves *downward* relative to the footwall.

18. ___ Most folds result from *compressional* stresses in the crust.

19. ___ Most major mountain belts are *fault-block* mountains.

20. ___ The two most common types of folds are anticlines and *haloclines.*

21. ___ In a reverse fault, the foot-wall moves *upward* relative to the hanging wall.

22. ___ The *Appalachians* resulted from collisions between North America, Europe, and northern Africa.

23. ___ *Normal* faulting is often the type that occurs at spreading centers, where plates are diverging.

24. ___ Most major episodes of mountain building have occurred along *divergent* plate boundaries.

25. ___ *Fault-block* mountains are common in the Basin and Range Province of the southwestern United States.

26. ___ *Terrane* refers to any crustal fragment whose geologic history is distinct from that of adjoining fragments.

27. ___ The *cap* rock of an oil trap is both porous and permeable.

Word choice. Complete each of the following statements by selecting the most appropriate response.

1. When a rock's shape and size are permanently altered by folding and flowing, the rock has under-gone [elastic/orogenetic/plastic] deformation.

2. Forces that pull apart Earth's crust are called [compressional/tensional] stresses.

3. Oil traps consist of a(n) [permeable/impermeable] cap rock located [above/below] a(n) [permeable/impermeable] reservoir rock.

4. The concept of a floating crust in gravitational balance is called [orogenesis/isostasy].

5. A fault with primarily vertical displacement is called a(n) [dip-slip/strike-slip/oblique-slip] fault.

6. Reverse faults are usually produced by [tensional/compressional] forces.

7. Changes in elevation caused by the addition or removal of weight from Earth's crust are called [isostatic/volcanic/orogenetic] adjustments.

8. Fault-block mountains are associated with [compressional/tensional] forces.

9. Volcanic arcs are associated with [diverging/subducting] tectonic plates.

10. Troughs or downfolds in rock are called [anticlines/synclines].

11. Most mountain building is associated with the [margins/central regions] of tectonic plates.

12. Grabens and horsts are associated with regions of tectonic plate [convergence/divergence].

13. Continental crust is [more/less] dense than oceanic crust.

14. Over time, erosion and isostatic adjustments will [bury/expose] a mountain's core.

15. The central regions of upwarped mountains are usually composed of metamorphic and [sedimentary/igneous] rocks.

16. The accretion of terranes [is/is not] associated with the development of the west coast of North America.

Written questions

1. Why don't anticlines always appear as hills, even though the rocks beneath the surface are folded upward?

2. Briefly describe the composition and formation of an accretionary wedge.

3. Describe the process responsible for the formation of Earth's major mountain systems.

Geologic Time

Geologic Time opens with a brief history of geology that spans from James Ussher (mid-1600s) to James Hutton (late-1700s) and Sir Charles Lyell (mid-1800s). The chapter continues with a discussion of the fundamental principles of relative dating, including the law of superposition, principle of original horizontality, principle of cross-cutting relationships, and the uses of inclusions and unconformities. How rock units in different localities can be correlated is also investigated. The types of fossils and their significance to understanding geologic time precedes a discussion of the conditions favoring preservation. Also examined is the use of fossils in correlating and dating rock units. Following an explanation of radioactivity, the fundamentals and importance of radiometric dating are presented. The chapter concludes with an examination of the geologic time scale.

Learning Objectives

After reading, studying, and discussing this chapter, you should be able to:

- Describe the doctrine of uniformitarianism.
- Explain the difference between absolute and relative dating.
- List the laws and principles used in relative dating.
- Discuss unconformities.
- Explain correlation of rock layers.
- Describe fossils, fossilization, and the uses of fossils.
- Explain radioactivity and radiometric dating.
- Describe the geologic time scale.

Chapter Summary

- The *doctrine of uniformitarianism*, one of the fundamental principles of modern geology put forth by *James Hutton* in his published work *Theory of the Earth* in the late 1700s, states that the physical, chemical, and biological laws that operate today have also operated in the geologic past. The idea is often summarized as "the present is the key to the past." Hutton argued that processes that appear to be slow-acting could, over long spans of time, produce effects that were just as great as those resulting from sudden catastrophic events. *Catastrophism*, on the other hand, states that Earth's landscapes have been developed primarily by great catastrophes. *Sir Charles Lyell* (mid-1800s) is given the most credit for advancing the basic principles of modern geology with the publication of the eleven editions of his great work, *Principles of Geology*.

- The two types of dates used by geologists to interpret Earth history are 1) *relative dates*, which put events in their *proper sequence of formation*, and 2) *absolute dates*, which pinpoint the *time in years* when an event took place.

• Relative dates can be established using the *law of superposition, principle of original horizontality, principle of cross-cutting relationships, inclusions,* and *unconformities.*

• *Correlation,* the matching up of two or more geologic phenomena in different areas, is used to develop a geologic time scale that applies to the whole Earth.

• *Fossils* are the remains or traces of prehistoric life. Some fossils form by *petrification, replacement,* or *carbonization.* Others are *impressions, casts,* or have been *preserved in amber.* Fossils can also be indirect evidence of prehistoric life such as *tracks* or *burrows.* The special conditions that favor preservation are *rapid burial* and the possession of *hard parts* such as shells, bones, or teeth.

• Fossils are used to *correlate* sedimentary rocks that are from different regions by using the rocks' distinctive fossil content and applying the *principle of fossil succession.* The principle of fossil succession, which is based on the work of *William Smith* in the late 1700s, states that fossil organisms succeed one another in a definite and determinable order, and therefore any time period can be recognized by its fossil content. The use of *index fossils,* those that are widespread geographically and are limited to a short span of geologic time, provides an important method for matching rocks of the same age.

• Each atom has a nucleus containing *protons* (positively charged particles) and *neutrons* (neutral particles). Orbiting the nucleus are negatively charged *electrons.* The *atomic number* of an atom is the number of protons in the nucleus. The *mass number* is the number of protons plus the number of neutrons in an atom's nucleus. *Isotopes* are variants of the same atom, but with a different number of neutrons, and hence a different mass number.

• *Radioactivity* is the spontaneous breaking apart (decay) of certain unstable atomic nuclei. Three common forms of radioactive decay are 1) emission of alpha particles from the nucleus, 2) emission of a beta particle (or electron) from the nucleus, and 3) capture of an electron by the nucleus.

• An unstable *radioactive isotope,* called the *parent,* will decay and form *daughter products.* The length of time for one-half of the nuclei of a radioactive isotope to decay is called the *half-life* of the isotope. If the half-life of the isotope is known, and the parent/daughter ratio can be measured, the age of a sample can be calculated. *Carbon-14,* the radioactive isotope of carbon that is absorbed by living matter, is used to date very recent events. Although the procedure is quite complex, radiometric dating has vindicated the ideas of Hutton, Darwin, and others who have inferred that geologic time must be immense.

• The *geologic time scale* divides Earth's history into units of varying magnitude. It is commonly presented in chart form, with the oldest time and event at the bottom and the youngest at the top. The principal subdivisions of the geologic time scale, called *eons,* include the *Hadean, Archean, Proterozoic* (together, these three eons are commonly referred to as the *Precambrian*), and, beginning about 570 million years ago, the *Phanerozoic.* The Phanerozoic (meaning "visible life") eon is divided into the following *eras: Paleozoic* ("ancient life"), *Mesozoic* ("middle life"), and *Cenozoic* ("recent life").

• The primary problem in assigning absolute dates to units of time is that *not all rocks can be dated radiometrically.* A sedimentary rock may contain particles of many ages that have been weathered from different rocks that formed at various times. One way geologists assign absolute dates to sedimentary rocks is to relate them to datable igneous masses, such as volcanic ash beds.

Chapter Outline

I. Historical notes
  A. Catastrophism
    1. Landscape developed by catastrophes
    2. James Ussher, mid-1600s, concluded Earth was only a few thousand years old
  B. Modern geology
    1. Uniformitarianism
      a. Fundamental principle of geology
      b. "The present is the key to the past"
    2. James Hutton
      a. *Theory of the Earth*
      b. Late 1700s
    3. Sir Charles Lyell
      a. Advanced modern geology
      b. *Principles of Geology*
      c. Mid-1800s
II. Relative dating
  A. Placing rocks and events in sequence
  B. Principles and rules of
    1. Law of superposition—oldest rocks are on the bottom
    2. Principle of original horizontality—sediment is deposited horizontally
    3. Principle of cross-cutting relationships—younger feature cuts through an older feature
    4. Inclusions—one rock contained within another—rock containing the inclusions is younger
    5. Unconformities
      a. An unconformity is a break in the rock record
      b. Types of unconformities
        1. Angular unconformity—tilted rocks are overlain by flat-lying rocks
        2. Disconformity—strata on either side are parallel
        3. Nonconformity
          a. Metamorphic or igneous rocks below
          b. Younger sedimentary rocks above
III. Correlation of rock layers
  A. Matching rocks of similar age in different regions
  B. Often relies upon fossils
IV. Fossils
  A. Remains or traces of prehistoric life

B. Types of fossils
    1. Petrified—cavities and pores are filled with precipitated mineral matter
    2. Formed by replacement—cell material is removed and replaced with mineral matter
    3. Mold—shell or other structure is buried and then dissolved by underground water
    4. Cast—hollow space of a mold is filled with mineral matter
    5. Carbonization—organic matter becomes a thin residue of carbon
    6. Impression—replica of the fossil's surface preserved in fine-grained sediment
    7. Preservation in amber—hardened resin of ancient trees surrounds an organism
    8. Indirect evidence includes
      a. Tracks
      b. Burrows
      c. Coprolites—fossil dung and stomach contents
      d. Gastroliths—stomach stones used to grind food by some extinct reptiles
  C. Conditions favoring preservation
    1. Rapid burial
    2. Possession of hard parts
  D. Fossils and correlation
    1. Principle of fossil succession
      a. Fossils succeed one another in a definite and determinable order
      b. Proposed by William Smith—late 1700s and early 1800s
    2. Index fossils
      a. Widespread geographically
      b. Existed for a short range of geologic time
V. Radioactivity and radiometric dating
  A. Atom
    1. Nucleus
      a. Protons—positively charged
      b. Neutrons
        1. Neutral charge
        2. Protons and electrons combined
    2. Orbiting the nucleus are electrons—negative electrical charges
    3. Atomic number
      a. An element's identifying number

b. Number of protons in the atom's nucleus

4. Mass number
   a. Number of protons plus (added to) the number of neutrons in an atom's nucleus
   b. Isotope
      1. Variant of the same parent atom
      2. Different number of neutrons
      3. Different mass number than the parent atom

B. Radioactivity
   1. Spontaneous breaking apart (decay) of atomic nuclei
   2. Radioactive decay
      a. Parent—an unstable isotope
      b. Daughter products—isotopes formed from the decay of a parent
      c. Types of radioactive decay
         1. Alpha emission
         2. Beta emission
         3. Electron capture

C. Radiometric dating
   1. Half-life—the time for one-half of the radioactive nuclei to decay
   2. Requires a closed system
   3. Cross-checks are used for accuracy
   4. Complex procedure
   5. Yields absolute dates

D. Carbon-14 dating
   1. Half-life of only 5730 years
   2. Used to date very recent events
   3. Carbon-14 produced in upper atmosphere
      a. Incorporated into carbon dioxide
      b. Absorbed by living matter
   4. Useful tool for anthropologists, archeologists, historians, and geologists who study very recent Earth history

E. Importance of radiometric dating
   1. Radiometric dating is a complex procedure that requires precise measurement
   2. Rocks from several localities have been dated at more than 3 billion years
   3. Confirms the idea that geologic time is immense

VI. Geologic time scale
   A. Divides geologic history into units
   B. Originally created using relative dates
   C. Subdivisions
      1. Eon
         a. Greatest expanse of time
         b. Names
            1. Phanerozoic ("visible life")—the most recent eon
            2. Proterozoic
            3. Archean
            4. Hadean—the oldest eon
      2. Era
         a. Subdivision of an eon
         b. Eras of the Phanerozoic eon
            1. Cenozoic ("recent life")
            2. Mesozoic ("middle life")
            3. Paleozoic ("ancient life")
      3. Eras are subdivided into periods
      4. Periods are subdivided into epochs
   D. Difficulties in dating the time scale
      1. Not all rocks are datable (sedimentary ages are rarely reliable)
      2. Materials are often used to bracket events and arrive at ages

Key Terms

Page numbers shown in ( ) refer to the textbook page where the term first appears.

absolute date (p. 264)
angular unconformity (p. 267)
catastrophism (p. 263)
Cenozoic era (p. 279)
conformable (p. 266)
correlation (p. 267)
cross-cutting relationships, principle of (p. 264)
disconformity (p. 267)
eon (p. 279)
epoch (p. 279)
era (p. 279)

fossil (p. 268)
fossil succession, principle of (p. 273)
geologic time scale (p. 277)
half-life (p. 274)
inclusions (p. 265)
index fossil (p. 273)
Mesozoic era (p. 279)
nonconformity (p. 267)
original horizontality, principle of (p. 264)
Paleozoic era (p. 279)

period (p. 279)
Phanerozoic eon (p. 279)
Precambrian (p. 279)
radioactivity (p. 274)
radiometric dating (p. 274)
relative dating (p. 264)
superposition, law of (p. 264)
unconformity (p. 266)
uniformitarianism (p. 263)

Vocabulary Review

Choosing from the list of key terms, furnish the most appropriate response for the following statements.

1. Simply stated, the doctrine of _____ states that the physical, chemical, and biological laws that operate today have also operated in the geologic past.

2. The _____ is the age of "recent life" on Earth.

3. To state specifically the point in history when something took place, for example, the extinction of the dinosaurs about 66 million years ago, is to use a type of date called a(n) _____.

4. The "matching-up" of rocks from similar ages but different regions is referred to as _____.

5. The time required for one-half of the nuclei in a radioactive sample to decay is called the _____ of the isotope.

6. The doctrine of _____ adheres to the idea that Earth's landscape was produced by sudden and often worldwide disasters of unknowable causes that no longer operate.

7. To place events in their proper sequence is to apply a type of dating technique called _____.

8. The Hadean, the Archean, and the Proterozoic eons are commonly referred to as the _____.

9. To assume that rock layers that are inclined have been moved into that position by crustal disturbances is to apply a principle of relative dating known as _____.

10. To state that "in an undeformed sequence of sedimentary rocks the oldest rock is at the bottom" is to use a basic principle of relative dating called the _____.

11. A(n) _____ is a break in the rock record during which deposition ceased, erosion removed previously formed rocks, and then deposition resumed.

12. A(n) _____ is the remains or trace of prehistoric life.

13. The spontaneous breaking apart of atomic nuclei is a process known as _____.

14. A(n) _____ is a subdivision of a geologic era.

15. Pieces of one rock unit contained within another are called _____.

16. The unit of geologic time known as the _____ is noted for its "ancient life."

17. A fossil of an organism that was widespread geographically but limited to a short span of geologic time is often referred to as a(n) _____.

18. The fact that fossils succeed one another in a definite and determinable order is known as the _____.

19. Using radioactive isotopes to calculate the ages of rocks and minerals is a procedure called _____.

20. The unit of geologic time called a(n) _____ represents the greatest expanse of time.

Comprehensive Review

1. Write a brief statement that compares the doctrine of catastrophism with the doctrine of uniformitarianism.

2. Dinosaurs became extinct about 66 million years ago, while large coal swamps flourished in North America about 300 million years ago. Using these facts, write relative and absolute date statements.

   a) Relative date:

   b) Absolute date:

3. Briefly state the following:

   a) Law of superposition:

   b) Principle of original horizontality:

   c) Principle of fossil succession:

4. Using Figure 10.1, select the letter of the diagram that illustrates each of the following types of unconformities. (Unconformities are indicated with a dark wavy line and the small "v" symbol represents igneous rock.)

   a) Nonconformity: ____

   b) Angular unconformity: ____

   c) Disconformity: ____

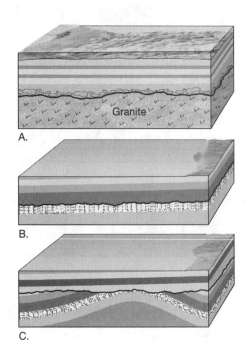

Figure 10.1

5. List the two special conditions that are necessary for an organism to be preserved as a fossil.

   1)

   2)

6. Examine the illustrations of the sedimentary rocks exposed along two separate cliffs in Figure 10.2. Different types of rocks are represented with different patterns. Write the letter of the rock layer at location 2 that correlates with the indicated rock layer at location 1.

   Location 1   Location 2

   a) layer b   correlates with   layer ____

   b) layer e   correlates with   layer ____

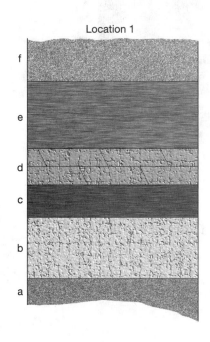
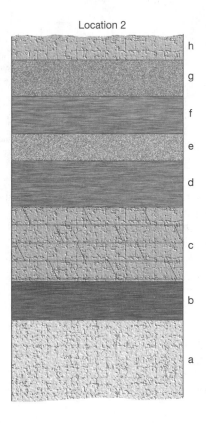

Figure 10.2

7. Figure 10.3 illustrates two types of fossilization. Write the letter of the photograph that is representative of each of the following types.

   a) Cast: ____

   b) Impression: ____

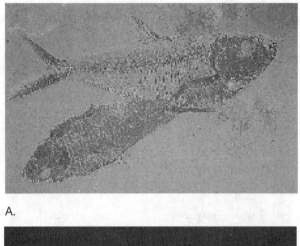

A.

B.

Figure 10.3

8. Explain the difference between an atom's atomic number and its mass number.

9. List three common types of radioactive decay.

   1)

   2)

   3)

10. Why can radiometric dating be considered reliable?

11. Briefly explain how carbon-14 is produced in the upper atmosphere.

12. What is the primary problem in assigning absolute dates to the units of time of the geologic time scale?

13. Using Figure 10.4, answer the following questions concerning the relative ages of the features illustrated. Also, when indicated, list the law or principle of relative dating you used to arrive at your answer.

    a) Dike B is _____ [older, younger] than fault B.

        Law or principle:

    b) The shale is _____ [older, younger] than the sandstone.

        Law or principle:

    c) Dike B is _____ [older, younger] then the batholith.

    d) The sandstone is _____ [older, younger] than Dike A.

    e) The conglomerate is _____[older, younger] than the shale.

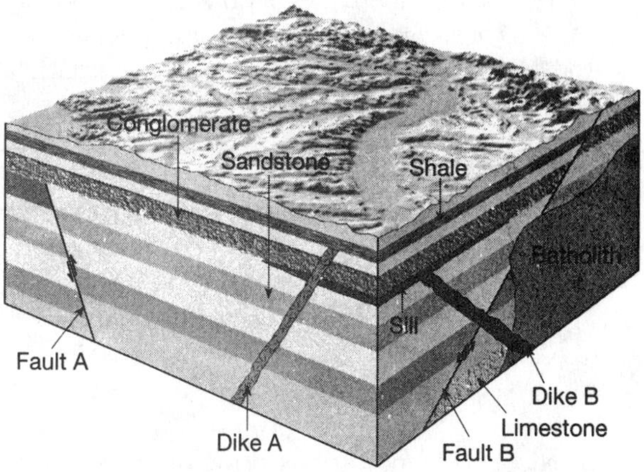

Figure 10.4

14. Assume a radioactive isotope has a half-life of 50,000 years. If the ratio of radioactive parent to stable daughter product is 1:3, how old is the rock containing the radioactive material?

Practice Test

Multiple choice. Choose the best answer for the following multiple choice questions.

1. "The physical, chemical, and biological laws that operate today have also operated in the geologic past" is a statement of the doctrine of _____.
    a) uniformitarianism   c) catastrophism   e) paleontology
    b) universal truth   d) unity

2. The task of matching up rocks of similar ages in different regions is known as _____.
    a) indexing   c) linking   e) superposition
    b) correlation   d) succession

3. Fossils that are widespread geographically and are limited to a short span of time are referred to as _____ fossils.
    a) key   c) succeeding   e) relative
    b) matching   d) index

4. A(n) _____ in the rock record represents a long period during which deposition ceased, erosion removed previously formed rocks, and then deposition resumed.
    a) absolute date   c) relative date   e) unconformity
    b) superposition   d) inclusion

5. The process by which atomic nuclei spontaneously break apart (decay) is termed _____.
   - a) ionization
   - b) fusion
   - c) reduction
   - d) nucleation
   - e) radioactivity

6. The doctrine of _____ implies that Earth's landscape has been developed by worldwide disasters over a short time span.
   - a) uniformitarianism
   - b) destruction
   - c) catastrophism
   - d) universal time
   - e) suddenness

7. The greatest expanses of geologic time are referred to as _____.
   - a) eras
   - b) periods
   - c) eons
   - d) epochs
   - e) systems

8. Nicolaus Steno proposed the most basic principle of relative dating, the law of _____.
   - a) superposition
   - b) secondary intrusion
   - c) sequential stacking
   - d) succession
   - e) suddenness

9. Which eon of the geologic time scale means "visible life"?
   - a) Phanerozoic
   - b) Archean
   - c) Hadean
   - d) Proterozoic

10. The doctrine of _____ was put forth by James Hutton in his monumental work *Theory of the Earth*.
    - a) destruction
    - b) catastrophism
    - c) suddenness
    - d) uniformitarianism
    - e) universal time

11. Rock layers that have been deposited essentially without interruption are said to be _____ strata.
    - a) included
    - b) conformable
    - c) abnormal
    - d) succeeding
    - e) sequential

12. The major units of the geologic time scale were delineated during the _____.
    - a) 1700s
    - b) 1800s
    - c) 1900s

13. The type of date that places events in proper sequence is referred to as a(n) _____ date.
    - a) sequential
    - b) secondary
    - c) relative
    - d) positional
    - e) temporary

14. The remains or traces of prehistoric life are known as _____.
    - a) indicators
    - b) fossils
    - c) replicas
    - d) paleotites
    - e) fissures

15. Between 1830 and 1872, _____ produced eleven editions of his great work *Principles of Geology*.
    - a) Archbishop James Ussher
    - b) James Hutton
    - c) Sir Charles Lyell
    - d) William Smith
    - e) Charles Darwin

16. Atoms with the same atomic number but different mass numbers are called _____.
    - a) protons
    - b) isotopes
    - c) ions
    - d) variants
    - e) nucleoids

17. A break that separates older metamorphic rocks from younger sedimentary rocks immediately above them is a type of unconformity called a(n) _____.
    a) disconformity
    b) nonconformity
    c) angular unconformity
    d) contact unconformity
    e) pseudoconformity

18. "Most layers of sediments are deposited in a horizontal position" is a statement of the principle of _____.
    a) original horizontality
    b) cross-cutting relationships
    c) fossil succession
    d) sediment
    e) cross-bedding

19. Which one of the following is NOT an era of the Phanerozoic eon?
    a) Proterozoic
    b) Mesozoic
    c) Paleozoic
    d) Cenozoic

20. Pieces of one rock unit contained within another are called _____.
    a) intrusions
    b) interbeds
    c) hosts
    d) conformities
    e) inclusions

21. Which one of the following is NOT a common type of radioactive decay?
    a) alpha particle emission
    b) nuclei ionization
    c) beta particle emission
    d) electron capture

22. Earth is about _____ years old.
    a) 4,000
    b) 4.6 million
    c) 5.8 million
    d) 4.6 billion
    e) 6.4 billion

23. A date that pinpoints the time in history when something took place is known as a(n) _____ date.
    a) relative
    b) exact
    c) known
    d) absolute
    e) factual

24. During radioactive decay, what will be the parent/daughter ratio after three half-lives?
    a) 1:3
    b) 1:4
    c) 1:5
    d) 1:6
    e) 1:7

25. The principle of fossil _____ states that fossil organisms succeed one another in a definite and determinable order.
    a) inclusions
    b) succession
    c) sequences
    d) superposition
    e) evolution

*True/false. For the following true/false questions, if a statement is not completely true, mark it false. For each false statement, change the **italicized** word to correct the statement.*

1. ___ The entire geologic time scale was created using *absolute* dates.

2. ___ Acceptance of the concept of uniformitarianism means the acceptance of a very *short* history for Earth.

3. ___ *Index* fossils are useful for correlating the rocks of one area with those of another.

4. ___ To state that a rock is 120 million years old is an example of applying a(n) *absolute* date.

5. ___ Rapid burial and the possession of *soft* parts are necessary conditions for fossilization.

6. ___ The rates of geologic processes have undoubtedly *varied* through time.

7. ___ All rocks *can* be dated radiometrically.

8. ___ Every element has a different number of *electrons* in the nucleus.

9. ___ *Catastrophists* believed that Earth was only a few thousand years old.

10. ___ When magma works its way into a rock and crystallizes, we can assume that the intrusion is *older* than the rock that has been intruded.

11. ___ During radioactive decay, as the percentage of radioactive parent atoms declines, the proportion of stable daughter atoms *decreases*.

12. ___ In addition to being important tools for correlation, fossils are also useful *environmental* indicators.

13. ___ Most layers of sediment are deposited in a *horizontal* position.

14. ___ Widespread occurrence of abundant, visible life marks the beginning of the *Phanerozoic* eon.

15. ___ The fossil record of those organisms with *hard* body parts that lived in areas of sedimentation is quite abundant.

16. ___ No place on Earth has a complete set of *conformable* strata.

17. ___ Among the major contributions of geology is the concept that Earth history is exceedingly *long*.

18. ___ The "present is the key to the past" summarizes the basic idea of *catastrophism*.

19. ___ *Index* fossils are important time indicators.

20. ___ An accurate radiometric date can only be obtained if the material remains in a *closed* system during the entire period since its formation.

Word choice. Complete each of the following statements by selecting the most appropriate response.

1. The most useful fossils for geologic dating are called [index/parent] fossils.

2. In the geologic time scale, [eons/eras/periods] represent the greatest expanses of time.

3. Isotopes of an element have different numbers of [electrons/neutrons/protons].

4. [James Ussher/James Hutton] put forth the fundamental principle of geology known as uniformitarianism.

5. Carbon-14 dating is used with [igneous/organic] materials.

6. Catastrophism states that the forces that shaped the major features of Earth were [the same as/different from] the current causes.

7. With most isotopes used in radiometric dating, the ratio of the original isotope to its [emitted particles/daughter product] is measured.

8. When rocks are deposited sequentially without interruption they are said to be [sequential/conformable].

9. The best index fossils formed from organisms that had [narrow/wide] geographic ranges and existed for [long/short] spans of time.

10. The rates of natural processes [do/do not] vary at different times and different places.

11. The entire geologic time scale was originally created using methods of [relative/absolute] dating.

12. Fossils would be best preserved in [coarse/fine] sediments.

13. Isotopes produced by the decay of radioactive elements are called the [parent/daughter] products.

14. Uniformitarianism requires that Earth be [older/younger] than the age required by catastrophism.

15. One of the radioactive decay products of uranium-238 is [nitrogen/carbon/radon], a colorless, odorless, invisible gas.

16. Inclusions are [younger/older] than the rock they are found in.

17. Rocks from several localities have been dated at more than [three/five] billion years.

18. Placing events in proper sequence is known as [relative/absolute] dating.

19. Angular unconformities are most likely to follow periods of [volcanism/folding and erosion].

20. [Igneous/Sedimentary] rocks are the least suitable for radiometric dating.

Written questions

    1. Explain the difference between radiometric and relative dating.

    2. List three laws or principles that are used to determine relative dates.

    3. How are fossils helpful in geologic investigations?

EARTH HISTORY:

A Brief Summary

11

Earth History: A Brief Summary opens with a presentation of the origin of Earth, beginning about 5 billion years ago with an enormous cloud of interstellar material. Also examined are Earth's primitive atmosphere and how its composition changed in response to the evolution of life. Earth's history is systematically detailed with a description of the geologic and biologic events that have occurred from the Precambrian, through the Paleozoic, Mesozoic, and Cenozoic Eras.

Learning Objectives

After reading, studying, and discussing this chapter, you should be able to:

- Describe the origin of Earth and the solar system in general.
- Discuss how Earth's atmosphere has evolved and changed through time.
- List the principal geologic and biologic events for each era of geologic time.

Chapter Summary

- The *nebular hypothesis* describes the formation of the solar system. The planets and sun began forming about 5 billion years ago from a large cloud of dust and gases composed of hydrogen and helium, with only a small percentage of all the other heavier elements. As the cloud contracted, it began to rotate and assume a disk shape. Material that was gravitationally pulled toward the center became the *protosun*. Within the rotating disk, small centers, called *protoplanets*, swept up more and more of the cloud's debris. Due to their high temperatures and weak gravitational fields, the inner planets (Mercury, Venus, Earth, and Mars) were unable to accumulate and retain many of the lighter components (hydrogen, helium, ammonia, methane, and water). However, because of the very cold temperatures existing far from the sun, the fragments from which the large outer planets (Jupiter, Saturn, Uranus, and Neptune) formed consisted of huge amounts of hydrogen and other light materials. These gaseous substances account for the comparatively large sizes and low densities of the outer planets.

- The decay of radioactive elements and heat released by colliding particles aided the melting of Earth's interior, allowing the denser elements, principally iron and nickel, to sink to its center. As a result of this *differentiation*, Earth's interior consists of shells or spheres of materials, each having rather distinct properties.

- Earth's primitive atmosphere probably consisted of the gases water vapor, carbon dioxide, nitrogen, and several trace gases that were released in volcanic emissions, a process called *outgassing*. The first life forms on Earth, probably *anaerobic bacteria*, did not need oxygen. As life evolved, plants, through the process of *photosynthesis*, used carbon dioxide and water and released oxygen into the atmosphere. Once the available iron on Earth was oxidized (combined with oxygen), substantial quantities of oxygen accumulated in the atmosphere. About 4 billion years into Earth's existence, the fossil record reveals abundant ocean-dwelling organisms that require oxygen to live.

• The *Precambrian* spans about 87 percent of Earth's history, beginning with the formation of Earth about 4.6 billion years ago and ending approximately 570 million years ago with the diversification of life that marks the start of the Paleozoic Era. It is the least understood span of Earth's history because most Precambrian rocks are buried from view. However, on each continent there is a "core area" of Precambrian rocks called the *shield*. The iron ore deposits of Precambrian age represent the time when oxygen became abundant in the atmosphere and combined with iron to form iron oxide. The most common middle Precambrian fossils are *stromatolites*. Microfossils of bacteria and blue-green algae, both primitive *prokaryotes* whose cells lack organized nuclei, have been found in chert, a hard, dense, chemical sedimentary rock in southern Africa (3.1 billion years of age) and near Lake Superior (1.7 billion years of age). *Eukaryotes*, with cells containing organized nuclei, are among billion-year-old fossils discovered in Australia. Plant fossils date from the middle Precambrian, but animal fossils came a bit later, in the late Precambrian. Many of these fossils are *trace fossils*, and not of the animals themselves.

• The *Paleozoic Era* extends from 570 million years ago to about 245 million years ago. The beginning of the Paleozoic is marked by the *appearance of the first life forms with hard parts* such as shells. Therefore, abundant Paleozoic fossils occur and a far more detailed record of Paleozoic events can be constructed. During the early Paleozoic (the Cambrian, Ordovician, and Silurian Periods), the vast southern continent of *Gondwanaland* existed. Seas inundated and receded from North America several times, leaving thick evaporite beds of rock salt and gypsum. Life in the early Paleozoic was restricted to the seas and consisted of several invertebrate groups. During the late Paleozoic (the Devonian, Mississippian, Pennsylvanian, and Permian Periods), ancestral North America collided with Africa to produce the original northern Appalachian Mountains, and the northern continent of *Laurasia* formed. By the close of the Paleozoic, all the continents had fused into the supercontinent of *Pangaea*. During most of the late Paleozoic, organisms diversified dramatically. Insects and plants moved onto the land, and amphibians evolved and diversified quickly. By the Pennsylvanian Period, large tropical swamps, which became the major coal deposits of today, extended across North America, Europe, and Siberia. At the close of the Paleozoic, altered climatic conditions caused one of the most dramatic biological declines in all of Earth's history.

• The *Mesozoic Era*, often called the *"age of dinosaurs,"* begins about 245 million years ago and ends approximately 66 million years ago. Early in the Mesozoic much of the land was above sea level. However, by the middle Mesozoic, seas invaded western North America. As Pangaea began to break up, the westward-moving North American plate began to override the Pacific plate, causing crustal deformation along the entire western margin of the continent. Organisms that had survived extinction at the end of the Paleozoic began to diversify in spectacular ways. *Gymnosperms* (cycads, conifers, and ginkgoes) quickly became the dominant trees of the Mesozoic because they could adapt to the drier climates. Reptiles quickly became the dominant land animals, with one group eventually becoming the birds. The most awesome of the Mesozoic reptiles were the *dinosaurs*. At the close of the Mesozoic, many reptile groups, including the dinosaurs, became extinct.

• The *Cenozoic Era*, or *"era of recent life,"* begins approximately 66 million years ago and continues today. It is *the time of mammals*, including humans. The widespread, less disturbed rock formations of the Cenozoic provide a rich geologic record. Most of North America was above sea level throughout the Cenozoic. Due to their different relations with tectonic plate boundaries, the eastern and western margins of the continent experienced contrasting events. The stable eastern margin was the site of abundant sedimentation as isostatic adjustment raised the Appalachians, causing the streams to erode with renewed vigor and deposit their sediment along the continental margin. In the West, building of the Rocky Mountains (the *Laramide orogeny*) was coming to an end, the Basin and Range Province was forming, and volcanic activity was extensive. The Cenozoic is often called *"the age of mammals"* because these animals replaced the reptiles as the dominant land life. Two groups of mammals, the *marsupials* and the *placentals*, evolved and expanded to dominate the era. One tendency was for some mammal groups to become very large. However, a wave of late *Pleistocene* extinctions rapidly eliminated these animals from the landscape. Some scientists believe that early humans hastened their decline by selectively hunting the larger animals. The Cenozoic could also be called the *"age of flowering plants."* As a source of food, flowering plants strongly influenced the evolution of both birds and herbivorous (plant-eating) mammals throughout the Cenozoic Era.

Chapter Outline

I. Origin of Earth
   A. Theory called the nebular hypothesis
      1. Bodies of the solar system evolved from an enormous cloud (nebula) composed of hydrogen and helium, with a small percentage of all the other heavier elements
      2. Sequence of events within the nebula
         a. Cloud began to contract about 5 billion years ago
         b. Small concentrations form the nuclei of the planets
         c. Protosun (sun in the making) forms
         d. Protoplanets accumulate more and more debris
         e. Inner planets (Mercury through Mars) heat and lose their lighter components
         f. Cold outer planets (Jupiter through Saturn) accumulated gases and other light materials
   B. Partial melting of Earth's early interior
      1. Heat comes from the decay of radioactive materials and collisions
      2. Iron and nickel sink to center
      3. Lighter components rise toward the surface
      4. Gaseous materials escape from the interior and form the primitive atmosphere

II. Earth's atmosphere
   A. Primitive atmosphere formed from volcanic gases
      1. A process called outgassing
      2. Water vapor, carbon dioxide, nitrogen, and several trace gases
      3. Very little free oxygen
   B. Water vapor condenses and forms primitive oceans as Earth cools
   C. Bacteria evolve

   D. Plants evolve and photosynthesis produces oxygen
   E. Oxygen content in the atmosphere increases
   F. By about 4 billion years after Earth formed, abundant ocean-dwelling organisms that require oxygen existed

III. Earth's history
   A. Precambrian Era
      1. 4.6 billion to 570 million years ago
      2. 87 percent of Earth's history
      3. Only sketchy knowledge
      4. Most Precambrian rocks devoid of fossils
      5. Precambrian rocks
         a. Most are buried from view
         b. Each continent has a "core area" of Precambrian rocks called a shield
         c. Contain extensive iron ore deposits
         d. Absent are fossil fuels
      6. Precambrian fossils
         a. Most common are stromatolites
            1. Material deposited by algae
            2. Common about 2 billion years ago
         b. Microfossils of bacteria and algae have been found in chert
            1. Southern Africa (3.1 billion years of age)
            2. Lake Superior area (1.7 billion years of age)
         c. Plant fossils date from the middle Precambrian
         d. Animal fossils date from the late Precambrian
         e. Diverse and multicelled organisms exist by the close of the Precambrian
   B. Paleozoic Era

1. 570 million years ago to about 245 million years ago
2. First life forms with hard parts
3. Abundant Paleozoic fossils
4. Early Paleozoic history
   a. Southern continent of Gondwanaland exists
   b. North America
      1. A barren lowland
      2. Seas move inland and recede several times and shallow marine basins evaporate leaving rock salt and gypsum deposits
      3. Taconic orogeny, a mountain building event, affects eastern North America
5. Early Paleozoic life
   a. Restricted to seas
   b. Vertebrates had not yet evolved
   c. Life consisted of several invertebrate groups
      1. Trilobites
      2. Brachiopods
      3. Cephalopods
   d. First organisms with hard parts, such as shells—perhaps for protection
6. Late Paleozoic history
   a. Supercontinent of Pangaea forms
   b. Several mountain belts formed during the movements of the continents
   c. World's climate becomes very seasonal, causing the dramatic extinction of many species
7. Late Paleozoic life
   a. Organisms diversified dramatically
   b. Land plants develop
   c. Fishes evolve into two groups of bony fish
      1. Lung fish
      2. Lobe-finned fish which become the amphibians
   d. Insects invade the land
   e. Amphibians diversify rapidly
   f. Extensive coal swamps develop
C. Mesozoic Era
   1. 245 million years ago to about 66 million years ago
   2. Often called the "age of dinosaurs"
   3. Mesozoic history
      a. Begins with much of the world's land above sea level
      b. Seas invade western North America
      c. Breakup of Pangaea begins forming the Atlantic ocean
      d. North American plate began to override the Pacific plate
      e. Mountains of western North America began forming
   4. Mesozoic life
      a. Survivors of the great Paleozoic extinction
      b. Gymnosperms become the dominant trees
      c. Reptiles (first true terrestrial animals) readily adapt to the dry Mesozoic climate
      d. Reptiles have shell-covered eggs that can be laid on the land
      e. Dinosaurs dominate
      f. One group of reptiles led to the birds
      g. Many reptile groups, along with many other animal groups, become extinct at the close of the Mesozoic
         1. One hypothesis is that a large asteroid or comet struck Earth
         2. Another possibility is extensive volcanism
D. Cenozoic Era
   1. 66 million years ago to the present
   2. Often called the "age of mammals"
   3. Smaller fraction of geologic time than either the Paleozoic or the Mesozoic
   4. North America
      a. Most of the continent was above sea level throughout the Cenozoic Era
      b. Many events of mountain building, volcanism, and earthquakes in the West
      c. Eastern North America
         1. Stable with abundant marine sedimentation
         2. Eroded Appalachians were raised by isostatic adjustments
      d. Western North America
         1. Building of the Rocky Mountains was coming to an end
         2. Large region is uplifted
            a. Basin and Range province formed

b. Re-elevates the Rockies
c. Rivers erode and form gorges (e.g., Grand Canyon and Black Canyon)
3. Volcanic activity is common
   a. Fissure eruptions form the Columbia Plateau
   b. Volcanoes form from northern California to the Canadian border
4. Coast Ranges form
5. Sierra Nevada become fault-block mountains
5. Cenozoic life
   a. Mammals replace reptiles as the dominant land animals
   b. Angiosperms (flowering plants with covered seeds) dominate the plant world

1. Strongly influenced the evolution of both birds and mammals
2. Food source for both birds and mammals
c. Two groups of mammals evolve after the reptilian extinctions at the close of the Mesozoic
   1. Marsupials
   2. Placentals
d. Mammals diversify quite rapidly and some groups become very large
   1. e.g., Hornless rhinoceros, which stood nearly 16 feet high
   2. Many large animals became extinct
e. Humans evolve

Key Terms

Page numbers shown in ( ) refer to the textbook page where the term first appears.

nebular hypothesis (p. 287)   shields (p. 289)   stromatolites (p. 290)
outgassing (p. 288)

Vocabulary Review

Choosing from the list of key terms, furnish the most appropriate response for the following statements.

1. On each continent, large "core areas" of Precambrian rocks, called _____, dominate the surface, mostly as deformed metamorphic rocks.

2. The proposal, known as the _____, suggests that approximately 5 billion years ago the bodies of our solar system condensed from an enormous cloud composed mostly of hydrogen and helium.

3. During the process called _____, gases that were formerly dissolved in molten rock are released.

4. The most common Precambrian fossils are _____.

Comprehensive Review

1. Figure 11.1 illustrates the sequence of events that scientists believe occurred during the formation of the solar system. Using the figure as a guide, describe what is happening in each of the four stages.

   Diagram A:

   Diagram B:

Diagram C:

Diagram D:

Figure 11.1

2. When compared to the inner planets, why do the outer planets (Jupiter, Saturn, Uranus, and Neptune) have such low densities?

3. Why is Earth's interior heterogeneous, consisting of shells or spheres of materials having different properties?

4. What was the source of the gases that formed Earth's original atmosphere? What were the gases that most likely made up the original atmosphere?

5. Where did the significant percentage of oxygen in Earth's present atmosphere come from?

6. Write the name of the geologic era during which each of the following events occurred.

   a) Humans develop: f) Dinosaurs dominant:

   b) First one-celled organisms: g) First shelled organisms:

   c) First birds: h) Mammals dominant:

   d) First fishes: i) First reptiles:

   e) Formation of Pangaea: j) Breakup of Pangaea:

7. Why is Precambrian time, which represents about 87 percent of Earth's history, not subdivided into detailed time divisions like the other eras?

8. What important event in animal evolution marks the beginning of the Paleozoic Era?

9. What are two hypotheses for the extinction of the dinosaurs, along with many other plant and animal groups, at the end of the Mesozoic Era?

   1)

   2)

10. Compare the geologic events that occurred along the eastern and western margins of North America during the Cenozoic Era.

11. Following the demise of most Mesozoic reptiles, what were the four principal directions that mammals followed in their subsequent development and specialization?

    1)

    2)

    3)

    4)

12. List the ages and briefly summarize the events of the following geologic eras.

    a) Cenozoic Era:

    b) Mesozoic Era:

    c) Paleozoic Era:

Practice Test

Multiple choice. Choose the best answer for the following multiple choice questions.

1. Life on Earth began during the _____ Era.
   a) Cenozoic
   b) Mesozoic
   c) Paleozoic
   d) Precambrian
   e) Evolutionary

2. Earth's original atmosphere consisted of gases expelled from within the planet during a process known as _____.
   a) exhalation
   b) expulsion
   c) venting
   d) intrusion
   e) outgassing

3. Large "core areas" of Precambrian rocks, called _____, exist on each continent.
   a) plateaus
   b) shields
   c) roots
   d) complexes
   e) rifts

4. Once the available metal _____ was oxidized, substantial quantities of oxygen began to accumulate in Earth's atmosphere.
   a) nickel
   b) copper
   c) magnesium
   d) lead
   e) iron

5. The proposal that suggests that the bodies of our solar system condensed from an enormous cloud is termed the _____ hypothesis.
   a) formation
   b) nebular
   c) protosystem
   d) nucleation
   e) big bang

6. The beginning of the Paleozoic Era is marked by the appearance of the first life forms with _____.
   a) legs
   b) cells
   c) fins
   d) hard parts
   e) lungs

7. The smallest divisions of the geologic time scale are referred to as _____.
   a) eras
   b) periods
   c) eons
   d) epochs
   e) systems

8. Some of the oldest microfossils are dated at about _____ years of age.
   a) 570 million
   b) 890 million
   c) 1.2 billion
   d) 2.8 billion
   e) 3.1 billion

9. Which era of the geologic time scale is often referred to as the "age of mammals"?
   a) Cenozoic
   b) Mesozoic
   c) Paleozoic
   d) Precambrian

10. Which one of the following was NOT a gas that geologists hypothesize existed in Earth's original atmosphere?
    a) water vapor
    b) carbon dioxide
    c) oxygen
    d) nitrogen

11. Planets in the making are referred to as _____.
    a) neoplanets
    b) exoplanets
    c) isoplanets
    d) protoplanets
    e) nebularplanets

12. Which era of the geologic time scale is often referred to as the "age of reptiles"?
    a) Cenozoic
    b) Mesozoic
    c) Paleozoic
    d) Precambrian

13. The supercontinent of Pangaea formed during the late _____.
    a) Cenozoic
    b) Mesozoic
    c) Paleozoic
    d) Precambrian

14. The petroleum traps of the Gulf Coast occur in _____ strata.
    a) Cenozoic
    b) Mesozoic
    c) Paleozoic
    d) Precambrian

15. Which one of the following events is NOT associated with the late Paleozoic Era?
    a) establishment of land plants
    b) evolution of amphibians
    c) evolution of bony fishes
    d) first flowering plants
    e) extensive coal swamps

16. As the first oceans formed on Earth, large amounts of the gas _____ became dissolved in the water.
    a) oxygen
    b) carbon dioxide
    c) sulfur oxide
    d) argon
    e) hydrogen

17. Which one of the following is notably absent in Precambrian rocks?
    a) oxidized iron
    b) metamorphic rocks
    c) silicate minerals
    d) fossil fuels
    e) igneous rocks

18. During their formation, the inner planets were unable to retain appreciable amounts of the lighter, gaseous components because of their high surface temperatures and weak _____ fields.
    a) gravitational
    b) nuclear
    c) radioactive
    d) magnetic
    e) isostatic

19. Which era of geologic time spans the largest percentage of Earth's history?
    a) Precambrian
    b) Mesozoic
    c) Paleozoic
    d) Cenozoic

20. Plants, employing a process called _____, are the major contributors of oxygen to the atmosphere.
    a) conversion
    b) oxidation
    c) hydrolysis
    d) photosynthesis
    e) calcification

21. The presence of elevated proportions of the element _____ in a thin layer of sediment deposited 65 million years ago supports the theory that a large asteroid or comet collided with Earth.
    a) iridium
    b) radon
    c) silicon
    d) chlorine
    e) carbon

22. Earth is about _____ years old.
    a) 4,000
    b) 4.6 million
    c) 5.8 million
    d) 4.6 billion
    e) 6.4 billion

23. Because the outer planets consist of huge amounts of _____ substances, they have large sizes and low densities.
    a) fluid   c) unknown   e) silicon
    b) solid   d) gaseous

24. One group of _____, exemplified by the fossil Archaeopteryx, led to the birds.
    a) reptiles   c) rodents   e) angiosperms
    b) large cats   d) fishes

25. The late _____ extinction was the greatest of at least five mass extinctions to occur over the past 600 million years.
    a) Cenozoic   c) Paleozoic
    b) Mesozoic   d) Precambrian

26. About 65 million years ago, during a time of mass extinction called the _____ (or KT) boundary, more than half of all plant and animal species died out.
    a) Cretaceous-Tertiary   c) Cretaceous-Triassic
    b) Carboniferous-Triassic   d) Carboniferous-Tertiary

27. Toward the end of the _____, a mountain building event, called the Laramide Orogeny, formed the middle and southern ranges of the Rocky Mountains.
    a) Cenozoic   c) Paleozoic
    b) Mesozoic   d) Precambrian

28. Which one of the following was NOT a characteristic of primitive mammals?
    a) small size   c) short legs
    b) large brain   d) flat, five-toed feet

29. With the perfection of the shelled egg, reptiles quickly became the dominant animals during the _____ Era.
    a) Cenozoic   c) Paleozoic
    b) Mesozoic   d) Precambrian

30. The beginning of the breakup of the supercontinent Pangaea occurred during the _____ Era.
    a) Cenozoic   c) Paleozoic
    b) Mesozoic   d) Precambrian

31. Many of the most ancient fossils are preserved in _____, a hard, dense, chemical sedimentary rock.
    a) chert   c) limestone   e) mudstone
    b) iron oxide   d) shale

32. _____ bacteria thrive in environments that lack free oxygen.
    a) Anaerobic   c) Nucleated   e) Photobiotic
    b) Proaerobic   d) Placental

*True/false. For the following true/false questions, if a statement is not completely true, mark it false. For each false statement, change the **italicized** word to correct the statement.*

1. ___ *Plants* are responsible for dramatically altering the composition of the entire planet's atmosphere.

2. ___ The enormous cloud from which the solar system formed was composed mostly of the gases hydrogen and *nitrogen*.

3. ___ The *Devonian* Period is often called the "age of fishes."

4. ___ During the *Mesozoic*, the entire western margin of North America was subjected to a continuous wave of deformation as the North American plate began to override the Pacific plate.

5. ___ Many large mammals were rapidly eliminated during a wave of late *Pleistocene* extinctions.

6. ___ The climate of the Mesozoic Era was *wetter* than that of the Paleozoic.

7. ___ The smallest division of geologic time is a(n) *period*.

8. ___ The dominant animals of the Mesozoic Era were *amphibians*.

9. ___ Melting of Earth's interior allowed the *denser* elements to sink to the center.

10. ___ The ancestral Appalachian Mountains formed during the late Paleozoic as North America and *Africa* collided.

11. ___ The first appearance of mammals was during the *Mesozoic* Era.

12. ___ In North America, the Canadian Shield encompasses slightly more than *seven* million square kilometers.

13. ___ The beginning of the *Cenozoic* Era is marked by the appearance of the first life forms with hard parts.

14. ___ Today, of the two groups of mammals, *marsupials* are found primarily in Australia.

15. ___ The *Mesozoic* Era is the least-understood span of Earth's history.

16. ___ By late Devonian time, *lobe-finned* fish had evolved into true air-breathing amphibians.

17. ___ Evidence indicates that some dinosaurs, unlike their present-day reptile relatives, were *warm* blooded.

18. ___ The Basin and Range Province of western North America formed during the *Mesozoic* Era.

19. ___ The most common Precambrian fossils are *cephalopods*.

20. ___ Most mammals, including humans, are *placental*.

Word choice. Complete each of the following statements by selecting the most appropriate response.

1. Most of the divisions of geologic time are based upon the [life forms/geologic formations] present at the time.

2. Rocks from the middle Precambrian contain most of Earth's [iron/gold] ore, mainly as the mineral hematite.

3. Life began to colonize the land during the [Paleozoic/Mesozoic] Era.

4. Earth's early atmosphere was composed chiefly of gases expelled from [the oceans/molten rock].

5. The predominant plant life of the Cenozoic Era are [angiosperms/gymnosperms].

6. In the early Paleozoic Era, North America and Europe are thought to have been [near/far from] the equator.

7. The [Paleozoic/Precambrian] encompasses most of Earth's history.

8. Grasses and accompanying herbivores developed during the [Quaternary/Tertiary] Period of the Cenozoic Era.

9. In the Precambrian, plant fossils occur [before/after] animal fossils.

10. By the close of the [Paleozoic/Mesozoic] Era all the continents had fused into the supercontinent of Pangaea.

11. The appearance of higher organisms with preservable hard parts occurred in the [Cambrian/Permian] Period.

12. A [drier/wetter] climate at the end of the Paleozoic Era is thought to have triggered the extinction of many species on land and sea.

13. Isolation caused by the separation of continents [increases/decreases] the diversity of animals in the world.

14. Most of North America was [above/below] sea level throughout the Cenozoic Era.

15. Volcanic activity was common in [eastern/western] North America during much of Cenozoic time.

Written questions

1. List the two hypothesis most frequently used to explain the extinction of the dinosaurs.

2. Why is so little known about the Precambrian?

3. Describe the evolutionary advancement that freed reptiles from needing to live part of their lives in water, as did the amphibians that preceded them.

4. List four traits that separate mammals from reptiles.

Ocean Waters and the Ocean Floor

12

Ocean Waters and the Ocean Floor introduces oceanography with a brief discussion of the extent and distribution of the world ocean. After examining the composition of seawater, the ocean's layered temperature and salinity structures are presented. The ocean floor section of the chapter begins with an examination of continental margins and continues with submarine canyons and associated turbidity currents, ocean basin features, and mid-ocean ridges. The origin of coral reefs and atolls is also presented. The chapter ends with a discussion of seafloor sediments and how they are used to study climatic changes.

Learning Objectives

After reading, studying, and discussing this chapter, you should be able to:

- Describe the extent and boundaries of the world ocean.
- Discuss the chemical composition of ocean water.
- Explain the ocean's layered temperature and salinity structures.
- Describe the relation between submarine canyons, turbidity currents, and turbidites.
- Describe the major features of the continental margin, ocean basin floor, and mid-ocean ridges.
- List the types of seafloor sediments.
- Describe how ocean floor sediments relate to climatic changes.

Chapter Summary

- *Oceanography* is a composite science that draws on the methods and knowledge of biology, chemistry, physics, and geology to study all aspects of the world ocean.

- *Earth is a planet dominated by oceans.* Seventy-one percent of Earth's area consists of oceans and marginal seas. In the Southern Hemisphere, often called the *water hemisphere*, about 81 percent of the surface is water. Of the three major oceans, Pacific, Atlantic, and Indian, the *Pacific Ocean is the largest*, contains slightly *more than half of the water* in the world ocean, and has the *greatest average depth*—3940 meters (12,900 feet).

- *Salinity* is the proportion of dissolved salts to pure water, usually expressed in parts per thousand (‰). The average salinity in the open ocean ranges from 35‰ to 37‰. The principal elements that contribute to the ocean's salinity are *chlorine* (55 percent) and *sodium* (31 percent). The primary *sources for the salts* in the ocean are *chemical weathering* of rocks on the continents and *outgassing* through volcanism. Outgassing is also considered to be the principal source of water in the oceans as well as in the atmosphere.

• Commercial products obtained from seawater include common salt (sodium chloride), the light-weight metal magnesium, and bromine, a gasoline additive also used in the manufacture of fireproofing material. Fresh water is derived from seawater by *desalination,* the removal of salts and other chemicals.

• In most regions, open oceans exhibit a *three-layered temperature and salinity structure.* Ocean water temperatures are warmest at the surface because of solar energy. The mixing of waves as well as the turbulence from currents can distribute this heat to a depth of about 450 meters or more. Beneath the sun-warmed zone of mixing, a layer of rapid temperature change, called the *thermocline*, occurs. Below the thermocline, in the deep zone, temperatures fall only a few more degrees. The changes in salinity with increasing depth correspond to the general three-layered temperature structure. In the low and middle latitudes, a surface zone of higher salinity is underlaid by a layer of rapidly decreasing salinity, called the *halocline*. Below the halocline, salinity changes are small.

• Ocean depths are determined using an *echo sounder*, a device carried by a ship that bounces sound off the ocean floor. The time it takes for the sound waves to make the round trip to the bottom and back to the ship is directly related to the depth. Continuous monitoring of the echoes are plotted to produce a profile of the ocean floor.

• The zones that collectively make up the *continental margin* include the *continental shelf* (a gently sloping, submerged surface extending from the shoreline toward the deep-ocean basin), *continental slope* (the true edge of the continent, which has a steep slope that leads from the continental shelf into deep water), and in regions where trenches do not exist, the steep continental slope merges into a gradual incline known as the *continental rise*. The continental rise consists of sediments that have moved downslope from the continental shelf to the deep-ocean floor.

• *Submarine canyons* are deep, steep-sided valleys that originate on the continental slope and may extend to depths of three kilometers. Some of these canyons appear to be the seaward extensions of river valleys. However, most information seems to favor the view that many submarine canyons have been excavated by *turbidity currents* (downslope movements of dense, sediment-laden water). *Turbidites*, sediments deposited by turbidity currents, are characterized by a decrease in sediment grain size from bottom to top, a phenomenon known as *graded bedding*.

• The *ocean basin floor* lies between the continental margin and the mid-oceanic ridge system. The features of the ocean basin floor include *deep-ocean trenches* (the deepest parts of the ocean, where moving crustal plates descend into the mantle), *abyssal plains* (the most level places on Earth, consisting of thick accumulations of sediments that were deposited atop the low, rough portions of the ocean floor by turbidity currents), and *seamounts* (isolated volcanic peaks on the ocean floor that originate near oceanic ridges or in association with volcanic hot spots).

• *Mid-ocean ridges*, the sites of seafloor spreading, are found in all major oceans and represent more than 20 percent of Earth's surface. These broad features are characterized by an elevated position, extensive faulting, and volcanic structures that have developed on newly formed oceanic crust. Most of the geologic activity associated with ridges occurs along a narrow region on the ridge crest, called the *rift zone*, where magma from the asthenosphere moves upward to create new slivers of oceanic crust.

• *Coral reefs*, which are confined largely to the warm, sunlit waters of the Pacific and Indian Oceans, are constructed over thousands of years primarily from the skeletal remains and secretions of corals and certain algae. Coral islands, called *atolls*, consist of a continuous or broken ring of coral reef surrounding a central lagoon. Atolls form from corals that grow on the flanks of sinking volcanic islands, where the corals continue to build the reef complex upward as the island sinks.

• *There are three broad categories of seafloor sediments. Terrigenous sediment* consists primarily of mineral grains that were weathered from continental rocks and transported to the ocean. *Biogenous sediment* consists of shells and skeletons of marine animals and plants. *Hydrogenous sediment* includes minerals that crystallize directly from seawater through various chemical reactions.

• Seafloor sediments are helpful when studying worldwide climatic changes because they often contain the remains of organisms that once lived near the sea surface. The numbers and types of these organisms change as the climate changes, and their remains in the sediments record these changes.

Chapter Outline

I. The world ocean
  A. Global extent of the world ocean
    1. 71 percent of Earth
    2. 61 percent of Northern Hemisphere
    3. 81 percent of Southern Hemisphere
  B. Size and depth of the major oceans
    1. Pacific by far the largest
    2. Pacific has greatest average depth
    3. Atlantic the shallowest
  C. Composition of seawater
    1. 3.5 percent (by weight) dissolved minerals
    2. Salinity
      a. Proportion of dissolved salts
      b. Measured in parts-per-thousand (‰)
      c. Salinity of open ocean: 33 percent to 37 percent
      d. Major constituents: Cl^- and Na^+
    3. Sources of salts
      a. Chemical weathering of rocks
      b. Outgassing—gases from volcanic eruptions
  D. Resources from seawater
    1. Commercial products from seawater
      a. Salt (sodium chloride)
      b. Magnesium—a light-weight metal
      c. Bromine
    2. Fresh water
      a. Obtained by the desalination of seawater
      b. High cost to produce
      c. Small total production
  E. The ocean's three-layered structure
    1. Temperature
      a. Warmest at surface
      b. Thermocline—rapid decrease in temperature with depth
      c. Below thermocline—little change in temperature
    2. Salinity
      a. In low and middle latitudes
        1. Higher at surface
        2. Fresh water is evaporated
      b. Halocline—rapid decrease in salinity with depth
      c. Below halocline—little change in salinity with depth
II. Earth beneath the sea
  A. Measuring depth
    1. Originally done with weighted lines
    2. Echo sounder
      a. Primary instrument for measuring depth
      b. Reflects sound from ocean floor
  B. Major topographic units of the sea floor
    1. Continental margin
      a. Continental shelf
        1. Flooded extension of the continent
        2. Shoreline toward ocean basin
        3. Gently sloping
        4. Submerged surface
      b. Continental slope
        1. Seaward edge of the shelf
        2. Steep gradient into deep water
        3. True edge of the continent
        4. Associated with continental slopes are
          a. Turbidity currents
            1. Dense, sediment-laden water
            2. Deposits are called turbidites
              a. Layered

b. Graded bedding—
coarse sediments
on the bottom
b. Submarine canyons are
excavated by turbidity
currents
c. Continental rise
1. Found where trenches
are absent
2. Thick accumulation
of sediment
3. At base of slope
4. Turbidity currents form
deep-sea fans
2. Ocean basin floor
a. Deep-ocean trench
1. Deepest part of ocean
2. Where plates plunge into
the mantle
b. Abyssal plain
1. Most level place on Earth
2. Thick accumulations
of sediment
3. Found in all oceans
c. Seamount
1. Isolated volcanic peak
2. Some form over hot
spots
3. Mid-ocean ridge
a. Site of seafloor spreading
b. Found in all major oceans
c. Rift zone
1. On ridge crest
2. Geologically active

3. Magma moves up from the
asthenosphere
4. New ocean crust forms here
C. Coral reefs and atolls
1. Coral reef
a. Built up over thousands of
years by
1. Remains and secretions of
corals
2. Certain algae
b. Thrives in warm waters
2. Atoll
a. Forms on the flanks of a sinking
volcanic island
b. Coral reef surrounding a lagoon
c. Formation explained by Darwin
D. Seafloor sediment
1. Thickness varies
2. Mud is the most common sediment
3. Types of sediments
a. Terrigenous sediment—weathered
continental rocks
b. Biogenous sediment
1. Shells, skeletons, and plants
2. Calcareous and siliceous oozes
c. Hydrogenous sediment—chemicals precipitate directly from seawater (e.g., manganese nodules)
4. Climatic change and seafloor
sediment
a. Life in the sea changes with the
climate
b. Remains of organisms in the sediments record the changes

Key Terms

Page numbers shown in ( ) refer to the textbook page where the term first appears.

abyssal plain (p. 325)	desalination (p. 317)	rift zone (p. 326)
atoll (p. 327)	echo sounder (p. 319)	salinity (p. 315)
biogenous sediment (p. 328)	graded bedding (p. 324)	seamount (p. 326)
continental margin (p. 322)	guyot (p. 326)	submarine canyon (p. 323)
continental rise (p. 323)	halocline (p. 318)	terrigenous sediment (p. 328)
continental shelf (p. 323)	hydrogenous sediment (p. 330)	thermocline (p. 318)
continental slope (p. 323)	manganese nodule (p. 330)	turbidite (p. 324)
coral reef (p. 327)	mid-ocean ridge (p. 326)	turbidity current (p. 324)
deep-ocean trench (p. 324)	oceanography (p. 314)	
deep-sea fan (p. 323)	outgassing (p. 316)	

Vocabulary Review

Choosing from the list of key terms, furnish the most appropriate response for the following statements.

1. _____ is the composite science that draws on the methods and knowledge of biology, chemistry, physics, and geology to study all aspects of the world ocean.

2. The proportion of dissolved salts to pure water, expressed in parts-per-thousand, is referred to as _____.

3. A long, narrow, deep ocean trough is known as a(n) _____.

4. The _____ is the gently sloping submerged surface of the continental margin that extends from the shoreline toward the deep-ocean basin.

5. The downslope movement of dense, sediment-laden water is known as a(n) _____.

6. Likely one of the most level places on Earth, a(n) _____ is an incredibly flat feature on the ocean floor.

7. The _____ is the site of seafloor spreading.

8. The principal source of water in the oceans as well as in the atmosphere is a process known as _____.

9. Sediment that consists primarily of mineral grains that were weathered from continental rocks and transported to the ocean is known as _____.

10. A(n) _____ consists of a continuous or broken ring of coral reef surrounding a central lagoon.

11. The _____ is the ocean layer where there is a rapid change in salinity.

12. A(n) _____ is constructed primarily from the skeletal remains and secretions of corals and certain algae.

13. An isolated volcanic peak found on the ocean floor is known as a(n) _____.

14. Sediment consisting of minerals that crystallize directly from seawater through various chemical reactions is called _____.

15. The layer of the ocean where there is a rapid temperature change is known as the _____.

16. Ocean sediment consisting of shells and skeletons of marine animals and plants is known as _____.

17. In regions where trenches do not exist, the steep continental slope merges into a more gradual incline known as the _____.

18. A(n) _____ is a submerged, flat-topped seamount.

19. Much of the geologic activity on the ocean floor occurs along a relatively narrow zone on the mid-ocean ridge crest known as the _____.

20. The _____ marks the seaward edge of the continental shelf.

21. The instrument used to measure ocean depths by transmitting sound waves toward the ocean floor is the _____.

22. The removal of salts and other chemicals from seawater is called _____.

23. A(n) _____ is a turbidity current deposit characterized by graded bedding.

24. A rounded lump of hydrogenous sediment composed of a complex mixture of minerals that forms on the floor of the ocean basin is referred to as a(n) _____.

Comprehensive Review

1. Briefly contrast the distribution of land and water in the Northern Hemisphere with that in the Southern Hemisphere.

2. Describe the three-layered structure normally found in the open ocean.

3. What are the principal elements that contribute to the ocean's salinity?

4. Using Figure 12.1, select the letter that identifies each of the following features of the continental margin.

   a) Deep-sea fan: ____   d) Continental shelf: ____

   b) Submarine canyon: ____   e) Continental rise: ____

   c) Continental slope: ____

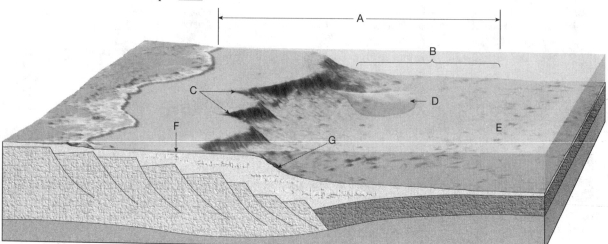

Figure 12.1

5. Describe abyssal plains and explain their origin. In Figure 12.1, letter _____ illustrates an abyssal plain.

6. What are the two primary sources for the salts in the ocean?

    1)

    2)

7. Explain the reason for the difference in the ocean's surface salinity between subtropical and equatorial regions.

8. What is the probable origin for submarine canyons that are not seaward extensions of river valleys?

9. List three water conditions that are necessary for the growth of coral.

    1)

    2)

    3)

10. Briefly describe the formation of an atoll.

11. List the three broad categories of seafloor sediments and briefly describe the origin of each.

    1)

    2)

    3)

12. Using Figure 12.2, select the letter that identifies each of the following ocean basin features.

    a) Deep-ocean trench: _____   d) Rift zone: _____

    b) Seamount: _____   e) Guyot: _____

    c) Mid-ocean ridge: _____

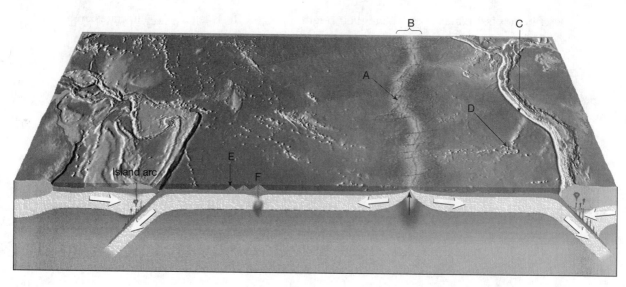

Figure 12.2

Practice Test

Multiple choice. Choose the best answer for the following multiple choice questions.

1. Approximately _____ percent of Earth's area is represented by oceans and marginal seas.
   - a) 50
   - b) 60
   - c) 70
   - d) 80
   - e) 90

2. Important mineral deposits, including large reservoirs of petroleum and natural gas, are associated with _____.
   - a) continental shelves
   - b) ocean trenches
   - c) mid-ocean ridges
   - d) rift zones

3. Seawater consists of about _____ percent (by weight) of dissolved mineral substances.
   - a) 1.5
   - b) 2.5
   - c) 3.5
   - d) 4.5
   - e) 5.5

4. The most prominent feature in the oceans, forming an almost continuous mountain range, are the _____.
   - a) deep-ocean trenches
   - b) mid-ocean ridges
   - c) seamounts
   - d) abyssal plains
   - e) guyots

5. The layer of rapid temperature change in the ocean is known as the _____.
   - a) halocline
   - b) centicline
   - c) thermocline
   - d) faricline
   - e) tempegrad

6. The largest of Earth's water bodies is the _____.
   - a) Pacific Ocean
   - b) Mediterranean Sea
   - c) Atlantic Ocean
   - d) Gulf of Mexico
   - e) Indian Ocean

7. The principal element that contributes to the ocean's salinity is _____.
   - a) sodium
   - b) potassium
   - c) bromine
   - d) calcium
   - e) chlorine

8. As a consequence of the distribution of land and water on Earth, the Southern Hemisphere is referred to as the _____ hemisphere.
   a) water
   b) open
   c) land
   d) mixed
   e) dry

9. Which of the following is NOT one of the three major topographic units of the ocean basins?
   a) continental margins
   b) ocean basin floor
   c) coastal plain
   d) mid-ocean ridges

10. The seaward edge of the continental shelf is marked by the _____.
    a) abyssal plain
    b) ocean trench
    c) continental slope
    d) mid-ocean ridge
    e) shoreline

11. Which process is thought to be the principal source of water in the oceans as well as in the atmosphere?
    a) sublimation
    b) outgassing
    c) deposition
    d) crystallization
    e) freezing

12. Sediment that consists primarily of mineral grains that were weathered from continental rocks and transported to the ocean is known as _____ sediment.
    a) biogenous
    b) hydrogenous
    c) terrigenous

13. The layer of rapid salinity change in the ocean is known as the _____.
    a) halocline
    b) saligrad
    c) thermocline
    d) salicline
    e) halograd

14. The _____ are likely the most level places on Earth.
    a) mid-ocean ridges
    b) abyssal plains
    c) deep-ocean trenches
    d) continental slopes
    e) guyots

15. Which one of the following is NOT a zone included in the continental margin?
    a) continental slope
    b) continental rise
    c) continental coastal plain
    d) continental shelf

16. Which of the following is NOT considered one of the three major oceans?
    a) Atlantic Ocean
    b) Arctic Ocean
    c) Pacific Ocean
    d) Indian Ocean

17. The fact that elements such as chlorine and bromine are more abundant in the ocean than in Earth's crust confirms the hypothesis that _____ is largely responsible for the present oceans.
    a) subduction
    b) convergence
    c) dissolving
    d) volcanism
    e) precipitation

18. Salinity variations in the open ocean normally range from 33‰ to _____.
    a) 37‰
    b) 45‰
    c) 60‰
    d) 67‰
    e) 83‰

19. Depths in the ocean are often measured using a(n) _____.
    a) altimeter
    b) laser
    c) submersible
    d) rope
    e) echo-sounder

20. Which one of the following water bodies has the highest salinity?
    a) Lake Michigan   c) Red Sea   e) Hudson Bay
    b) Baltic Sea   d) Gulf of Mexico

21. Deep, steep-sided valleys that originate on the continental slope and may extend to depths of 3 km are referred to as _____.
    a) slope canyons   c) rifts   e) submarine canyons
    b) abyssal plains   d) trenches

22. The most abundant salt in the sea is _____.
    a) calcium chloride   d) sodium chloride
    b) magnesium chloride   e) sodium fluoride
    c) potassium chloride

23. Where trenches do not exist, the steep continental slope merges into a more gradual incline known as the continental _____.
    a) abyss   c) end   e) run
    b) rise   d) coast

24. The deepest parts of the ocean are long, narrow features known as deep-ocean _____.
    a) ridges   c) scars   e) rifts
    b) trenches   d) holes

25. Isolated volcanic peaks on the ocean floor are known as _____.
    a) ridges   c) nodules   e) rifts
    b) abyssals   d) seamounts

26. Downslope movements of dense, sediment-laden water are known as _____ currents.
    a) avalanche   c) density   e) longshore
    b) turbidity   d) sediment

27. Which one of the following is NOT a water condition necessary for coral growth?
    a) warm   c) deep
    b) sunlit   d) clear

28. The speed of sound in water is about _____ meters per second.
    a) 500   c) 1500   e) 2500
    b) 1000   d) 2000

*True/false. For the following true/false questions, if a statement is not completely true, mark it false. For each false statement, change the **italicized** word to correct the statement.*

1. ____ Ocean temperature and salinity vary with depth and generally follow a *three-layered* structure in the open ocean.

2. ____ Because it has many shallow adjacent seas, the *Pacific* Ocean is the shallowest of the three major oceans.

3. ____ The true edge of the continent is the continental *slope*.

4. ____ In the ocean, *high* salinities are found where evaporation is high.

5. ____ Turbidity currents originate along the continental *rise*.

6. ___ *Low* ocean surface salinities occur where large quantities of fresh water are supplied by rivers.

7. ___ Deep-ocean *trenches* in the open ocean are often paralleled by volcanic island arcs.

8. ___ In the ocean where heavy precipitation occurs, *higher* surface salinities prevail.

9. ___ The most common sediment covering the deep-ocean floor is *sand*.

10. ___ Below the thermocline, temperatures *fall* only a few more degrees.

11. ___ Much of the geologic activity along mid-ocean ridges occurs in the *rift* zone.

12. ___ The continental *shelf* varies greatly in width from one continent to another.

13. ___ Deep-ocean trenches are the sites where moving crustal plates plunge back into the *mantle*.

14. ___ Mid-ocean ridges exist in *all* major oceans.

15. ___ The Northern Hemisphere can be referred to as the *land* hemisphere.

16. ___ Atolls owe their existence to the *sudden* sinking of oceanic crust.

Word choice. Complete each of the following statements by selecting the most appropriate response.

1. [Runoff/Evaporation] increases the salinity of sea water.

2. The continental shelf is underlaid by [oceanic/continental] crust.

3. The shallow surface layer of the ocean is called the [mixed/transition] zone.

4. The greatest number of seamounts have been identified in the [Pacific/Atlantic/Indian] Ocean.

5. The proportion of land to water is [larger/smaller] in the Northern Hemisphere than it is in the Southern Hemisphere.

6. A coral atoll forms during the [subsidence/uplift] of a seamount.

7. In low and middle latitudes, the salinity of the ocean's surface layer is [greater/less] than the deep-zone salinity.

8. During the Pleistocene epoch (Ice Age) sea level was [higher/lower] than it is today.

9. Turbidity currents often form [deep-sea fans/guyots] on the flat ocean floor.

10. The largest ocean is the [Atlantic/Pacific/Indian].

11. [Biogenous/Terrigenous] sediments are most helpful in determining past climates.

12. Variations in the salinity of ocean water are primarily the consequence of changes in the [salt/water] content of the solution.

13. Most of the geologic activity associated with a mid-ocean ridge occurs along the [rift/trench] zone.

14. Deep-ocean trenches are the sites where crustal plates plunge into Earth's [mantle/core].

15. The Atlantic Ocean has [more/less] extensive abyssal plains than the Pacific.

16. Coral reefs are built by [living organisms/chemical deposition].

17. Most seafloor sediment consists of [mud/sand].

18. Shells and skeletons of marine animals and plants form [biogenous/hydrogenous] sediment.

19. Mid-ocean ridges form along [divergent/convergent] plate boundaries.

Written questions

1. Describe the term *salinity*. What is the average salinity of the ocean?

2. What are the three major topographic units of the ocean basins?

3. Relate deep-ocean-trenches and mid-ocean ridges to seafloor spreading.

The Restless Ocean

13

The Restless Ocean opens with a discussion of ocean surface currents and their importance. Deep-ocean circulation is also briefly examined. Tides are analyzed, along with a detailed look at wave mechanics and wave erosion. Also investigated are the shoreline features and erosional problems that are caused by wave action and rising sea levels. The chapter concludes with a discussion of emergent and submergent coastal regions.

Learning Objectives

After reading, studying, and discussing this chapter, you should be able to:

- List the factors that influence surface ocean currents.
- Discuss the importance of surface ocean currents.
- Describe deep-ocean circulation.
- Discuss the factors that influence tides.
- Describe wave characteristics and types.
- Describe wave erosion and the features produced by wave erosion.
- Discuss shoreline erosional problems and solutions.
- Explain the differences between an emergent and submergent coast.

Chapter Summary

- *Surface ocean currents* are parts of huge, slowly moving, circular whirls, or *gyres*, that begin near the equator in each ocean. *Wind is the driving force* for the ocean's surface currents. Where wind is in contact with the ocean, it passes energy to the water through friction and causes the surface layer to move. The most significant factor other than wind that influences the movement of ocean waters is the *Coriolis effect*, the deflective force of Earth's rotation which causes free-moving objects to be deflected to the right in the Northern Hemisphere and to the left in the Southern Hemisphere. Because of the Coriolis effect, surface currents form clockwise gyres in the Northern Hemisphere and counterclockwise gyres in the Southern Hemisphere. Winds may also produce vertical water movements. *Upwelling* is the rising of cold water from deeper layers to replace warmer surface water. It is most characteristic along the eastern shores of the oceans.

- Ocean currents are important in navigation and travel and for the effect that they have on climates. The moderating effect of poleward-moving warm ocean currents during the winter in middle latitudes is well known.

- In contrast to surface currents, *deep-ocean circulation* is governed by gravity and driven by *density* differences. The two factors that are most significant in creating a dense mass of water are *temperature* and *salinity*.

• *Tides*, the daily rise and fall in the elevation of the ocean surface at a specific location, are caused by the *gravitational attraction* of the moon and, to a lesser extent, by the sun. Near the times of new and full moons, the sun and moon are aligned and their gravitational forces are added together to produce especially high and low tides. These are called the *spring tides*. Conversely, at about the times of the first and third quarters of the moon, when the gravitational forces of the moon and sun are at right angles, the daily *tidal range* is less. These are called *neap tides*. The three types of tides are 1) *semidiurnal*, with two high and two low tides each tidal day and relatively small differences in the high and low water heights, 2) *diurnal*, with a single high and low water height each tidal day, and 3) *mixed*, which have two high and two low waters each day and are characterized by a large inequality in high water heights, low water heights, or both.

• The three factors that influence the *height, wavelength*, and *period* of a wave are 1) *wind speed*, 2) *length of time the wind has blown*, and 3) *fetch*, the distance that the wind has traveled across the open water.

• The two types of wind-generated waves are 1) *waves of oscillation*, which are waves in the open sea in which the wave form advances as the water particles move in circular orbits, and 2) *waves of translation*, the turbulent advance of water formed near the shore as waves of oscillation collapse, or *break*, and form *surf.*

• Wave erosion is caused by *wave impact, pressure*, and *abrasion* (the sawing and grinding action of water armed with rock fragments). The bending of waves is called *wave refraction*. Due to refraction, wave impact is concentrated against the sides and ends of headlands.

• Most waves reach the shore at an angle. The uprush and backwash of water from each breaking wave moves the sediment in a zigzag pattern along the beach. This movement, called *beach drift*, can transport sand and pebbles hundreds or even thousands of meters each day. Oblique waves also produce *longshore currents* within the surf zone that flow parallel to the shore.

• Features produced by *shoreline erosion* include *wave-cut cliffs* (which originate from the cutting action of the surf against the base of coastal land), *wave-cut platforms* (relatively flat, benchlike surfaces left behind by receding cliffs), *sea arches* (formed when a headland is eroded and two caves from opposite sides unite), and *sea stacks* (formed when the roof of a sea arch collapses).

• Some of the features formed when sediment is moved by beach drift and longshore currents are *spits* (elongated ridges of sand that project from the land into the mouth of an adjacent bay), *baymouth bars* (sand bars that completely cross a bay), and *tombolos* (ridges of sand that connect an island to the mainland or to another island).

• Local factors that influence shoreline erosion are 1) the proximity of a coast to sediment-laden rivers, 2) the degree of tectonic activity, 3) the topography and composition of the land, 4) prevailing winds and weather patterns, and 5) the configuration of the coastline and nearshore areas.

• Three basic responses to shoreline erosion problems are 1) building *structures* such as *groins* (short walls built at a right angle to the shore to trap moving sand), *breakwaters* (structures built parallel to the shoreline to protect boats from the force of large breaking waves), and *seawalls* (barriers constructed to prevent waves from reaching the area behind the wall) to hold the shoreline in place, 2) *beach nourishment*, which involves the addition of sand to replenish eroding beaches, and 3) *relocation of* buildings away from the beach.

• Because of basic geological differences, the nature of shoreline erosion problems along America's Pacific and Atlantic Coasts is very different. Much of the development along the Atlantic and Gulf Coasts has occurred on barrier islands which receive the full force of major storms. A majority of the

Pacific Coast is characterized by narrow beaches backed by steep cliffs and mountain ranges. A major problem facing the Pacific shoreline is a narrowing of beaches because the natural flow of materials to the coast has been interrupted by dams built for irrigation and flood control.

• One frequently used classification of coasts is based upon changes that have occurred with respect to sea level. Emergent coasts, often with wave-cut cliffs and wave-cut platforms above sea level, develop either because an area experiences uplift or as a result of a drop in sea level. Conversely, submergent coasts, with their drowned river mouths, called *estuaries,* are created when sea level rises or the land adjacent to the sea subsides.

Chapter Outline

I. Ocean water movements
    A. Surface currents
        1. Huge, slowly moving gyres
        2. Generated by wind
        3. Related to atmospheric circulation
        4. Deflected by the Coriolis effect
            a. To the right in the Northern Hemisphere
            b. To the left in the Southern Hemisphere
        5. Importance of surface currents
            a. Navigation
            b. Influence climates
    B. Upwelling
        1. Vertical water movement
        2. Along eastern shores of oceans
        3. Caused by
            a. Coriolis effect
            b. Surface water moving from the shore
    C. Deep-water circulation
        1. Governed by gravity
        2 Driven by density differences
        3. Factors creating a dense mass of water
            a. Temperature—cold water is dense
            b. Salinity—density increases with increasing salinity
        4. Called thermohaline circulation
        5. Dense water masses are created in
            a. Arctic regions
            b. Antarctic regions
    D. Tides
        1. Changes in elevation of the ocean surface
        2. Caused by the gravitational forces of the
            a. Moon, and to a lesser extent the
            b. Sun
        3. Tidal heights
            a. Spring tide
                1. During new and full moons
                2. Gravitational forces added together
                3. Especially high and low tides
                4. Large daily tidal range
            b. Neap tide
                1. First and third quarters of the moon
                2. Gravitational forces are offset
                3. Daily tidal range is least
        4. Other influencing factors
            a. Shape of the coastline
            b. Configuration of the ocean basin
        5 Types of tides
            a. Semidiurnal tide
                1. Two high and two low tides each tidal day
                2. Little difference in the high and low water heights
                3. Occur along the Atlantic coast of the U.S.
            b. Diurnal tide
                1. A single high and low water height each tidal day
                2. Occur along the northern shore of the Gulf of Mexico
            c. Mixed tide
                1. Two high and two low waters each day
                2. Large inequality in high water heights, low water heights, or both
                3. Prevalent along the Pacific coast of the U.S.
        6. Tidal currents

a. Horizontal flow accompanying tides

b. Types of tidal currents

   1. Flood current—advances into the coastal zone

   2. Ebb current—seaward-moving water

E. Wind-generated waves

  1. Derive their energy and motion from wind

  2. Parts

    a. Crest

    b. Trough

  3. Measurements of a wave

    a. Wave height—the distance between a trough and a crest

    b. Wavelength—the horizontal distance between crests

    c. Wave period—the time interval between crests

  4. Height, length, and period depend upon

    a. Wind speed

    b. Length of time wind blows

    c. Fetch—the distance wind travels

  5. Types of waves

    a. Wave of oscillation

      1. In open sea

      2. Shape moves forward

    b. Wave of translation

      1. Wave breaks along the shore

      2. Water advances up the shore

      3. Forms surf

  6. Wave erosion is caused by

    a. Wave impact and pressure

    b. Abrasion by rock fragments

  7. Wave refraction

    a. Bending of a wave along a coast

    b. Wave arrives parallel to shore

    c. Results

      1. Wave energy directed against headland

      2. Wave erosion straightens an irregular shoreline

  8. Moving sand along the beach

    a. Beach drift—sediment moves in a zigzag pattern

    b. Longshore current

      1. Current in surf zone

      2. Flows parallel to coast

II. Shoreline features

A. Features caused by wave erosion

  1. Wave-cut cliff

  2. Wave-cut platform

  3. Associated with headlands

    a. Sea arch

    b. Sea stack

B. Related to beach drift and longshore currents

  1. Spit—a ridge of sand extending from the land into a bay with an end that often hooks landward

  2. Baymouth bar—a sand bar that completely crosses a bay

  3. Tombolo—connects an island to the mainland

C. Barrier island

  1. Mainly along the Atlantic and Gulf coasts

  2. Parallels the coast

  3. 3 to 30 kilometers offshore

  4. Originates in several ways

D. Result of shoreline erosion and deposition is to eventually produce a straighter coast

III. Shoreline erosion problems

A. Influenced by the local factors

  1. Proximity to sediment-laden rivers

  2. Degree of tectonic activity

  3. Topography and composition of the land

  4. Prevailing wind and weather patterns

  5. Configuration of the coastline

B. Three basic responses to erosion problems

  1. Building structures

    a. Types of structures

      1. Groin

        a. Barrier built at a right angle to the beach

        b. Traps sand

      2. Breakwater

        a. Barrier built offshore and parallel to the coast

        b. Protects boats from the force of large breaking waves

      3. Seawall

        a. Barrier parallel to shore

        b. Stops waves from reaching the beach areas behind the wall

    b. Often not effective

  2. Addition of sand to replenish the beaches

a. Called beach nourishment
b. Not a permanent solution
3. Relocation of buildings away from beach
C. Atlantic and Pacific Coasts
1. Shoreline erosion problems are different along the opposite coasts
2. Atlantic Coast
a. Broad, gently sloping coastal plains
b. Development occurs mainly on the barrier islands
1. Face open ocean
2. Receive full force of storms
3. Pacific Coast
a. Relatively narrow beaches backed by steep cliffs and mountain ranges
b. Major problem is the narrowing of the beaches

1. Sediment for beaches is interrupted by dams and reservoirs
2. Rapid erosion along beaches
IV. Emergent and submergent coasts
A. Emergent coast
1. Caused by
a. Uplift of an area, or
b. A drop in sea level
2. Features of an emergent coast
a. Wave-cut cliffs
b. Wave-cut platforms
B. Submergent coast
1. Caused by
a. Land adjacent to sea subsides, or
b. Sea level rises
2. Features of a submergent coast
a. Highly irregular shoreline
b. Estuaries—drowned river mouths

Key Terms

Page numbers shown in ( ) refer to the textbook page where the term first appears.

abrasion (p. 345)	longshore current (p. 346)	tidal flat (p. 340)
barrier island (p. 348)	neap tide (p. 340)	tide (p. 338)
baymouth bar (p. 347)	sea arch (p. 347)	tombolo (p. 348)
beach drift (p. 346)	sea stack (p. 347)	upwelling (p. 336)
beach nourishment (p. 355)	seawall (p. 354)	wave-cut cliff (p. 346)
breakwater (p. 354)	spit (p. 347)	wave-cut platform (p. 347)
Coriolis effect (p. 335)	spring tide (p. 340)	wave height (p. 341)
emergent coast (p. 356)	submergent coast (p. 357)	wavelength (p. 341)
estuary (p. 357)	surf (p. 345)	wave of oscillation (p. 342)
fetch (p. 341)	thermohaline circulation (p. 337)	wave of translation (p. 345)
groin (p. 352)	tidal current (p. 340)	wave period (p. 341)
gyre (p. 334)	tidal delta (p. 340)	wave refraction (p. 345)

Vocabulary Review

Choosing from the list of key terms, furnish the most appropriate response for the following statements.

1. An elongated ridge of sand that projects from the land into the mouth of an adjacent bay is called a(n) _____.

2. _____ is the movement of sediment in a zigzag pattern along a beach.

3. The large circular surface current pattern found in each ocean is call a(n) _____.

4. The _____ is the vertical distance between the trough and crest of a wave.

5. Deep-ocean circulation is called _____.

6. Turbulent water created by breaking waves is called _____.

7. The deflective force of Earth's rotation on all free-moving objects is known as the _____.

8. A wave in open water is called a(n) _____.

9. The horizontal distance separating successive wave crests is the _____.

10. A(n) _____ is a low ridge of sand that parallels a coast at a distance from 3 to 30 kilometers offshore.

11. The horizontal flow of water accompanying the rise and fall of a tide is called a(n) _____.

12. When a wave of oscillation breaks along a shore it becomes a type of wave called a(n) _____.

13. The rising of colder water from deeper ocean layers to replace warmer surface water is called _____.

14. A(n) _____ is a ridge of sand that connects an island either to the mainland or to another island.

15. The time interval between the successive passage of wave crests is known as the _____.

16. When a sea arch collapses it leaves an isolated remnant called a(n) _____.

17. A coast that develops either because the area experiences uplift or a drop in sea level is a(n) _____.

18. A(n) _____ is a drowned river mouth.

19. The area affected by alternating tidal currents is known as a(n) _____.

20. The bending of a wave is called _____.

21. _____ is the term that describes the distance that wind has traveled across the open water.

22. A near-shore current that flows parallel to the shore is called a(n) _____.

23. A(n) _____ is a relatively flat, benchlike surface left behind by a receding wave-cut cliff.

24. The _____ is the daily change in the elevation of the ocean surface.

25. A(n) _____ is a barrier built at a right angle to the beach to trap sand that is moving parallel to the shore.

26. The highest tidal variation that occurs during the times of new and full moons is called a(n) _____.

27. A(n) _____ forms when a rapidly moving tidal current emerges from a narrow inlet and slows, depositing its load of sediment.

Comprehensive Review

1. Briefly describe the influence that the Coriolis effect has on the movement of ocean waters.

2. Using Figure 13.1, select the letter that identifies each of the following surface currents.

   a) Gulf Stream: ____   d) West Wind Drift: ____   g) North Pacific Drift: ____

   b) Peruvian:____   e) Benguela:____   h) Equatorial:____

   c) Canaries:____   f) Labrador:____

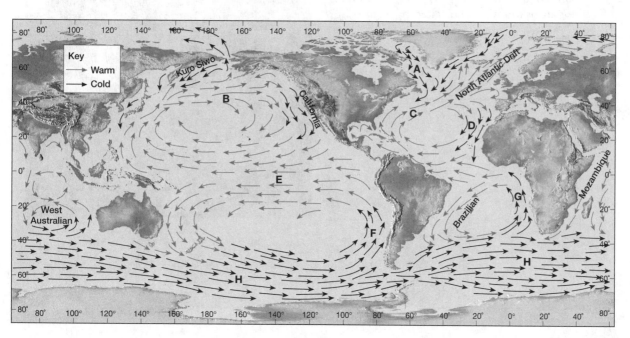

Figure 13.1

3. Describe the effect cold ocean currents have on climates.

4. In addition to the gravitational influence of the sun and moon, what are two other factors that influence the tide?

   1)

   2)

5. Using Figure 13.2, select the letter of the diagram that illustrates the relation of the sun, moon, and Earth for each of the following types of tides.

    a) Neap tide: _____

    b) Spring tide: _____

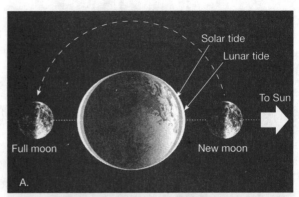

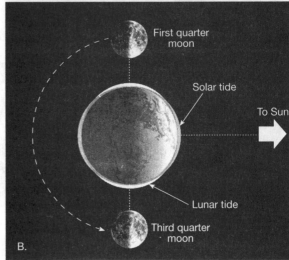

Figure 13.2

6. Using Figure 13.3, write the proper term for the part of the wave illustrated by each of the following letters.

    a) Letter A: _____

    b) Letter B: _____

    c) Letter C: _____

    d) Letter D: _____

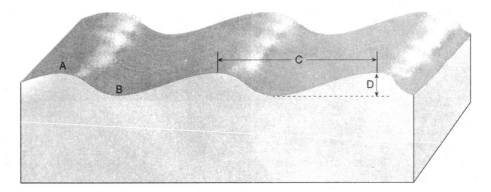

Figure 13.3

7. List the three factors that influence the height, length, and period of open water waves.

    1)

    2)

    3)

8. What is the difference between a wave of oscillation and a wave of translation?

9. What two factors are most significant in creating a dense mass of ocean water?

    1)

    2)

10. Where are the two major regions where dense water masses are created?

    1)

    2)

11. Describe the effect that wave refraction has on the energy of waves in a bay.

12. List two mechanisms which are responsible for transporting sediment along a coast.

    1)

    2)

13. Figure 13.4 illustrates the changes that take place through time along an initially irregular, relatively stable coastline. Using Figure 13.4, select the letter that illustrates each of the following features.

    a) Baymouth bar: ____

    b) Spit: ____

    c) Sea stack: ____

    d) Estuary: ____

    e) Wave-cut cliff: ____

    f) Tombolo: ____

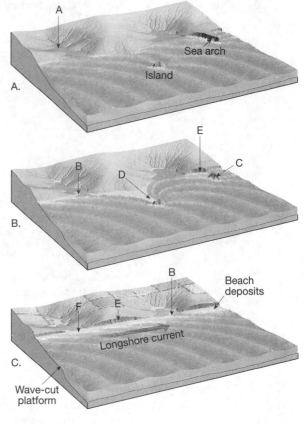

Figure 13.4

14. What are two ways that barrier islands may originate?

    1)

    2)

15. What are three factors that influence the type and degree of shoreline erosion?

    1)

    2)

    3)

16. Briefly describe the conditions that could result in the formation of the following types of coasts. Also list one feature you would expect to find associated with each type.

    a) Emergent coast:

        Feature: _____

b) Submergent coast:

　　Feature: _____

17. Briefly describe the high and low water characteristics of each of the three types of tides.

　　a) Semidiurnal tide:

　　b) Diurnal tide:

　　c) Mixed tide:

Practice Test

Multiple choice. Choose the best answer for the following multiple choice questions.

1. The movement of particles of sediment in a zigzag pattern along a beach is known as _____.
   - a) longshore transportation
   - b) beach drift
   - c) turbidity current
   - d) beach abrasion
   - e) fetch

2. A barrier built at a right angle to the beach to trap sand that is moving parallel to the shore is known as a(n) _____.
   - a) seawall
   - b) groin
   - c) pier
   - d) stack
   - e) headland

3. A ridge of sand that connects an island to the mainland or to another island is called a _____.
   - a) baymouth bar
   - b) sea arch
   - c) sea stack
   - d) tombolo
   - e) spit

4. Where the atmosphere and ocean are in contact, energy is passed from the moving air to the water through _____.
   - a) friction
   - b) radiation
   - c) oscillation
   - d) collision
   - e) translation

5. Waves in the open sea are called waves of _____.
   - a) translation
   - b) motion
   - c) modulation
   - d) combination
   - e) oscillation

6. Which one of the following is NOT a west coast desert influenced by cold offshore waters?
   - a) Namib
   - b) Gobi
   - c) Atacama

7. Exposed wave-cut cliffs and platforms are frequently found along _____ coasts.
   - a) submergent
   - b) emergent

8. The densest water in all of the oceans is formed in _____ waters.
   a) Equatorial   c) North Pacific   e) Indian Ocean
   b) Antarctic   d) North Atlantic

9. The energy and motion of most waves is derived from _____.
   a) currents   c) gravity   e) earthquakes
   b) tides   d) wind

10. Other than wind, the most significant factor that influences the movement of ocean water is the
    _____.
    a) translation force   c) force of gravity   e) ocean slope
    b) Coriolis effect   d) density effect

11. Which one of the following features occurs more frequently along the Atlantic and Gulf Coasts
    than along the Pacific Coast?
    a) wave-cut cliffs   c) sea stacks
    b) barrier islands   d) wave-cut platforms

12. Tidal currents that advance into the coastal zone as the tide rises are called _____
    currents.
    a) density   c) flow   e) ebb
    b) turbidity   d) longshore

13. Equatorial currents derive their energy principally from the _____ winds.
    a) easterly   c) monsoon   e) trade
    b) westerly   d) chinook

14. In the open sea, the movement of water particles in a wave becomes negligible at a depth equal
    to about _____ the distance from crest to trough.
    a) one-fourth   c) one-half   e) three-fourths
    b) one-third   d) two-thirds

15. Deep-ocean circulation is referred to as _____ circulation.
    a) density   c) ocean-floor   e) abyssal
    b) subsurface   d) thermohaline

16. When a wave breaks it becomes a wave of _____.
    a) translation   c) modulation   e) oscillation
    b) motion   d) combination

17. The areas affected by alternating tidal currents are known as tidal _____.
    a) swamps   c) flats   e) estuaries
    b) marshes   d) platforms

18. Daily changes in the elevation of the ocean surface are called _____.
    a) currents   c) density currents   e) estuaries
    b) waves   d) tides

19. The large central area of the North Atlantic which has no well-defined currents is known as the
    _____ Sea.
    a) Red   c) Sargasso   e) Central
    b) Baltic   d) Central Atlantic

20. Highly irregular coasts with numerous estuaries frequently characterize a(n) _____ coast.
    a) submergent
    b) emergent

21. The horizontal distance separating wave crests is known as the _____.
    a) wavelength   c) wave height   e) oscillation
    b) fetch   d) wave period

22. The largest daily tidal ranges are associated with _____ tides.
    a) spring   c) neap   e) lunar
    b) ebb   d) flow

23. The time interval between the passage of successive wave crests is called the _____.
    a) wave height   c) wave period   e) oscillation
    b) wavelength   d) fetch

24. A wave begins to "feel bottom" at a water depth equal to about _____ its wavelength.
    a) one-fourth   c) one-half   e) three-fourths
    b) one-third   d) two-thirds

25. Massive barriers built to prevent waves from reaching the shore are called _____.
    a) groins   c) stacks   e) piers
    b) seawalls   d) barrier islands

26. The greatest difference between high and low water heights occurs during a _____ tide.
    a) spring   c) neap   e) ebb
    b) flow   d) back

27. Which one of the following types of tides is most common along the Atlantic coast of the United States?
    a) mixed tide   c) semidiurnal tide   e) diurnal tides
    b) ebb tide   d) flow tide

*True/false. For the following true/false questions, if a statement is not completely true, mark it false. For each false statement, change the **italicized** word to correct the statement.*

1. ___ Turbulent water currents within the surf zone that flow parallel to the shore are called *longshore* currents.

2. ___ Evaporation and/or freezing tend to make the water *less* salty.

3. ___ *Reach* is the distance that wind has traveled across the open water.

4. ___ The two most significant factors in creating a dense mass of water are temperature and *salinity*.

5. ___ In the open sea, it is the wave *form* that moves forward, not the water itself.

6. ___ In the Southern Hemisphere, ocean currents form *counterclockwise* gyres.

7. ___ Groins have proven to be *unsatisfactory* solutions to shoreline erosion problems.

8. ___ The tide-generating potential of the sun is slightly less than *half* that of the moon.

9. ___ Two caves on opposite sides of a headland can unite to form a sea *cliff.*

10. ___ *Cold* ocean currents often have a dramatic impact on the climate of tropical deserts along the west coasts of continents.

11. ___ Ocean currents transfer *excess* heat from the tropics to the heat-deficient polar regions.

12. ___ Wind speeds less than *three* kilometers per hour generally do not produce stable, progressive waves.

13. ___ In the Northern Hemisphere, the Coriolis effect causes ocean currents to be deflected to the *left.*

14. ___ Turbulent water created by breaking waves is called *drift.*

15. ___ The lowest daily tidal ranges are associated with *neap* tides.

Word choice. Complete each of the following statements by selecting the most appropriate response.

1. In deep water, the motion of a water particle is best described as [circular/zigzag/vertical].

2. Increasing the salinity of sea water will cause its density to [increase/decrease].

3. The greatest amount of shoreline erosion is caused by [currents/tides/waves].

4. The Coriolis effect causes ocean currents to be [clockwise/counterclockwise] in the Northern Hemisphere and [clockwise/counterclockwise] in the Southern Hemisphere.

5. Daily tidal changes are the result of Earth's [rotation/revolution].

6. The presence of wave-cut terraces above sea level is one line of evidence that the area is a(n) [emergent/submergent] coast.

7. Recent research indicates that sea level is [falling/rising].

8. Ocean currents affect climate by storing and transporting [heat/plankton/water].

9. The Pacific Coast of North America is undergoing [uplift/subsidence].

10. The largest thermohaline currents begin in [equatorial/polar] regions.

11. When waves break they become waves of [oscillation/translation].

12. If Earth had an ocean with uniform depth covering its surface, the moon would produce [one/two/three] tidal bulges.

13. Waves are created by [gravity/tides/wind].

14. A breakwater will often cause the beach behind it to [gain/lose] sand.

15. Damming rivers has caused west coast beaches to [widen/narrow].

16. Generally, if the tidal range is less than [eight/eighteen] meters, or if narrow, enclosed bays are absent, tidal power development is uneconomical.

17. The zigzag movement of sediment along a beach is called [beach/longshore] drift.

18. The horizontal flow of water associated with rising and falling tides is called a tidal [current/fetch].

19. Beaches downcurrent from groins will usually [gain/lose] sand.

20. [Barrier/Tombolo] islands consist of low ridges of sand that parallel coastlines.

21. A highly irregular coastline is indicative of a(n) [emergent/submergent] coast.

22. Narrowing of beaches results in the cliffs behind them being [more/less] vulnerable to erosion.

23. [Reach/Fetch/Period] is the distance that wind has blown over the water.

24. The bending of waves is called [oscillation/refraction].

25. Upwelling is caused by [gravity/tides/winds].

Written questions

1. Describe coastal upwelling and the effect it has on fish populations.

2. Briefly describe thermohaline circulation.

3. What is the cause of ocean tides?

4. Describe the motion of a water particle in a wave of oscillation and in a wave of translation.

Composition, Structure, and Temperature

14

Composition, Structure, and Temperature introduces the subject of meteorology by presenting its definition, noting the differences between weather and climate, and listing the elements of weather. Included is a discussion of the major gases (nitrogen and oxygen) and variable components (water vapor, dust, and ozone) of the atmosphere. The structure and extent of the atmosphere are also examined.

Atmospheric heating begins by examining Earth's motions. The variables that control the quantity of solar radiation intercepted by a particular place are discussed. This is followed by a detailed investigation of the seasons. Following an examination of radiant energy and the common mechanisms of heat transfer is a discussion of atmospheric heating by solar and terrestrial radiation. Various temperature measurements are also explained. The factors that cause variations in temperature, such as differential heating of land and water, altitude, geographic position, cloud cover, and albedo, are investigated. The chapter concludes with a short description of the global distribution of Earth's surface temperatures.

Learning Objectives

After reading, studying, and discussing this chapter, you should be able to:

- Describe the science of meteorology.
- Explain the difference between weather and climate.
- List the most important elements of weather and climate.
- List the major and variable components of air.
- Describe the extent and structure of the atmosphere.
- Describe how the atmosphere is heated.
- Explain the causes of the seasons.
- List the factors that cause temperature to vary from place to place.
- Describe the general distribution of global surface temperatures.

Chapter Summary

- *Weather* is the state of the atmosphere at a particular place for a short period of time. *Climate*, on the other hand, is a generalization of the weather conditions of a place over a long period of time.

- The most important *elements*, those quantities or properties that are measured regularly, of weather and climate are 1) air *temperature*, 2) *humidity*, 3) type and amount of *cloudiness*, 4) type and amount of *precipitation*, 5) air *pressure*, and 6) the speed and direction of the *wind*.

- If water vapor, dust, ozone, and other variable components of the atmosphere were removed, clean, dry air would be composed almost entirely of *nitrogen* (N), about 78 percent of the atmosphere by volume, and *oxygen* (O_2), about 21 percent. *Carbon dioxide* (CO_2), although present only in minute amounts (0.035 percent), is important because it has the ability to absorb heat radiated by Earth and thus helps keep the atmosphere warm. Among the variable components of air, *water vapor* is very important because it is the source of all clouds and precipitation and, like carbon dioxide, it is also a heat absorber. *Ozone* (O_3), the triatomic form of oxygen, is concentrated in the 10- to 50-kilometer altitude range of the atmosphere; it is important to life because of its ability to absorb potentially harmful ultraviolet radiation from the sun.

• Because the atmosphere gradually thins with increasing altitude, it has no sharp upper boundary but simply blends into outer space. Based on temperature, the atmosphere is divided vertically into four layers. The *troposphere* is the lowermost layer. In the troposphere, temperature usually decreases with increasing altitude. This *environmental lapse rate* is variable, but averages about 6.5°C per kilometer (3.5°F per 1000 feet). Essentially all important weather phenomena occur in the troposphere. Above the troposphere is the *stratosphere*, which exhibits warming because of absorption of ultraviolet radiation by ozone. In the *mesosphere*, temperatures again decrease. Upward from the mesosphere is the *thermosphere*, a layer with only a minute fraction of the atmosphere's mass and no well-defined upper limit.

• The two principal motions of Earth are 1) *rotation*, the spinning of Earth about its axis, which produces the daily cycle of daylight and darkness, and 2) *revolution*, the movement of Earth in its orbit around the sun.

• *Several factors act together to cause the seasons. Earth's axis is inclined 23 ½° from the perpendicular to the plane of its orbit around the sun and remains pointed in the same direction* (toward the North Star) as Earth journeys around the sun. As a consequence, Earth's orientation to the sun continually changes. The yearly fluctuations in the angle of the sun and length of daylight brought about by Earth's changing orientation to the sun cause seasons. The four days that have special seasonal significance are the *summer solstice, winter solstice, autumnal equinox,* and the *spring equinox.*

• The three mechanisms of heat transfer are 1) *conduction*, the transfer of heat through matter by molecular activity, 2) *convection*, the transfer of heat by the movement of a mass or substance from one place to another, and 3) *radiation*, the transfer of heat by electromagnetic waves.

• *Electromagnetic radiation* is energy emitted in the form of rays, or waves, called electromagnetic waves. All radiation is capable of transmitting energy through the vacuum of space. One of the most important differences between electromagnetic waves is their *wavelengths*, which range from very long *radio waves* to very short *gamma rays*. *Visible light* is the only portion of the electromagnetic spectrum we can see. Some of the basic laws that govern radiation as it heats the atmosphere are 1) all objects emit radiant energy; 2) hotter objects radiate more total energy than do colder objects; 3) the hotter the radiating body, the shorter the wavelengths of maximum radiation; and 4) objects that are good absorbers of radiation are good emitters as well.

• *The atmosphere is largely transparent to incoming solar radiation.* Approximately 50 percent of the incoming solar radiation is ultimately absorbed at Earth's surface, while the remainder is either *absorbed, scattered,* or *reflected* back to space by the atmosphere. The fraction of the total radiation that is reflected by a surface is called its *albedo.*

• The general drop in temperature with increasing altitude in the troposphere supports the fact that *the atmosphere is heated from the ground up.* The solar energy, primarily in the form of the shorter wavelengths, that penetrates through the atmosphere is ultimately absorbed at Earth's surface. Earth releases the absorbed radiation in the form of long-wave radiation. The atmospheric absorption of this long-wave *terrestrial radiation*, primarily by water vapor and carbon dioxide, is responsible for heating the atmosphere. This very important phenomenon has been termed the *greenhouse effect.*

• Various temperature measurements of a place may include the *daily mean temperature, daily range, monthly mean, annual mean,* and *annual temperature.* Human comfort and the sensation of temperature the human body feels is influenced by several factors including *air temperature, relative humidity, wind,* and *solar radiation.*

• The factors that cause temperature to vary from place to place, also called the *controls of temperature,* are 1) differences in the *receipt of solar radiation*—the greatest single cause, 2) the unequal *heating and cooling* of *land and water,* in which land heats more rapidly and to higher temperatures than water and cools more rapidly and to lower temperatures than water, 3) *altitude,* 4) *geographic position,* 5) *cloud cover,* and 6) the *albedo* of the surface.

• Temperature distribution is shown on a map by using *isotherms,* which are lines that connect equal temperatures.

Chapter Outline

I. Weather and climate
    A. Weather
       1. Weather measured over a short
          period of time
       2. Constantly changing
    B. Climate
       1. Climate measured over a long period
          of time
       2. Generalized, composite of weather
    C. Elements of weather and climate
       1. Properties that are measured
          regularly
       2. Most important elements
          a. Temperature
          b. Humidity
          c. Cloudiness
          d. Precipitation
          e. Air pressure
          f. Wind speed and direction
II. Composition of the atmosphere
    A. Air is a mixture of discrete gases
    B. Major components of clean, dry air
       1. Nitrogen (N)—78 percent
       2. Oxygen (O_2)—21 percent
       3. Argon and other gases
       4. Carbon dioxide (CO_2)—0.035
          percent—absorbs heat energy
          from Earth
    C. Variable components of air
       1. Water vapor
          a. Up to about 4 percent of the
             air's volume
          b. Forms clouds and precipitation
          c. Absorbs heat energy from Earth

       2. Dust
          a. Includes pollen and spores
          b. Water vapor condenses on
          c. Reflects sunlight
          d. Helps color sunrise and sunset
       3. Ozone
          a. Three-atom oxygen (O_3)
          b. Distribution not uniform
          c. Concentrated between 10 to 50
             kilometers above the surface
          d. Absorbs harmful UV radiation
          e. Human activity is depleting ozone
             by adding chlorofluorocarbons
             (CFCs)
III. Structure of the atmosphere
    A. Pressure changes
       1. Pressure is the weight of the air
          above
       2. Average sea level pressure
          a. Slightly more than 1000 millibars
          b. About 14.7 pounds per square
             inch
       3. Pressure decreases with altitude
          a. One-half of the atmosphere is
             below 3.5 miles (5.6 km)
          b. 90 percent of the atmosphere is
             below 10 miles (16 km)
    B. Atmospheric layers based on temperature
       1. Troposphere
          a. Bottom layer
          b. Temperature decreases
             with altitude—called the
             environmental lapse rate
             1. 6.5°C per kilometer (average)

2. 3.5°F per 1000 feet (average)
  c. Thickness varies—average height is about 12 km
  d. Outer boundary is named the tropopause
2. Stratosphere
  a. About 12 km to 50 km
  b. Temperature increases at top
  c. Outer boundary is named the stratopause
3. Mesosphere
  a. About 50 km to 80 km
  b. Temperature decreases
  c. Outer boundary is named the mesopause
4. Thermosphere
  a. No well-defined upper limit
  b. Fraction of atmosphere's mass
  c. Gases moving at high speeds

IV. Earth-sun relations
  A. Earth motions
    1. Rotates on its axis
    2. Revolves around the sun
  B. Seasons
    1. Result of
      a. Changing sun angle
      b. Changing length of daylight
    2. Caused by Earth's changing orientation to the sun
      a. Axis is inclined 23½°
      b. Axis is always pointed in the same direction
    3. Special days (Northern Hemisphere)
      a. Summer solstice
        1. June 21-22
        2. Sun's vertical rays are located at the Tropic of Cancer (23½°N latitude)
      b. Winter solstice
        1. December 21-22
        2. Sun's vertical rays are located at the Tropic of Capricorn (23½°S latitude)
      c. Autumnal equinox
        1. September 22-23
        2. Sun's vertical rays are located at the Equator (0° latitude)
      d. Spring equinox
        1. March 21-22
        2. Sun's vertical rays are located at the Equator (0° latitude

V. Atmospheric heating
  A. Heat is always transferred from warmer to cooler objects
  B. Mechanisms of heat transfer
    1. Conduction through molecular activity
    2. Convection
      a. Movement of a mass
      b. Usually vertical motions
      c. Horizontal motions are called advection
    3. Radiation (electromagnetic radiaion)
      a. Velocity: 186,000 miles per second in a vacuum
      b. Consists of different wavelengths
        1. Gamma (very short waves)
        2. X-rays
        3. Ultraviolet (UV)
        4. Visible
        5. Infrared
        6. Microwaves
        7. Radio (longest waves)
      c. Governed by basic laws
  C. Incoming solar radiation
    1. Atmosphere is largely transparent to incoming solar radiation
    2. Atmospheric effects
      a. Scattering
      b. Reflection—albedo (percent reflected)
      c. Absorption
    3. Most visible radiation reaches the surface
    4. 50 percent absorbed at Earth's surface
  D. Radiation from Earth's surface
    1. Earth reradiates radiation at the longer wavelengths
    2. Longer terrestrial radiation is absorbed by
      a. Carbon dioxide and
      b. Water vapor
    3. Lower atmosphere is heated from Earth's surface

VI. Temperature measurement
  A. Daily maximum and minimum
  B. Other measurements
    1. Daily mean temperature
    2. Daily range
    3. Monthly mean
    4. Annual mean
    5. Annual temperature range

C. Human perception of temperature
 1. Anything that influences the rate of heat loss from the body also influences the sensation of temperature
 2. Important factors are
  a. Air temperature
  b. Relative humidity
  c. Wind
  d. Solar radiation

VII. Controls of temperature
 A. Cause temperature to vary
 B. Receipt of solar radiation is the most important control
 C. Other important controls
  1. Differential heating of land and water
   a. Land heats more rapidly than water
   b. Land gets hotter than water
   c. Land cools faster than water
   d. Land gets cooler than water
  2. Altitude
  3. Geographic position
  4. Cloud cover

  5. Surface albedo
VIII. World distribution of temperature
 A. Temperature maps
  1. Isotherm—a line connecting places of equal temperature
  2. Temperatures adjusted to sea level
  3. January and July used for analysis because they represent the temperature extremes
 B. Global temperature patterns
  1. Temperature decreases poleward from the tropics
  2. Isotherms exhibit a latitudinal shift with the seasons
  3. Warmest and coldest temperatures occur over land
  4. In the Southern Hemisphere
   a. Isotherms are straighter
   b. Isotherms are more stable
  5. Isotherms show ocean currents
  6. Annual temperature range
   a. Small near equator
   b. Increases with an increase in latitude
   c. Greatest over continental locations

Key Terms

Page numbers shown in ( ) refer to the textbook page where the term first appears.

air (p. 363)
albedo (p. 377)
autumnal equinox (p. 371)
circle of illumination (p. 369)
climate (p. 363)
conduction (p. 375)
convection (p. 375)
electromagnetic radiation (p. 375)
element (of weather and climate) (p. 363)

environmental lapse rate (p. 367)
greenhouse effect (p. 379)
inclination of the axis (p. 371)
infrared (p. 376)
isotherm (p. 383)
mesosphere (p. 367)
radiation (p. 375)
revolution (p. 361)
rotation (p. 369)
spring equinox (p. 372)

stratosphere (p. 367)
summer solstice (p. 371)
thermosphere (p. 368)
Tropic of Cancer (p. 371)
Tropic of Capricorn (p. 371)
troposphere (p. 367)
ultraviolet (p. 376)
visible light (p. 376)
weather (p. 363)
winter solstice (p. 371)

Vocabulary Review

Choosing from the list of key terms, furnish the most appropriate response for the following statements.

1. The bottom layer of the atmosphere, where temperature decreases with an increase in altitude, is the _____.

2. _____ is a word used to denote the state of the atmosphere at a particular place for a short period of time.

3. _____ is the spinning of Earth about its axis.

4. The outermost layer of the atmosphere is the _____.

5. _____ might be best described as an aggregate or composite of weather.

6. A line on a map connecting points of equal temperature is called a(n) _____.

7. The _____ is located at 23½ degrees north latitude.

8. A quantity or property of weather and climate that is measured regularly is termed a(n) _____.

9. _____ is the transfer of heat by the movement of a mass or substance from one place to another, usually vertically.

10. The fact that Earth's axis is not perpendicular to the plane of its orbit is referred to as the _____.

11. In the Northern Hemisphere, the time when the vertical rays of the sun strike the equator moving southward is known as the _____.

12. The temperature decrease with increasing altitude in the troposphere is called the _____.

13. The _____ is located at 23½ degrees south latitude.

14. The great circle that separates daylight from darkness is the _____.

15. In the atmospheric layer known as the _____, temperatures decrease with height until approximately 80 kilometers above the surface.

16. The transfer of energy through space by electromagnetic waves is known as _____.

17. The movement of Earth in its orbit around the sun is referred to as _____.

18. _____ is the transfer of heat through matter by molecular activity.

19. The atmospheric layer immediately above the tropopause is the _____.

20. Although we cannot see it, we detect _____ radiation as heat.

21. In the Southern Hemisphere, the time when the vertical rays of the sun are striking the Tropic of Cancer is known as the _____.

22. Intense exposure to _____ radiation is responsible for sunburn.

23. The percentage of the total radiation that is reflected by a surface is called its _____.

24. The _____ refers to the mechanism responsible for heating the atmosphere.

25. In the Northern Hemisphere, the time when the vertical rays of the sun are striking the Tropic of Cancer is known as the _____.

Comprehensive Review

1. Distinguish between the terms *weather* and *climate*.

2. Using Figure 14.1, select the letter that indicates each of the following parts of Earth's atmosphere.

   a) Mesosphere: ____

   b) Tropopause: ____

   c) Stratosphere: ____

   d) Troposphere: ____

   e) Stratopause: ____

   f) Thermosphere: ____

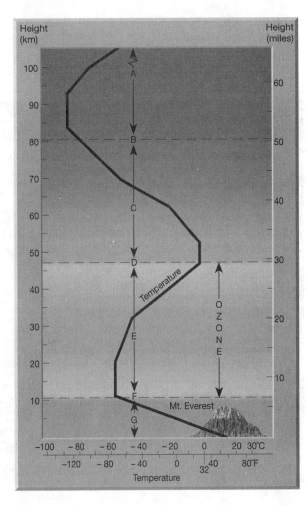

Figure 14.1

3. List and describe the two principal motions of Earth.

   1)

   2)

4. List the six most important elements of weather and climate.

   1)   4)

   2)   5)

   3)   6)

5. List two of the meteorological roles of atmospheric dust.

   1)

   2)

6. What are the two ways that the seasonal variation in the altitude of the sun affect the amount of solar energy received at Earth's surface?

   1)

   2)

7. Using Figure 14.2, list the Northern Hemisphere season, date, and latitude of the vertical sun for each of the four lettered positions.

   a) Position A:

   b) Position B:

   c) Position C:

   d) Position D:

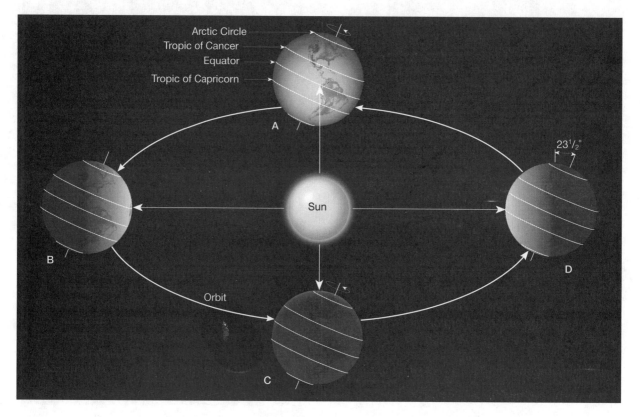

Figure 14.2

8. What are the two most abundant gases in clean, dry air? What are their percentages?

9. How does ozone differ from oxygen? What function does atmospheric ozone serve, especially for those of us on Earth?

10. In an average situation, approximately what percentage of incoming solar radiation is

   a) Absorbed by the atmosphere and clouds: _____ percent

   b) Absorbed by Earth's surface: _____ percent

   c) Reflected by clouds: _____ percent

11. Beginning with the fact that more incoming solar radiation is absorbed by Earth's surface than by the atmosphere, describe the mechanism responsible for atmospheric heating.

12. Which two atmospheric gases are the principal absorbers of heat energy?

13. List and briefly describe the three mechanisms of heat transfer.

1)

2)

3)

14. Briefly explain how each of the following temperatures are determined.

a) Daily mean temperature:

b) Annual mean temperature:

c) Annual temperature range:

15. What is the effect of each of the following on temperature?

a) Differential heating of land and water:

b) Altitude:

c) Geographic position:

d) Cloud cover:

16. Why are the months of January and July often selected for the analysis of global temperature patterns?

Practice Test

Multiple choice. Choose the best answer for the following multiple choice questions.

1. Which one of the following is NOT generally considered a major element of weather and climate?
   a) ocean currents   c) air temperature   e) wind speed
   b) humidity   d) air pressure

2. A form of oxygen that combines three oxygen atoms into each molecule is called _____.
   a) trioxygen   c) oxulene   e) azide
   b) oxyurid   d) ozone

3. Oxygen and ozone are efficient absorbers of incoming _____ radiation.
    a) ultraviolet   c) radio   e) gamma
    b) infrared   d) visible

4. The generalization of atmospheric conditions over a long period of time is referred to as _____.
    a) meteorology   c) global synthesis   e) element analysis
    b) climate   d) weather

5. On June 21, the vertical rays of the sun strike a line of latitude known as the _____.
    a) equator   c) Tropic of Capricorn e) Antarctic circle
    b) Arctic circle   d) Tropic of Cancer

6. The red and orange colors of sunset and sunrise are the result of _____ in the atmosphere.
    a) moisture   c) insects   e) argon
    b) dust   d) pollen

7. The bottom layer of the atmosphere in which we live is called the _____.
    a) mesosphere   c) troposphere   e) stratosphere
    b) thermosphere   d) hemisphere

8. Which one of the following forms of radiation has the longest wavelength?
    a) radio waves   c) blue light   e) infrared
    b) ultraviolet   d) yellow light

9. Ozone in the atmosphere is concentrated in a layer called the _____.
    a) troposphere   c) mesosphere   e) hemisphere
    b) stratosphere   d) thermosphere

10. Which one of the following is NOT a mechanism of heat transfer?
    a) conduction   c) radiation
    b) condensation   d) convection

11. The temperature decrease with altitude in the troposphere is called the _____ lapse rate.
    a) environmental   c) ascending   e) incremental
    b) altitude   d) atmospheric

12. On December 21, the vertical rays of the sun strike a line of latitude known as the _____.
    a) equator   c) Tropic of Capricorn e) Antarctic circle
    b) Arctic circle   d) Tropic of Cancer

13. Which one of the following gases has the ability to absorb heat energy radiated by Earth?
    a) oxygen   c) ozone   e) carbon dioxide
    b) helium   d) nitrogen

14. The chemicals contributing to the depletion of atmospheric ozone are referred to as _____.
    a) CCCs   c) FCDs   e) CFDs
    b) CFCs   d) PCBs

15. Approximately what percentage of the solar energy that strikes the top of the atmosphere reaches Earth's surface?
    a) 30 percent   c) 50 percent   e) 70 percent
    b) 40 percent   d) 60 percent

16. Which one of the following is NOT an important control of air temperature?
    a) heating of land and water   d) geographic position
    b) altitude   e) ocean currents
    c) wind

17. The two principal atmospheric absorbers of terrestrial radiation are carbon dioxide and _____.
    a) oxygen   c) nitrogen   e) helium
    b) water vapor   d) ozone

18. When water changes from one state to another, _____ heat is stored or released.
    a) latent   c) fusion   e) elemental
    b) vapor   d) reduced

19. Which one of the following yearly fluctuations is NOT responsible for the seasons?
    a) length of daylight   c) Earth-sun distance
    b) sun angle

20. The value of the normal lapse rate is _____C degrees per kilometer.
    a) 3.5   c) 5.5   e) 7.5
    b) 4.5   d) 6.5

21. Ninety percent of the atmosphere lies below _____ miles.
    a) 5   c) 15   e) 25
    b) 10   d) 20

22. In the Southern Hemisphere, the greatest number of daylight hours occurs on _____.
    a) June 21   c) March 21   e) September 21
    b) December 21   d) April 21

23. On the equinoxes, the vertical rays of the sun strike a line of latitude known as the _____.
    a) equator   c) Tropic of Capricorn   e) Antarctic circle
    b) Arctic circle   d) Tropic of Cancer

24. The amount of energy received at Earth's surface is controlled by the _____ of the sun.
    a) temperature   c) distance   e) rotation
    b) altitude   d) color

25. Temperature distribution is shown on a map using lines called _____ that connect places of equal temperature.
    a) isotemps   c) isobars
    b) equagrads   d) isotherms

26. Which form of radiation do we detect as heat?
    a) radio waves   c) red light   e) infrared
    b) ultraviolet   d) yellow light

27. Combustion of fossil fuels adds vast quantities of the gas _____ to the atmosphere.
    a) oxygen   c) ozone   e) carbon dioxide
    b) water vapor   d) nitrogen

28. The annual temperature range will be greatest for a place located _____.
    a) at the pole   d) at high altitude
    b) on the equator   e) near the coast
    c) in the interior of a continent

29. Which one of the following surfaces has the highest albedo?
    a) concrete   c) soil   e) forest
    b) snow   d) water

30. The most abundant gas in clean, dry air is _____.
    a) oxygen   c) nitrogen   e) ozone
    b) argon   d) carbon dioxide

*True/false. For the following true/false questions, if a statement is not completely true, mark it false. For each false statement, change the **italicized** word to correct the statement.*

1. ___ In the Northern Hemisphere, December 21 is referred to as the *winter* solstice.

2. ___ *Climate* is a word used to denote the state of the atmosphere at a particular place over a short period of time.

3. ___ The *higher* the solar angle or altitude of the sun, the more spread out and less intense is the solar radiation that reaches Earth's surface.

4. ___ On the *equinoxes*, all places on Earth receive 12 hours of daylight.

5. ___ Atmospheric ozone is an absorber of potentially harmful *ultraviolet* radiation from the sun.

6. ___ The transfer of heat through matter by molecular activity is called *radiation*.

7. ___ An international agreement known as the *Montreal Protocol* established a schedule for eliminating CFCs from general use.

8. ___ The most important difference among electromagnetic waves is their *wavelength*.

9. ___ If Earth's axis were not *inclined*, we would have no seasons.

10. ___ The atmosphere is divided into four layers on the basis of *composition*.

11. ___ The *higher* the solar angle or altitude of the sun, the less distance solar radiation must penetrate through the atmosphere to Earth's surface.

12. ___ Essentially all important weather occurs in the *stratosphere*.

13. ___ The line separating the dark half of Earth from the lighted half is called the circle of *daylight*.

14. ___ In the Southern Hemisphere, June 21 is referred to as the *summer* solstice.

15. ___ There is a *sharp* boundary between the atmosphere and outer space.

16. ___ The hotter the radiating body, the *shorter* the wavelength of maximum radiation.

17. ___ Earth's two principal motions are rotation and *revolution*.

18. ___ In the United States, between 1970 and the early 1990s the yearly output of primary pollutants *increased*.

19. ___ Heating an object causes its atoms to move *slowly*.

20. ___ The percentage of the total radiation that is reflected by an object is called its *albedo*.

21. ___ The principal source for atmospheric heat is *terrestrial* radiation.

22. ___ Land heats *more* rapidly than water.

23. ___ Compared to the Northern Hemisphere, the Southern Hemisphere has a *greater* annual temperature variation.

24. ___ One category of pollutants, the *primary* pollutants, are emitted directly from identifiable sources.

25. ___ The *visible* wavelengths of radiation from the sun reach Earth's surface.

Word choice. Complete each of the following statements by selecting the most appropriate response.

1. The composition of air [is constant/varies].

2. On June 21 the vertical rays of the sun strike the Tropic of [Cancer/Capricorn].

3. Temperature [increases/decreases] with an increase in altitude in the troposphere.

4. The changes in distance between Earth and the sun throughout the year have [major/insignificant] influence on weather and climate.

5. Hotter objects radiate [more/less] total energy per unit area than colder objects.

6. The average value of the environmental lapse rate is [3.5/5.2]°F per 1000 feet.

7. Temperature variations are [greater/smaller] over large bodies of water than they are over land.

8. Carbon dioxide in the air has the ability to absorb [water/heat energy].

9. The seasonal north-south migration of isotherms is [greater/less] over the continents than over the oceans.

10. Air pressure [increases/decreases] as altitude increases.

11. Earth's axis is inclined [53.2/23.5] degrees from the perpendicular.

12. When water vapor changes from one state of matter to another it absorbs or releases [conduction/latent] heat.

13. Ozone in the atmosphere absorbs the potentially harmful [ultraviolet/infrared] radiation from the sun.

14. On the date of an equinox, the vertical rays of the sun strike the [equator/tropics].

15. The generalization of the weather conditions of a place over a long period of time is called [weather/climate].

16. All other factors being the same, the greatest daily temperature range would occur in a(n) [arid/humid] region.

17. The atmosphere is divided into four principal layers on the basis of [composition/temperature].

18. The greatest annual extremes of temperature occur over [land/water].

19. The upper boundary of the atmosphere [is/is not] sharply defined.

20. The hotter the object, the [longer/shorter] the wavelength of maximum radiation.

21. Annual temperature ranges are [greatest/least] near the equator.

22. The only mechanism of heat transfer that can transmit heat through the relative emptiness of space is [conduction/radiation].

23. Increased cloud cover will [increase/decrease] the rate of nighttime cooling.

24. Most of the radiation that heats the atmosphere directly is emitted by [the sun/Earth].

25. The transfer of heat by the movement of a mass or substance from one place to another is called [conduction/convection].

26. The specific heat for water is [greater/less] than the specific heat for land.

Written questions

1. What is the relation between CFCs and the ozone problem?

2. What causes the amount of solar energy reaching places on Earth's surface to vary with the seasons?

3. Briefly describe how Earth's atmosphere is heated.

Moisture

15

Moisture begins with an examination of the processes and energy requirements involved in the changes of state of water, including the ability of water to store and release latent heat. After presenting the various methods used to express humidity, the relations between temperature, water vapor content, relative humidity, and dew-point temperature are explored. Also investigated are the adiabatic process and condensation aloft. Following a detailed explanation of the conditions that influence the stability of air, the three processes that lift air are described. Included in the chapter are the classification of clouds as well as the formation and types of fog. The chapter concludes with a discussion of the processes involved in the formation of precipitation and the various forms of precipitation.

Learning Objectives

After reading, studying, and discussing this chapter, you should be able to:

- List the processes that cause water to change from one of state of matter to another.
- Explain saturation, vapor pressure, specific humidity, relative humidity, and dew point.
- Describe how relative humidity is determined.
- Explain the basic cloud-forming process.
- Describe stable and unstable air.
- List the three processes that initiate the vertical movement of air.
- Discuss the conditions necessary for condensation.
- List the criteria used to classify clouds.
- Describe the formation of fog.
- Discuss the formation and forms of precipitation.

Chapter Summary

- *Water vapor*, an odorless, colorless gas, changes from one state of matter (solid, liquid, or gas) to another at the temperatures and pressures experienced near Earth's surface. The processes involved in *changing the state of matter* of water are *evaporation, condensation, melting, freezing, sublimation,* and *deposition.*

- *Humidity* is the general term used to describe the amount of water vapor in the air. *Relative humidity,* the *ratio* (expressed as a percent) of the air's water vapor *content* to its water vapor *capacity* at a given temperature, is the most familiar term used to describe humidity. The water vapor capacity of air is temperature dependent, with warm air having a much greater capacity than cold air.

- *Relative humidity can be changed in two ways.* The first is by *adding or subtracting water vapor.* The second is by *changing the air's temperature.* When air is cooled, its relative humidity increases. Air is said to be *saturated* when it contains the maximum quantity of water vapor that it can hold at any given temperature and pressure. *Dew point* is the temperature to which air would have to be cooled in order to reach saturation.

• The cooling of air as it rises and expands due to successively lower pressure is the basic cloud-forming process. Temperature changes in air brought about by expanding or compressing the air are called *adiabatic temperature changes*. Unsaturated air warms by compression and cools by expansion at the rather constant rate of $10°C$ per 1000 meters of altitude change, a figure called the *dry adiabatic rate*. If air rises high enough it will cool sufficiently to cause condensation and form a cloud. From this point on, air that continues to rise will cool at the *wet adiabatic rate* which varies from $5°C$ to $9°C$ per 1000 meters of ascent. The difference in the wet and dry adiabatic rates is caused by the condensing water vapor releasing *latent heat*, thereby reducing the rate at which the air cools.

• The *stability of air* is determined by examining the temperature of the atmosphere at various altitudes. Air is said to be *unstable* when the *environmental lapse rate* (the rate of temperature decrease with increasing altitude in the troposphere) is greater than the *dry adiabatic rate*. Stated differently, a column of air is unstable when the air near the bottom is significantly warmer (less dense) than the air aloft.

• Three mechanisms that can initiate the vertical movement of air are 1) *orographic lifting*, which occurs when elevated terrains, such as mountains, act as barriers to the flow of air; 2) *frontal wedging*, when cool air acts as a barrier over which warmer, less dense air rises; and 3) *convergence*, which happens when air flows together and a general upward movement of air occurs.

• *For condensation to occur, air must be saturated*. Saturation takes place either when air is cooled to its dew point, which most commonly happens, or when water vapor is added to the air. There must also be a surface on which the water vapor may condense. In cloud and fog formation, tiny particles called *condensation nuclei* serve this purpose.

• *Clouds* are classified on the basis of their *appearance* and *height*. The three basic forms are *cirrus* (high, white, thin, wispy fibers), *cumulus* (globular, individual cloud masses), and *stratus* (sheets or layers that cover much or all of the sky). The four categories based on height are *high clouds* (bases normally above 6000 meters), *middle clouds* (from 2000 to 6000 meters), *low clouds* (below 2000 meters), and *clouds of vertical development*.

• *Fog* is defined as a cloud with its base at or very near the ground. Fogs form when air is cooled below its dew point or when enough water vapor is added to the air to bring about saturation. Various types of fog include *advection fog*, *radiation fog*, *upslope fog*, *steam fog*, and *frontal* (or *precipitation*), *fog*.

• For *precipitation* to form, millions of cloud droplets must somehow join together into large drops. Two mechanisms for the formation of precipitation have been proposed. First, in clouds where the temperatures are below freezing, ice crystals form and fall as snowflakes. At lower altitudes the snowflakes melt and become raindrops before they reach the ground. In the second mechanism, large droplets form in warm clouds that contain large *hygroscopic* ("water seeking") *nuclei*, such as salt particles. As these big droplets descend, they collide and join with smaller water droplets. After many collisions the droplets are large enough to fall to the ground as rain.

• The forms of precipitation include *rain, snow, sleet, hail,* and *rime.*

Chapter Outline ▬▬▬▬▬▬▬▬▬▬▬▬▬▬▬▬▬▬▬▬▬▬▬

I. Changes of state of water
   A. Three states of matter
      1. Solid
      2. Liquid
      3. Gas
   B. To change state, heat must be
      1. Absorbed, or
      2. Released
   C. Heat energy
      1. Measured in calories—one calorie is the heat necessary to raise the temperature of one gram of water one degree Celsius
      2. Latent heat
         a. Stored or hidden heat
         b. Not derived from temperature change
         c. Important in atmospheric processes
   D. Processes
      1. Evaporation
         a. Liquid is changed to gas
         b. 600 calories per gram of water are added—called latent heat of vaporization
      2. Condensation
         a. Water vapor (gas) is changed to a liquid
         b. Heat energy is released—called latent heat of condensation
      3. Melting
         a. Solid is changed to a liquid
         b. 80 calories per gram of water are added—called latent heat of melting
      4. Freezing
         a. Liquid is changed to a solid
         b. Heat is released—called latent heat of fusion
      5. Sublimation
         a. Solid is changed directly to a gas (e.g., ice cubes shrinking in a freezer)
         b. 680 calories per gram of water are added
      6. Deposition
         a. Water vapor (gas) changed to a solid (e.g., frost in a freezer compartment)
         b. Heat is released

II. Humidity
   A. Amount of water vapor in the air
      1. Water vapor adds pressure (called vapor pressure) to the air
      2. Saturated air is air that is filled with water vapor to capacity
      3. Capacity is temperature dependent—warm air has a much greater capacity
   B. Measurements of humidity
      1. Specific humidity
         a. Quantity of water vapor in a given mass of air
         b. Often measured in grams per kilogram
      2. Relative humidity
         a. Ratio of the air's actual water vapor content to its potential water vapor capacity, at a given temperature
         b. Expressed as a percent
         c. Saturated air
            1. Content equals capacity
            2. Has a 100 percent relative humidity
         d. Relative humidity can be changed in two ways
            1. Add or subtract moisture to the air
               a. Adding moisture raises the relative humidity
               b. Removing moisture lowers the relative humidity
            2. Changing the air temperature
               a. Lowering the temperature raises the relative humidity
               b. Raising the temperature lowers the relative humidity
         e. Dew point
            1. Temperature at which the air is saturated and the relative humidity is 100 percent
            2. Cooling the air below the dew point causes condensation
               a. e.g., Cloud formation
               b. Water vapor requires a surface to condense on
         f. Instruments used to measure humidity

1. Psychrometer
   a. Compares temperatures of
      1. Wet-bulb thermometer, and
      2. Dry-bulb thermometer
   b. If the air is saturated (100 percent relative humidity) then both thermometers read the same temperature
   c. The greater the difference between the thermometer readings, the lower the relative humidity
2. Hygrometer—reads the humidity directly

III. Adiabatic heating/cooling
A. Adiabatic temperature changes occur when
   1. Air is compressed
      a. Motion of air molecules increases
      b. Air will warm
      c. Descending air is compressed due to increasing air pressure
   2. Air expands
      a. Air parcel does work on the surrounding air
      b. Air will cool
      c. Rising air will expand due to decreasing air pressure
B. Adiabatic rates
   1. Dry adiabatic rate
      a. Unsaturated air
      b. Rising air expands and cools at 1°C per 100 meters
      c. Descending air is compressed and warms at 1°C per 100 meters
   2. Wet adiabatic rate
      a. Commences at condensation level
      b. Air has reached the dew point
      c. Condensation is occurring and latent heat is being liberated
      d. Heat released by the condensing water reduces the rate of cooling
      e. Rate varies from 0.5°C to 0.9°C per 100 meters

IV. Stability of air
A. Two types of stability
   1. Stable air
      a. Resists vertical displacement
         1. Cooler than surrounding air
         2. Denser than surrounding air
         3. Wants to sink
      b. No adiabatic cooling
      c. Stability occurs when the environmental lapse rate is less than the wet adiabatic rate
      d. Often results in widespread clouds with little vertical thickness
      e. Precipitation, if any, is light to moderate
   2. Unstable air
      a. Acts like a hot air balloon
      b. Rising air
         1. Warmer than surrounding air
         2. Less dense than surrounding air
         3. Continues to rise until it reaches an altitude with the same temperature
      c. Adiabatic cooling
      d. Air is unstable when the environmental lapse rate is greater than the dry adiabatic rate
      e. Clouds are often towering
      f. Often results in heavy precipitation
      g. Conditional instability occurs when the atmosphere is stable for an unsaturated parcel of air but unstable for a saturated parcel
B. Determines to a large degree
   1. Type of clouds that develop
   2. Intensity of the precipitation

V. Processes that lift air
A. Orographic lifting
   1. Elevated terrains act as barriers
   2. Result can be a rainshadow desert
B. Frontal wedging
   1. Cool air acts as a barrier to warm air
   2. Fronts are part of the storm systems called middle-latitude cyclones
C. Convergence where the air is flowing together and rising

VI. Condensation and cloud formation
A. Condensation
   1. Water vapor in the air changes to a liquid and forms dew, fog, or clouds
   2. Water vapor requires a surface to condense on
      a. Possible condensation surfaces on the ground can be the grass, a car window, etc.
      b. Possible condensation surfaces in the atmosphere are tiny bits of particulate matter
         1. Called condensation nuclei

2. Dust, smoke, etc.
3. Ocean salt crystals which serve as hygroscopic ("water seeking") nuclei

B. Clouds
1. Made of millions and millions of
   a. Minute water droplets, or
   b. Tiny crystals of ice
2. Classification based on
   a. Form (three basic forms)
      1. Cirrus—high, white, thin
      2. Cumulus
         a. Globular cloud masses
         b. Often associated with fair weather
      3. Stratus
         a. Sheets or layers
         b. Cover much or all of the sky
   b. Height
      1. High clouds
         a. Above 6000 meters
         b. Types
            1. Cirrus
            2. Cirrostratus
            3. Cirrocumulus
      2. Middle clouds
         a. 2000 to 6000 meters
         b. Types (alto as part of the name)
            1. Altocumulus
            2. Altostratus
      3. Low clouds
         a. Below 2000 meters
         b. Types
            1. Stratus
            2. Stratocumulus
            3. Nimbostratus (nimbus means "rainy")
      4. Clouds of vertical development
         a. From low to high altitudes
         b. Called cumulonimbus
         c. Often produce
            1. Rain showers
            2. Thunderstorms

VII. Fog
A. Considered an atmospheric hazard
B. Cloud with its base at or near the ground
C. Most fogs form because of
   1. Radiation cooling, or
   2. Movement of air over a cold surface
D. Types of fog

1. Fogs caused by cooling
   a. Advection fog—warm, moist air moves over a cool surface
   b. Radiation fog
      1. Earth's surface cools rapidly
      2. Forms during cool, clear, calm nights
   c. Upslope fog
      1. Humid air moves up a slope
      2. Adiabatic cooling occurs
2. Evaporation fogs
   a. Steam fog
      1. Cool air moves over warm water and moisture is added to the air
      2. Water has a steaming appearnce
   b. Frontal fog, or precipitation fog
      1. Forms during frontal wedging when warm air lifted over colder air
      2. Rain evaporates to form fog

VIII. Precipitation
A. Cloud droplets
   1. Less than 10 micrometers in diameter
   2. Fall incredibly slow
B. Formation of precipitation
   1. Bergeron process
      a. Temperature in the cloud is below freezing
      b. Ice crystals collect water vapor
      c. Large snowflakes form and
         1. Fall to the ground as snow, or
         2. Melt on their descent and form rain
   2. Collision-coalescence process
      a. Warm clouds
      b. Large hygroscopic condensation nuclei
      c. Large droplets form
      d. Droplets collide with other droplets during their descent
C. Forms of precipitation
   1. Rain and drizzle
      a. Rain—droplets have at least a 0.5 mm diameter
      b. Drizzle—droplets have less than a 0.5 mm diameter
   2. Snow—ice crystals, or aggregates of ice crystals
   3. Sleet and glaze
      a. Sleet

1. Wintertime phenomenon
2. Small particles of ice
3. Occurs when
   a. Warmer air overlies colder air
   b. Rain freezes as it falls
   b. Glaze, or freezing rain—impact with a solid causes freezing
   c. Hail
      1. Hard rounded pellets
         a. Concentric shells
         b. Most diameters range from 1 to 5 cm
      2. Formation
         a. Occurs in large cumulonimbus clouds with violent up- and downdrafts
         b. Layers of freezing rain are caught in up-and-down drafts in the cloud
         c. Pellets fall to the ground when they become too heavy

d. Rime
   1. Forms on cold surfaces
   2. Freezing of
      a. Supercooled fog, or
      b. Cloud droplets
D. Measuring precipitation
   1. Rain
      a. Easiest form to measure
      b. Measuring instruments
         1. Standard rain gauge
            a. Uses a funnel to collect and conduct rain
            b. Cylindrical measuring tube measures rainfall in centimeters or inches
         2. Recording gauge
   2. Snow has two measurements
      a. Depth
      b. Water equivalent
         1. General ratio is 10 snow units to 1 water unit
         2. Varies widely

Key Terms

Page numbers shown in ( ) refer to the textbook page where the term first appears.

absolute instability (p. 398)
absolute stability (p. 398)
adiabatic temperature change (p. 396)
advection fog (p. 408)
Bergeron process (p. 411)
calorie (p. 388)
cirrus (p. 404)
cloud (p. 404)
cloud of vertical development (p. 407)
collision-coalescence process (p. 411)
condensation (p. 389)
condensation nuclei (p. 404)
conditional instability (p. 398)
convergence (p. 402)
cumulus (p. 404)
deposition (p. 389)

dew point (p. 391)
dry adiabatic rate (p. 396)
evaporation (p. 389)
fog (p. 408)
freezing (p. 389)
freezing nuclei (p. 411)
frontal fog (p. 410)
frontal wedging (p. 402)
glaze (p. 413)
hail (p. 413)
high cloud (p. 405)
humidity (p. 389)
hygrometer (p. 394)
hygroscopic nuclei (p. 404)
latent heat (p. 388)
low cloud (p. 405)
melting (p. 389)
orographic lifting (p. 402)
precipitation fog (p. 410)

psychrometer (p. 393)
radiation fog (p. 409)
rain (p. 412)
rainshadow desert (p. 402)
relative humidity (p. 390)
rime (p. 414)
saturation (p. 389)
sleet (p. 413)
snow (p. 413)
specific humidity (p. 390)
steam fog (p. 409)
stratus (p. 404)
sublimation (p. 389)
supercooled (p. 411)
upslope fog (p. 409)
vapor pressure (p. 390)
wet adiabatic rate (p. 396)

Vocabulary Review

Choosing from the list of key terms, furnish the most appropriate response for the following statements.

1. The general term for the amount of water vapor in air is _____.

2. The process whereby water vapor changes to the liquid state is called _____.

3. _____ is the ratio of the air's actual water vapor content to its potential water vapor capacity at a given temperature.

4. _____ results when elevated terrains, such as mountains, act as barriers to flowing air.

5. The part of total atmospheric pressure attributable to water vapor content is called the _____.

6. A(n) _____ results when air is compressed or allowed to expand.

7. _____ refers to the energy stored or released during a change of state.

8. A(n) _____ is best described as visible aggregates of minute water droplets or tiny crystals of ice suspended in the air.

9. Frozen or semifrozen rain formed when raindrops freeze as they pass through a layer of cold air is called _____.

10. The term _____ refers to the maximum possible quantity of water vapor that the air can hold at any given temperature and pressure.

11. The rate of adiabatic temperature change in saturated air is the _____.

12. _____ is the conversion of a solid directly to a gas, without passing through the liquid state.

13. Sheets or layers of clouds that cover much or all of the sky are called _____ clouds.

14. The weight of water vapor per weight of a chosen mass of air (e.g., grams per kilogram), including the water vapor, is called _____.

15. _____ occurs when cool air acts as a barrier over which warmer, less dense air rises.

16. In the atmosphere, tiny bits of particulate matter, known as _____, serve as surfaces for water vapor condensation.

17. The process by which a solid is changed to a liquid is referred to as _____.

18. In meteorology, the term _____ is restricted to drops of water that fall from a cloud and have a diameter of at least 0.5 millimeter.

19. The temperature to which air would have to be cooled to reach saturation is the _____.

20. _____ is a deposit of ice crystals formed by the freezing of supercooled fog or cloud droplets on objects whose surface temperature is below freezing.

21. The conversion of a vapor directly to a solid is called _____.

22. Nearly spherical ice pellets which have concentric layers and are formed by the successive freezing of layers of water are referred to as _____.

23. The adiabatic rate of cooling or heating which applies only to unsaturated air is the _____.

24. Water absorbing particles, such as salt, are termed _____.

25. The process of converting a liquid to a gas is called _____.

26. _____ is a type of fog that forms on cool, clear, calm nights, when Earth's surface cools rapidly by radiation.

27. A cloud with its base at or very near Earth's surface is referred to as _____.

28. A dry area on the lee side of a mountain that forms as air descends and is warmed by compression, is called a(n) _____.

29. The _____, an instrument used to measure relative humidity, consists of two identical thermometers mounted side by side.

30. Whenever air masses flow together, _____ is said to occur.

31. A(n) _____, the common unit of heat energy, is the amount of heat required to raise the temperature of 1 gram of water 1°C.

32. Water in the liquid state below 0°C is referred to as _____ water.

33. The _____ is an instrument used to measure relative humidity that can be read directly, without the use of tables.

Comprehensive Review

1. Using Figure 15.1, select the letter that indicates each of the following processes.

a) Freezing: ____   d) Melting: ____

b) Evaporation: ____   e) Sublimation: ____

c) Deposition: ____   f) Condensation: ____

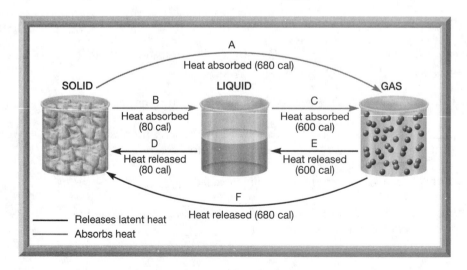

Figure 15.1

2. List two different ways that the relative humidity of a parcel of air can be increased.

    1)

    2)

3. Using Table 15.1 in the textbook, determine the relative humidity of a kilogram of air that has the following temperature and water vapor content.

    a) Temperature: 25°C; Water vapor content: 5 grams/kilogram

    b) Temperature: 41°F; Water vapor content: 3 grams/kilogram

4. Using Table 15.1 in the textbook, determine the dew point of a kilogram of air that has the following temperature and relative humidity.

    a) Temperature: 25°C; Relative humidity: 50 percent

    b) Temperature: 59°F; Relative humidity: 20 percent

5. Explain the principle of the psychrometer for measuring relative humidity.

6. Using Table 15.2 in the textbook, determine the relative humidity of a parcel of air that yielded the following psychrometer readings:

    a) Dry-bulb temperature: 24°C; Wet-bulb temperature: 16°C

7. Why is the wet adiabatic rate of cooling of air less than the dry adiabatic rate?

8. What is the difference between stable and unstable air?

9. Briefly describe each of the following processes that lift air.

    a) Orographic lifting:

    b) Frontal wedging:

    c) Convergence:

10. What two criteria are used as the basis for cloud classification?

    1)

    2)

11. Write the name of the cloud described by each of the following statements.

   a) Low, layered clouds with a uniform fog-like appearance that frequently cover much of the sky:

   b) High, thin, delicate clouds that sometimes appear as hooked filaments called "mares' tails":

   c) Dense, fluffy, globular masses found between 2000 and 6000 meters above the surface:

12. Using Figure 15.2, select the letter of the photograph that illustrates the following cloud types.

   a) Cumulus: _____   b) Cirrus: _____

A.

B.   *Figure 15.2*

13. Briefly describe the circumstances responsible for the formation of each of the following types of fog.

    a) Steam fog:

    b) Frontal fog:

14. List and briefly describe the two mechanisms that have been proposed to explain how precipitation forms.

    1)

    2)

15. Briefly describe the circumstances that result in the formation of each of the following forms of precipitation.

    a) Sleet:

    b) Hail:

Practice Test

Multiple choice. Choose the best answer for the following multiple choice questions.

1. The ratio of the air's water vapor content to its capacity at that same temperature is the _____.
   - a) vapor pressure
   - b) specific humidity
   - c) wet adiabatic rate
   - d) relative humidity
   - e) dry adiabatic rate

2. The term _____ means "rainy cloud."
   - a) ferrous
   - b) nimbus
   - c) stratus
   - d) cirrus
   - e) cumulus

3. Which one of the following changes from one state of matter to another at the temperatures and pressures experienced at Earth's surface?
   - a) oxygen
   - b) water
   - c) carbon dioxide
   - d) nitrogen
   - e) methane

4. Which of the following is NOT a necessary condition for condensation?
   - a) high altitude
   - b) dew-point temperature reached
   - c) surfaces
   - d) saturation
   - e) water vapor

5. Which one of the following processes involves the greatest quantity of heat energy?
   - a) melting
   - b) sublimation
   - c) freezing
   - d) evaporation
   - e) condensation

6. Which one of the following is NOT a process that lifts air?
   - a) convergence
   - b) orographic lifting
   - c) divergence
   - d) frontal wedging

7. Which form of precipitation is likely to occur when a layer of warm air with temperatures above freezing overlies a subfreezing layer near the ground?
    a) rain   c) snow   e) drizzle
    b) sleet   d) hail

8. Air that has reached its water vapor capacity is said to be _____.
    a) dry   c) soaked   e) saturated
    b) unstable   d) stable

9. Which clouds are best described as sheets or layers that cover much or all of the sky?
    a) cirrus   c) stratus   e) cumulonimbus
    b) cumulus   d) altocumulus

10. The temperature to which air would have to be cooled to reach saturation is called the _____ point.
    a) vapor   c) adiabatic   e) critical
    b) dew   d) sublimation

11. When the environmental lapse rate is less than the dry adiabatic rate, a parcel of air will be _____.
    a) stable   b) unstable

12. Which one of the following refers to the energy that is stored or released during a change of state of water?
    a) caloric heat   c) latent heat   e) evaporation heat
    b) ultraviolet heat   d) geothermal heat

13. The process where cool air acts as a barrier over which warmer, less dense air rises is called _____.
    a) divergence   c) orographic lifting   e) subduction
    b) frontal wedging   d) collision

14. Relative humidity is typically highest at _____.
    a) sunrise   c) late afternoon   e) midnight
    b) noon   d) early evening

15. With which cloud type is hail most associated?
    a) cirrus   c) cumulonimbus   e) cumulus
    b) stratus   d) nimbostratus

16. The wet adiabatic rate of cooling is less than the dry rate because _____.
    a) wet air is unsaturated   d) wet air is more common
    b) of the release of latent heat   e) of the dew point
    c) dry air is less dense

17. The conditions that favor the formation of radiation fog are _____.
    a) warm, cloudy, calm nights   d) cool, clear, calm nights
    b) cool, cloudy, windy nights   e) warm, cloudy, windy nights
    c) warm, clear, windy nights

18. The process responsible for deposits of white frost or hoar frost is _____.
    a) sublimation   c) freezing   e) deposition
    b) melting   d) evaporation

19. When the environmental lapse rate is greater than the dry adiabatic rate, a parcel of air will be
    _____.
    a) stable
    b) unstable

20. Compared to clouds, fogs are _____.
    a) colder
    b) thicker
    c) of a different composition
    d) at lower altitudes
    e) drier

21. Which clouds are high, white, and thin?
    a) cirrus
    b) cumulus
    c) stratus
    d) nimbostratus
    e) cumulonimbus

22. When air expands or contracts it will experience _____ temperature changes.
    a) volume
    b) radiation
    c) external
    d) adiabatic
    e) unnecessary

23. Which one of the following is NOT a process that has been proposed to explain the formation of precipitation?
    a) atmospheric subduction process
    b) collision-coalescence process
    c) Bergeron process

24. The conversion of a solid directly to a gas, without passing through the liquid state, is called
    _____.
    a) sublimation
    b) melting
    c) freezing
    d) evaporation
    e) deposition

25. Weather-producing fronts are parts of the storm systems called _____.
    a) mid-latitude cyclones
    b) hurricanes
    c) tornadoes
    d) tropical storms
    e) typhoons

26. Which one of the following is NOT produced by condensation?
    a) dew
    b) smog
    c) fog
    d) clouds

27. Which one of the following is a deposit of ice crystals formed by the freezing of supercooled fog or cloud droplets on objects whose surface temperature is below freezing?
    a) hail
    b) sleet
    c) snow
    d) rime
    e) glaze

28. The dry adiabatic rate is _____.
    a) 3.0°C/100 meters
    b) 0.5°C/meter
    c) 1°C/100 meters
    d) 1°C/1000 meters
    e) 0.5°C/1000 meters

29. That part of total atmospheric pressure that can be attributed to the water vapor content is called
    _____ pressure.
    a) moisture
    b) isostatic
    c) vapor
    d) gas
    e) water

30. Which one of the following is NOT a type of fog?
    a) white fog
    b) radiation fog
    c) upslope fog
    d) steam fog
    e) frontal fog

*True/false. For the following true/false questions, if a statement is not completely true, mark it false. For each false statement, change the **italicized** word to correct the statement.*

1. ___ At the *dew-point* temperature, the air is both saturated and has a 100 percent relative humidity.

2. ___ A *dyne* is the amount of heat required to raise the temperature of 1 gram of water 1°C.

3. ___ Clouds are classified on the basis of their form and *color*.

4. ___ When water vapor condenses, it *releases* heat energy.

5. ___ When air expands, it will *warm*.

6. ___ When air masses flow together, *convergence* is said to occur.

7. ___ Assuming equal volumes, a *greater* quantity of water vapor is required to saturate warm air than cold air.

8. ___ Clouds associated with *unstable* air are towering and usually accompanied by heavy precipitation.

9. ___ The principle of the psychrometer relies on the fact that the amount of cooling of the *dry-bulb* thermometer is directly related to the dryness of the air.

10. ___ Above the ground, tiny bits of particulate matter serve as *condensation* nuclei.

11. ___ The water vapor capacity of air is *temperature*-dependent.

12. ___ The *dry* adiabatic rate varies from about 5°C/1000 meters in moist air to 9°C/1000 meters in dry air.

13. ___ When the water vapor content remains constant, a(n) *increase* in temperature results in an increase in relative humidity.

14. ___ A raindrop large enough to fall to the ground contains roughly one *million* times more water than a single cloud droplet.

15. ___ The energy absorbed by water molecules during evaporation is referred to as latent heat of *fusion*.

16. ___ When the surface temperature is about 4°C or higher, snowflakes usually melt before they reach the ground and continue their descent as rain.

17. ___ *Sleet* is defined as a cloud with its base at or very near the ground.

18. ___ Although highly variable, a general snow/water ratio of *ten* units of snow to one unit of water is often used when exact information is not available.

Word choice. Complete each of the following statements by selecting the most appropriate response.

1. When water vapor condenses, it [absorbs/releases] heat.

2. The process of raindrop formation that relies on the fact that "giant" cloud droplets fall more rapidly than smaller droplets is referred to as the [collision-coalescence/Bergeron/evaporation] process.

3. The primary difference between a fog and a cloud is in the [droplet size/altitude/dew point].

4. The ratio of water vapor in the air to the air's water vapor capacity at the same temperature is called the [specific/relative/absolute] humidity.

5. Freezing nuclei are [more/less] common in the atmosphere than condensation nuclei.

6. [Hail/Sleet/Glaze] is liquid raindrops that freeze as they fall through a layer of sub-freezing air.

7. [More/Less] water vapor is required to saturate warm air than cold air.

8. Air that has a tendency to rise on its own is referred to as [stable/unstable] air.

9. As the relative humidity increases, the air's dew point [increases/decreases/remains the same].

10. Sinking air [compresses/expands] and therefore [warms/cools] adiabatically.

11. When the environmental lapse rate is greater than the dry adiabatic rate, the air exhibits [absolute instability/conditional instability/absolute stability].

12. When water vapor evaporates, it [absorbs/releases] heat.

13. [Cirrus/Stratus/Cumulus] clouds are best described as sheets or layers that cover much or all of the sky.

14. Cirrus clouds are most often composed of [ice crystals/water droplets].

15. The two ways of increasing relative humidity are by [increasing/decreasing] the temperature of the air and/or by [increasing/decreasing] the amount of water vapor in the air.

16. The wet adiabatic rate is [higher/lower] than the dry adiabatic rate due to the release of [residual/latent/kelvin] heat during [condensation/evaporation].

17. Ice crystals that grow by "absorbing" water vapor that is evaporating from liquid cloud droplets describes the [collision-coalescence/Bergeron/evaporation] process for the formation of precipitation.

18. When using a psychrometer, the greater the difference between the dry and wet bulb temperatures, the [higher/lower] the relative humidity.

19. Whenever air masses flow together, [convergence/divergence] occurs and the result is a general [upward/downward] flow of air.

20. Orographic lifting [increases, decreases] the amount of rainfall on the windward sides of mountains.

Written questions

1. Describe how clouds are classified.

2. Explain why air cools as it rises through the atmosphere.

3. By comparing the environmental lapse rate to the adiabatic rate, describe when air will be stable and when it will be unstable.

4. Of the three mechanisms that lift air, which is most prevalent in your location? Explain how this mechanism is responsible for the precipitation of your area.

Air Pressure and Wind

16

Air Pressure and Wind opens with a description of the units and instruments used for measuring atmospheric pressure. A definition of wind is followed by an analysis of the factors that affect it—pressure gradient force, Coriolis effect, and friction. A discussion of cyclones and anticyclones includes associated movements of air and weather patterns. Also presented are the general, global patterns of pressure and wind. A more detailed discussion of atmospheric circulation in the mid-latitudes is followed by descriptions of several local winds. Wind measurement and instruments are also briefly mentioned. The chapter closes with an examination of the relations between global precipitation and Earth's pressure and wind belts.

Learning Objectives

After reading, studying, and discussing this chapter, you should be able to:

- Describe air pressure, how it is measured, and how it changes with altitude.
- Explain how the pressure gradient force, Coriolis effect, and friction influence wind.
- Describe the movements of air associated with the two types of pressure centers.
- Describe the idealized global patterns of pressure and wind.
- Discuss the general atmospheric circulation in the mid-latitudes.
- List the names and causes of the major local winds.
- Describe the global distribution of precipitation.

Chapter Summary

- *Air has weight*: at sea level it exerts a pressure of 1 kilogram per square centimeter (14.7 pounds per square inch). *Air pressure* is the force exerted by the weight of air above. With increasing altitude there is less air above to exert a force, and thus air pressure decreases with altitude, rapidly at first, then much more slowly. The unit used by meteorologists to measure atmospheric pressure is the *millibar*. *Standard sea level pressure* is expressed as 1013.2 millibars. *Isobars* are lines on a weather map that connect places of equal air pressure.

- A *mercury barometer* measures air pressure using a column of mercury in a glass tube that is sealed at one end and inverted in a dish of mercury. As air pressure increases, the mercury in the tube rises; conversely, when air pressure decreases, so does the height of the column of mercury. A mercury barometer measures atmospheric pressure in *"inches of mercury"*—the height of the column of mercury in the barometer. Standard atmospheric pressure at sea level equals 29.92 inches of mercury. *Aneroid* ("without liquid") *barometers* consist of partially-evacuated metal chambers that compress as air pressure increases and expand as pressure decreases.

• *Wind* is the horizontal flow of air from areas of higher pressure to areas of lower pressure. Winds are controlled by the following combination of forces: 1) the *pressure gradient force* (amount of pressure change over a given distance), 2) *Coriolis effect* (deflective force of Earth's rotation—to the right in the Northern Hemisphere and to the left in the Southern Hemisphere), and 3) *friction* with Earth's surface (slows the movement of air and alters wind direction). Upper-air winds (the most prominent feature being the *jet streams*) generally flow parallel to the isobars and are called *geostrophic winds*.

• The two types of pressure centers are 1) *cyclones*, or *lows* (centers of low pressure) and 2) *anticyclones*, or *highs* (high-pressure centers). In the Northern Hemisphere, winds around a low (cyclone) are counterclockwise and inward. Around a high (anticyclone) they are clockwise and outward. In the Southern Hemisphere, the Coriolis effect causes winds to be clockwise around a low and counterclockwise around a high. Since air rises and cools adiabatically in a low pressure system, cloudy conditions and precipitation are often associated with their passage. In a high pressure system, descending air is compressed and warmed; therefore, cloud formation and precipitation are unlikely in an anticyclone, and "fair" weather is usually expected.

• Earth's *global pressure zones* include the *equatorial low*, *subtropical high*, *subpolar low*, and *polar high*. The *global surface winds* associated with these pressure zones are the *trade winds*, *westerlies*, and *polar easterlies*.

• Particularly in the Northern Hemisphere, large seasonal temperature differences over continents disrupt the idealized, or zonal, global patterns of pressure and wind. In winter, large, cold landmasses develop a seasonal high-pressure system from which surface air flow is directed off the land. In summer, landmasses are heated and a low-pressure system develops over them, which permits air to flow onto the land. These seasonal changes in wind direction are known as *monsoons*.

• In the middle latitudes, between 30 and 60 degrees latitude, the general west-to-east flow of the westerlies is interrupted by the migration of cyclones and anticyclones. The paths taken by these cyclonic and anticyclonic systems is closely correlated to upper-level air flow and the polar *jet stream*. The average position of the polar jet stream, and hence the paths of cyclonic systems, migrates equatorward with the approach of winter and poleward as summer nears.

• *Local winds* are small-scale winds produced by a locally generated pressure gradient. Local winds include *sea* and *land breezes* (formed along a coast because of daily pressure differences over land and water), *valley* and *mountain breezes* (daily wind similar to sea and land breezes except in a mountainous area where the air along slopes heats differently than the air at the same elevation over the valley floor), *chinook* and *Santa Ana winds* (warm, dry winds created when air descends the leeward side of a mountain and warms by compression).

• The two basic wind measurements are *direction* and *speed*. Winds are always labeled by the direction *from* which they blow. Wind speed is measured using a *cup anemometer*. The instrument most commonly used to measure wind direction is the *wind vane*. When the wind consistently blows more often from one direction than from any other, it is termed a *prevailing wind*.

• In general, regions influenced by high pressure, with its associated subsidence and divergent winds, experience relatively dry conditions. Conversely, regions under the influence of low pressure and its converging winds and ascending air receive ample precipitation. In addition to latitudinal variations in precipitation related to the global pattern of pressure and temperatures, the distribution of land and water also influences Earth's precipitation pattern.

Chapter Outline

I. Atmospheric pressure
   A. Force exerted by the weight of the air above
   B. Weight of the air at sea level
      1. 14.7 pounds per square inch
      2. 1 kilogram per square centimeter
   C. Decreases with increasing altitude
   D. Units of measurement
      1. Millibar (mb)—standard sea level pressure is 1013.2 mb
      2. Inches of mercury—standard sea level pressure is 29.92 inches of mercury
   E. Instruments for measuring
      1. Barometer
         a. Mercury barometer
            1. Invented by Torricelli in 1643
            2. Uses a glass tube filled with mercury
         b. Aneroid barometer
            1. "Without liquid"
            2. Uses an expanding chamber
      2. Barograph (continuously records the air pressure)

II. Wind
   A. Horizontal movement of air
      1. Out of areas of high pressure
      2. Into areas of low pressure
   B. Controls of wind
      1. Pressure gradient force
         a. Isobars—lines of equal air pressure
         b. Pressure gradient—pressure change over distance
      2. Coriolis effect
         a. Apparent deflection in the wind direction due to Earth's rotation
         b. Deflection is
            1. To the right in the Northern Hemisphere
            2. To left in the Southern Hemisphere
      3. Friction with Earth's surface
         a. Only important near the surface
         b. Acts to slow the air's movement
   C. Upper air winds
      1. Generally blow parallel to isobars—called geostrophic winds
      2. Jet stream
         a. "River" of air
         b. High altitude
         c. High velocity (120-240 kilometers per hour)

III. Cyclones and anticyclones
   A. Cyclone
      1. A center of low pressure
      2. Pressure decreases toward the center
      3. Winds associated with
         a. In the Northern Hemisphere
            1. Inward (convergence)
            2. Counterclockwise
         b. In the Southern Hemisphere
            1. Inward (convergence)
            2. Clockwise
      4. Associated with rising air
      5. Often bring clouds and precipitation
   B. Anticyclone
      1. A center of high pressure
      2. Pressure increases toward the center
      3. Winds associated with
         a. In the Northern Hemisphere
            1. Outward (divergence)
            2. Clockwise
         b. In the Southern Hemisphere
            1. Outward (divergence)
            2. Counterclockwise
      4. Associated with subsiding air
      5. Usually bring "fair" weather

IV. General atmospheric circulation
   A. Underlying cause is unequal surface heating
   B. On the rotating Earth there are three pairs of atmospheric cells that redistribute the heat
   C. Idealized global circulation
      1. Equatorial low pressure zone
         a. Rising air
         b. Abundant precipitation
      2. Subtropical high pressure zone
         a. Subsiding, stable, dry air
         b. Near 30° latitude
         c. Location of great deserts
         d. Air traveling equatorward from the subtropical high produces the trade winds

e. Air traveling poleward from the subtropical high produces the westerly winds

3. Subpolar low pressure zone
   a. Warm and cool winds interact
   b. Polar front—an area of storms

4. Polar high pressure zone
   a. Cold, subsiding air
   b. Air spreads equatorward and produces polar easterly winds
   c. Polar easterlies collide with the westerlies along the polar front

D. Influence of continents
   1. Seasonal temperature differences disrupt the
      a. Global pressure patterns
      b. Global wind patterns
   2. Influence is most obvious in the Northern Hemisphere
   3. Monsoon
      a. Seasonal change in wind direction
      b. Occur over continents
         1. During warm months
            a. Air flows onto land
            b. Warm, moist air from the ocean
         2. Winter months
            a. Air flows off the land
            b. Dry, continental air

V. Circulation in the mid-latitudes
   A. Complex
   B. Occurs in the zone of the westerlies
   C. Air flow is interrupted by cyclones
      1. Cells move west to east in the Northern Hemisphere
      2. Create anticyclonic and cyclonic flow
      3. Paths of the cyclones and anticyclones are associated with the upper-level airflow

VI. Local winds
   A. Produced from temperature differences
   B. Small scale winds
   C. Types
      1. Sea and land breezes
      2. Valley and mountain breezes
      3. Chinook and Santa Ana winds

VII. Wind measurement
   A. Two basic measurements
      1. Direction
      2. Speed
   B. Direction
      1. Winds are labeled from where they originate (e.g., North wind—blows from the north toward the south)
      2. Instrument for measuring wind direction is the wind vane
      3. Direction indicated by either
         a. Compass points (N, NE, etc.)
         b. Scale of 0° to 360°
      4. Prevailing wind comes more often from one direction
   C. Speed—often measured with a cup anemometer
   D. Changes in wind direction
      1. Associated with locations of
         a. Cyclones
         b. Anticyclones
      2. Often bring changes in
         a. Temperature
         b. Moisture conditions

VIII. Global distribution of precipitation
   A. Relatively complex pattern
   B. Related to global wind and pressure patterns
      1. High pressure regions
         a. Subsiding air
         b. Divergent winds
         c. Dry conditions
         d. e.g., Sahara and Kalahari deserts
      2. Low pressure regions
         a. Ascending air
         b. Converging winds
         c. Ample precipitation
         d. e.g., Amazon and Congo basins
   C. Related to distribution of land and water
      1. Large landmasses in the middle latitudes often have less precipitation toward their centers
      2. Mountain barriers also alter precipitation patterns
         a. Windward slopes receive abundant rainfall from orographic lifting
         b. Leeward slopes are usually deficient in moisture

Key Terms

Page numbers shown in ( ) refer to the textbook page where the term first appears.

aneroid barometer (p. 421)   geostrophic wind (p. 423)   pressure tendency (p. 428)
anticyclone (high) (p. 424)   isobar (p. 421)   prevailing wind (p. 433)
barograph (p. 421)   jet stream (p. 423)   Santa Ana (p. 432)
barometric tendency (p. 428) land breeze (p. 431)   sea breeze (p. 431)
chinook (p. 432)   mercury barometer (p. 420)   subpolar low (p. 429)
convergence (p. 426)   monsoon (p. 430)   subtropical high (p. 429)
Coriolis effect (p. 422)   mountain breeze (p. 432)   trade winds (p. 429)
cup anemometer (p. 433)   polar easterlies (p. 429)   valley breeze (p. 432)
cyclone (low) (p. 424)   polar front (p. 429)   westerlies (p. 429)
divergence (p. 426)   polar high (p. 429)   wind (p. 421)
equatorial low (p. 429)   pressure gradient (p. 421)   wind vane (p. 432)

Vocabulary Review

Choosing from the list of key terms, furnish the most appropriate response for the following statements.

1. The stormy belt separating the westerlies from the polar easterlies is known as the _____.

2. A(n) _____ is a line on a weather map that connects places of equal air pressure.

3. A(n) _____ is a seasonal reversal of wind direction associated with large continents, especially Asia.

4. The deflective force of Earth's rotation on all free-moving objects is called the _____.

5. Air that flows horizontally with respect to Earth's surface is referred to as _____.

6. _____, a useful aid in short-range weather prediction, refers to the nature of the change in atmospheric pressure over the past several hours.

7. The _____ is an instrument used for measuring air pressure that consists of evacuated metal chambers that change shape as pressure changes.

8. When the wind consistently blows more often from one direction than from any other, it is termed a(n) _____.

9. The instrument most commonly used to determine wind direction is the _____.

10. A center of low atmospheric pressure is called a(n) _____.

11. _____ is the condition that exists when the distribution of winds within a given area results in a net horizontal inflow of air into the area.

12. A(n) _____ is a local wind blowing from land toward the water during the night in coastal areas.

13. The _____ is a belt of low pressure lying near the equator and between the subtropical highs.

14. A wind blowing down the leeward side of a mountain and warming by compression is called a(n) _____.

15. A(n) _____ is a local wind blowing from the sea during the afternoon in coastal areas.

16. The _____ is an instrument used to measure wind speed.

17. The amount of pressure change over a given distance is referred to as the _____.

18. A(n) _____ is a swift (120 to 240 km/hour), high-altitude wind.

19. A center of high atmospheric pressure is called a(n) _____.

20. The _____ is the pressure zone located at about the latitude of the Arctic and Antarctic circles.

21. The _____ are global winds that blow from the polar high toward the subpolar low.

22. A(n) _____ is an instrument that continuously records air pressure changes.

23. _____ is the condition that exists when the distribution of winds within a given area results in a net horizontal outflow of air from the region.

24. The _____ is a region of several semipermanent anticyclonic centers characterized by subsidence and divergence located roughly between latitudes 25° and 35°.

25. The _____ are the dominant west-to-east winds that characterize the regions on the poleward sides of the subtropical highs.

26. The _____ are wind belts located on the equatorward sides of the subtropical highs.

Comprehensive Review

1. Which element of weather is measured by each of the following instruments? Briefly describe the principle of each instrument.

    a) Mercury barometer:

    b) Wind vane:

    c) Aneroid barometer:

    d) Cup anemometer:

    e) Barograph:

2. List and describe the three forces that control the wind.

    1)

    2)

    3)

3. On Figure 16.1, complete each of the map-views of the four pressure center diagrams by labeling the isobars with appropriate pressures and drawing several wind arrows to indicate surface air movement.

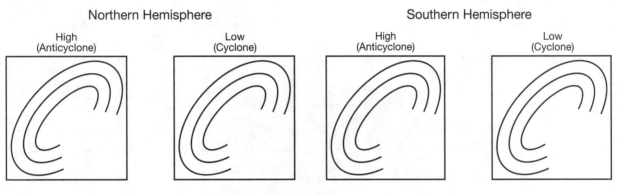

Figure 16.1

4. Referring to Figure 16.1, describe the general surface air flow associated with each of the following pressure systems.

    a) Northern Hemisphere low (cyclone):

    b) Southern Hemisphere high (anticyclone):

5. Figure 16.2 illustrates side-views of the air movements in a cyclone and an anticyclone. Using Figure 16.2, select the letter of the diagram that illustrates each of the following.

    a) Cyclone: ____

    b) Convergence aloft: ____

    c) Surface convergence: ____

    d) Anticyclone: ____

    e) Divergence aloft: ____

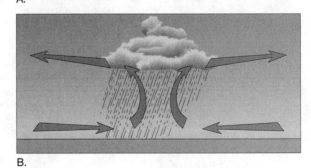

Figure 16.2

6. Why is "fair" weather usually associated with the approach of a high pressure system (anticyclone)?

7. On Figure 16.3, write the names of the global pressure zones indicated by letters A through E; and the wind belts indicated by letters F through K.

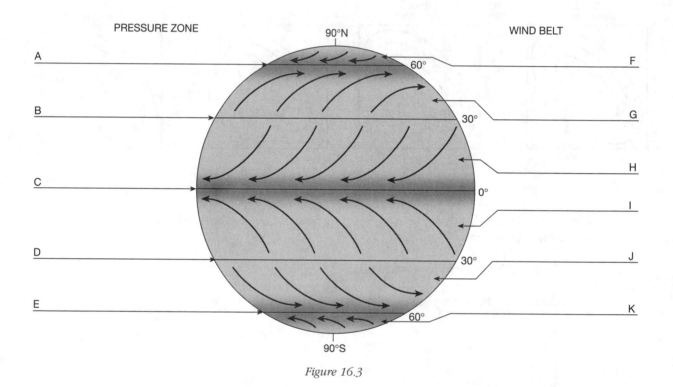

Figure 16.3

8. Write the name of the global pressure zone best described by each of the following statements.

   a) Warm and cold winds interact to produce stormy weather: _____

   b) Subsiding, dry air with extensive arid and semiarid regions: _____

   c) Warm, rising air marked by abundant precipitation: _____

   d) Cold, subsiding, polar air: _____

9. Briefly describe each of the following local winds.

   a) Chinook:

   b) Sea breeze:

10. If a wind is indicated with the following degree, from what compass direction is it blowing?

    a) 45 degrees: _____

    b) 270 degrees: _____

    c) 180 degrees: _____

Practice Test

Multiple choice. Choose the best answer for the following multiple choice questions.

1. The only truly continuous pressure belt on Earth is the _____.
   a) Northern Hemisphere subtropical high   d) Southern Hemisphere subpolar low
   b) equatorial low   e) Northern Hemisphere subpolar low
   c) Southern Hemisphere subtropical high

2. The force exerted by the weight of the air above is called _____.
   a) air pressure   c) the Coriolis effect   e) divergence
   b) convergence   d) the pressure gradient

3. Which one of the following is NOT a force that controls wind? _____
   a) magnetic force   c) pressure gradient force
   b) Coriolis effect   d) friction

4. Variations in air pressure from place to place are the principal cause of _____.
   a) snow   c) wind   e) hail
   b) rain   d) clouds

5. Fair weather can usually be expected with the approach of a(n) _____.
   a) cyclone   b) anticyclone

6. Centers of low pressure are called _____.
   a) anticyclones   c) cyclones   e) domes
   b) air masses   d) jet streams

7. In the winter, large landmasses, particularly Asia, develop a seasonal _____.
   a) high-pressure system   c) low-pressure system   e) cyclonic circulation
   b) trade wind system   d) chinook

8. Wind with a direction of 225 degrees would be a _____ wind.
   a) west   c) south   e) northwest
   b) northeast   d) southwest

9. The general movement of low pressure centers across the United States is from _____.
   a) north to south   c) east to west   e) northeast to southwest
   b) south to north   d) west to east

10. Near the equator, rising air is associated with a pressure zone known as the _____.
   a) equatorial high   c) equatorial low   e) subtropical low
   b) tropical high   d) tropical low

11. Standard sea level pressure is _____.
    a) 1020.6 millibars
    b) 15.8 pounds per square inch
    c) 29.92 inches of mercury
    d) 3.7 kilograms per square centimeter
    e) 100 kilopascals

12. A sea breeze is most intense _____.
    a) during mid- to late afternoon
    b) in the late morning
    c) late in the evening
    d) at midnight
    e) at sunrise

13. High-altitude, high-velocity "rivers" of air are referred to as _____.
    a) cyclones
    b) tornadoes
    c) anticyclones
    d) air currents
    e) jet streams

14. The deflection of wind due to the Coriolis effect is strongest at _____.
    a) the equator
    b) the mid-latitudes
    c) midnight
    d) the poles
    e) sunrise

15. Seasonal changes in wind direction associated with large landmasses and adjacent water bodies are called _____.
    a) chinooks
    b) geostrophic winds
    c) jet streams
    d) trade winds
    e) monsoons

16. Pressure data on a map are shown using lines that connect places of equal air pressure called _____.
    a) isotherms
    b) millibars
    c) aneroids
    d) kilograms
    e) isobars

17. Air is subsiding in the center of a(n) _____.
    a) low pressure system
    b) jet stream
    c) surface convergence area
    d) high pressure system
    e) chinook

18. A(n) _____ is an instrument that continuously records pressure changes.
    a) mercury barometer
    b) barograph
    c) aneroid barometer
    d) sling psychrometer
    e) cup anemometer

19. The surface winds between the subtropical high and equatorial low pressure zones are the _____.
    a) polar easterlies
    b) sea breezes
    c) trade winds
    d) monsoon winds
    e) westerlies

20. Which one of the following does NOT describe the surface air movement in a Northern Hemisphere low? _____
    a) inward
    b) counterclockwise
    c) net upward movement
    d) divergent

21. Winds that blow parallel to isobars are called _____.
    a) monsoon winds
    b) geostrophic winds
    c) sea breezes
    d) chinooks
    e) trade winds

22. Chinooks are warm, dry winds that commonly occur on the _____ slopes of the Rockies.
    a) northern
    b) eastern
    c) southern
    d) western

23. The westerlies and polar easterlies converge in a stormy region known as the _____.
    a) polar front   c) subtropical front   e) polar low
    b) equatorial low   d) subpolar high

24. In the mid-latitudes, the movement of cyclonic and anticyclonic systems is steered by the
    _____.
    a) polar easterlies   c) trade winds   e) ocean circulation
    b) upper-level air flow   d) pressure gradient

25. The surface winds between the subtropical high and subpolar low pressure zones are the
    _____.
    a) polar easterlies   c) trade winds   e) westerlies
    b) sea breezes   d) monsoon winds

*True/false. For the following true/false questions, if a statement is not completely true, mark it false. For each false statement, change the **italicized** word to correct the statement.*

1. ___ The ultimate driving force of wind is *solar* energy.

2. ___ The greater the pressure differences between two places, the *slower* the wind speed.

3. ___ Centers of high pressure are called *anticyclones.*

4. ___ In the Northern Hemisphere, the Coriolis effect causes the path of wind to be curved to the *left.*

5. ___ The *aneroid* barometer consists of metal chambers that change shape as air pressure changes.

6. ___ A *rising* barometric tendency often means that fair weather can be expected.

7. ___ Upper air flow is often nearly *parallel* to the isobars.

8. ___ Pressure decreases from the outer isobars toward the center in a(n) *anticyclone.*

9. ___ In India, the winter monsoon is dominated by *wet* continental air.

10. ___ A steep pressure gradient indicates *strong* winds.

11. ___ A *low* pressure system often brings cloudiness and precipitation.

12. ___ As wind speed increases, deflection by the Coriolis effect becomes *less.*

13. ___ The pressure zone of subsiding, dry air which encircles the globe near 30 degrees latitude, north and south, is the *mid-latitude* high.

14. ___ Wind direction is modified by both the Coriolis effect and *friction.*

15. ___ In both hemispheres, an anticyclone is associated with *divergence* of the air aloft.

16. ___ In the Northern Hemisphere, more cyclones are generated during the *warmer* months when the temperature gradient is greatest.

17. ___ The rainiest regions of Earth are associated with *low* pressure systems.

Word choice. Complete each of the following statements by selecting the most appropriate response.

1. Wind flows from areas of [higher/lower] air pressure to regions of [higher/lower] air pressure.

2. Fair weather is predicted when the air pressure is [falling/rising].

3. During summers, continents will tend to have [higher/lower] air pressures than adjacent oceans.

4. The Coriolis force is [strongest/weakest] at the equator.

5. Divergence aloft [diminishes/maintains] a surface low-pressure center.

6. The amount of pressure change occurring over a given distance is indicated by the [angle/spacing] of isobars.

7. Cyclones usually move from [south-to-north/west-to-east] across the United States.

8. In the Southern Hemisphere, moving objects are deflected to the [right/left] by the Coriolis force.

9. Jet streams travel in a [west-to-east/east-to-west] direction.

10. Clouds and precipitation are most often associated with [high/low] pressure systems.

11. The Coriolis effect [does/does not] change the wind speed.

12. The prevailing winds in the mid-latitudes are [easterlies/westerlies].

13. The greater the pressure gradient, the [higher/lower] the wind speed.

14. Sea breezes usually [raise/lower] temperatures along a coast.

15. Friction has the greatest effect on winds in the [upper/lower] atmosphere.

16. In general, high latitudes receive [more/less] precipitation than low latitudes.

17. Chinook and Santa Ana winds are warmed by [compression/expansion] of the air.

Written questions

1. How does the Coriolis effect influence the movement of air?

2. Briefly describe the weather that is usually associated with 1) a cyclone and 2) an anticyclone.

3. Describe the motions of air at the surface and aloft in a Northern Hemisphere cyclone.

Weather Patterns and Severe Storms

17

Weather Patterns and Severe Storms begins with a discussion of air masses, their source regions, and a description of the weather associated with each air mass type. Following a detailed examination of warm fronts and cold fronts is an in-depth discussion of the evolution of the middle-latitude cyclone and its idealized weather patterns. The chapter concludes with investigations of thunderstorms, tornadoes, and hurricanes.

Learning Objectives

After reading, studying, and discussing this chapter, you should be able to:

- Explain what an air mass is.
- Describe how air masses are classified.
- Describe the general weather associated with each air mass type.
- Discuss the differences between warm fronts and cold fronts.
- Describe the primary mid-latitude weather producing systems.
- List the atmospheric conditions that produce thunderstorms, tornadoes, and hurricanes.

Chapter Summary

- An *air mass* is a large body of air, usually 1600 kilometers (1000 miles) or more across, which is characterized by a *homogeneity of temperature and moisture* at any given altitude. When this air moves out of its region of origin, called the *source region*, it will carry these temperatures and moisture conditions elsewhere, perhaps eventually affecting a large portion of a continent.

- Air masses are classified according to 1) the nature of the surface in the source region and 2) the latitude of the source region. *Continental* (*c*) designates an air mass of land origin, with the air likely to be dry; whereas a *maritime* (*m*) air mass originates over water, and therefore will be humid. *Polar* (*P*) air masses originate in high latitudes and are cold. *Tropical* (*T*) air masses form in low latitudes and are warm. According to this classification scheme, the *four basic types of air masses* are *continental polar* (cP), *continental tropical* (cT), *maritime polar* (mP), and *maritime tropical* (mT). Continental polar (cP) and maritime tropical (mT) air masses influence the weather of North America most, especially east of the Rocky Mountains. Maritime tropical air is the source of much, if not most, of the precipitation received in the eastern two-thirds of the United States.

- *Fronts* are boundaries that separate air masses of different densities, one warmer and often higher in moisture content than the other. A *warm front* occurs when the surface position of the front moves so that warm air occupies territory formerly covered by cooler air. Along a warm front, a warm air mass overrides a retreating mass of cooler air. As the warm air ascends, it cools adiabatically to produce clouds and frequently light-to-moderate precipitation over a large area. A *cold front* forms where cold air is actively advancing into a region occupied by warmer air. Cold fronts are about twice as steep and move more rapidly than warm fronts. Because of these two differences, precipitation along a cold front is more intense and of shorter duration than precipitation associated with a warm front. A *stationary front* forms when the air flow on both sides of a front is almost parallel to the position of the front and the surface front does not move. An *occluded front* occurs when an active cold front overtakes a warm front and the warm air is forced aloft.

• The primary weather producers in the middle latitudes are *large centers of low pressure* that generally travel from *west to east*, called *middle-latitude cyclones*. Middle-latitude cyclones originate along a front where air masses (typically continental polar to the north and maritime tropical to the south) are moving in opposite directions. The frontal surface often takes on a wave shape, which becomes more pronounced as the cyclone matures, and a cold and a warm front evolve. Eventually the cold front catches up to the warm front, a process called *occlusion*, and the storm comes to an end. Middle-latitude cyclones, which are often the *bearers of stormy weather*, last from a few days to a week, have a *counterclockwise circulation* pattern in the Northern Hemisphere, and an *inward flow of air* toward their centers. Most middle-latitude cyclones have a *cold front and frequently a warm front* extending from the central area of low pressure. *Convergence and forceful lifting along the fronts* initiate cloud development and frequently cause precipitation. As a middle-latitude cyclone with its associated fronts passes over a region, it often brings with it abrupt changes in the weather. The particular weather experienced by an area depends on the path of the cyclone.

• *Thunderstorms* are caused by the upward movement of warm, moist, unstable air, triggered by a number of different processes. They are associated with *cumulonimbus clouds* that generate heavy rainfall, thunder, lightning, and occasionally hail. *Tornadoes*, destructive, local storms of short duration, are violent windstorms associated with severe thunderstorms that take the form of a rotating column of air that extends downward from a cumulonimbus cloud. Tornadoes are most often spawned along the cold front of a middle-latitude cyclone, most frequently during the spring months. *Hurricanes*, the greatest storms on Earth, are tropical cyclones with wind speeds in excess of 119 kilometers (74 miles) per hour. These complex tropical disturbances develop over tropical ocean waters and are fueled by the latent heat liberated when huge quantities of water vapor condense. Hurricanes form most often in late summer when ocean-surface temperatures reach 27°C (80°F) or higher and thus are able to provide the necessary heat and moisture to the air. Hurricanes diminish in intensity whenever they 1) move over cool ocean water that cannot supply adequate heat and moisture, 2) move onto land, or 3) reach a location where large-scale flow aloft is unfavorable.

Chapter Outline

I. Air masses
    A. Characteristics
       1. Large body of air
          a. 1600 km (1000 mi.) or more across
          b. Several kilometers thick
       2. Similar temperature at any given altitude
       3. Similar moisture at any given altitude
       4. Move and affect large part of a continent
    B. Source region—the place where an air mass acquires its properties
    C. Classification of air an mass
       1. Two criteria are used to classify
          a. By the nature of the surface in the source region
             1. Continental (c)
                a. Form over land
                b. Likely to be dry
             2. Maritime (m)
                a. Originate over water

            b. Humid air
   b. By the latitude of the source
      region
      1. Polar (P)
         a. High latitudes
         b. Cold
      2. Tropical (T)
         a. Low latitudes
         b. Warm
   2. Four basic types of air masses
      a. Continental polar (cP)
      b. Continental tropical (cT)
      c. Maritime polar (mP)
      d. Maritime tropical (mT)
D. Air masses and weather
   1. cP and mT air masses are the most
      important in North America, espe-
      cially east of the Rockies
   2. North America (east of the Rocky
      Mountains)
      a. Continental polar (cP)
         1. From northern Canada and
            interior of Alaska
            a. Winter—brings cold, dry air
            b. Summer—brings cool relief
         2. Responsible for lake-effect
            snows
            a. cP air mass crosses the
               Great Lakes
            b. Air picks up moisture from
               the lakes
            c. Snow occurs on the leeward
               shores of the lakes
      b. Maritime tropical (mT)
         1. From the Gulf of Mexico and
            the Atlantic Ocean
         2. Warm, moist, unstable air
         3. Brings precipitation to the
            eastern United States
   3. Continental tropical (cT)
      a. Southwest and Mexico
      b. Hot, dry
      c. Seldom important outside the
         source region
   4. Maritime polar (mP)
      a. Brings precipitation to the western
         mountains
      b. Occasional influence in the north-
         eastern United States causes the
         "Northeaster" in New England
         with its cold temperatures and
         snow

II. Fronts
   A. Boundary that separates air masses
      1. Air masses retain their identities
      2. Warmer, less dense air forced aloft
      3. Cooler, denser air acts as wedge
   B. Types of fronts
      1. Warm front
         a. Warm air replaces cooler air
         b. Shown on map with semicircles
         c. Small slope (1:200)
         d. Clouds become lower as the front
            nears
         e. Slow rate of advance
         f. Light-to-moderate precipitation
         g. Gradual temperature increase with
            the passage of the front
      2. Cold front
         a. Cold air replaces warm air
         b. Shown on a map with triangles
         c. Twice as steep as warm fronts
            (1:100)
         d. Advances faster than a warm front
         e. Weather more violent than a
            warm front
            1. Intensity of precipitation is
               greater
            2. Duration of precipitation is
               shorter
         f. Weather behind the front is domi-
            nated by
            1. Cold air mass
            2. Subsiding air
            3. Clearing conditions
      3. Stationary front
         a. Flow of air on both sides of the
            front is almost parallel
         b. Surface position of the front does
            not move
      4. Occluded front
         a. Active cold front overtakes a
            warm front
         b. Cold air wedges warm air upward
         c. Weather is often complex
         d. Precipitation is associated with
            warm air being forced aloft

III. Middle-latitude cyclones
   A. Primary weather producer in the
      middle-latitudes
   B. Life cycle
      1. Originate along a front where air
         masses are moving parallel to the
         front in opposite directions

a. Continental polar (cP) air is often north of the front

b. Maritime tropical (mT) air is often south of the front

2. Frontal surface takes on a wave shape with low pressure centered at the apex of the wave

3. Warm front and cold front form

4. Cold front catches up to warm front and produces an occlusion

5. Warm sector is forced aloft

6. Fronts discontinue

7. Storm comes to an end

C. General characteristics

1. Large center of low pressure
   a. Counterclockwise air circulation
   b. Air flows inward toward center

2. Travel west to east guided by the westerlies

3. Last a few days to more than a week

4. Extending from the center of the low are a
   a. Cold front, and
   b. Frequently a warm front

5. Convergence and forceful lifting cause
   a. Cloud development
   b. Abundant precipitation

D. Idealized weather

1. Middle-latitude cyclones move eastward across the United States
   a. First signs of their approach are in the west
   b. Require two to four days to pass

2. Largest weather contrasts occur in the spring

3. Changes in weather associated with the passage of a middle-latitude cyclone
   a. Changes depend on the path of the storm
   b. Weather associated with fronts
      1. Warm front
         a. Clouds become lower and thicker
         b. Light precipitation
            1. Perhaps over large area
            2. Perhaps prolonged duration
         c. After the passage of a warm front
            1. Winds become more southerly

2. Warmer temperature (mT air mass)

2. Cold front
   a. Wall of dark clouds
   b. Heavy precipitation
      1. Narrow band along the front
      2. Short duration
   c. After the passage of a cold front
      1. Wind becomes north to northwest
      2. Drop in temperature as a cP air mass moves in
      3. Clearing skies

E. Role of airflow aloft

1. Cyclones and anticyclones
   a. Generated by upper-level air flow
   b. Maintained by upper-level air flow

2. Cyclone
   a. Low pressure system
   b. Surface convergence
   c. Outflow (divergence) aloft sustains the low pressure

3. Anticyclone
   a. High pressure system
   b. Associated with cyclones
   c. Surface divergence
   d. Convergence aloft

IV. Severe weather types

A. Thunderstorm

1. Features
   a. Cumulonimbus clouds
   b. Heavy rainfall
   c. Lightning
   d. Occasional hail

2. Occurrence
   a. 2000 in progress at any one time
   b. 100,000 per year in the United States
   c. Most frequent in
      1. Florida
      2. Eastern Gulf Coast region

3. Stages of development
   a. All thunderstorms require
      1. Warm air
      2. Moist air
      3. Instability (lifting)
         a. High surface temperatures
         b. Most common in the afternoon
   b. Require continual supply of warm air

1. Each surge causes air to rise higher
2. Updrafts and downdrafts form

c. Eventually precipitation forms
   1. Most active stage
   2. Gusty winds, lightning, hail
   3. Heavy precipitation

d. Cooling effect of precipitation marks the end of thunderstorm activity

B. Tornado
1. Local storm of short duration
2. Features
   a. Violent windstorm
   b. Rotating column of air
   c. Extends down from a cumulonimbus cloud
   d. Low pressures inside causes the air to rush into
   e. Winds approach 480 km (300 miles) per hour
3. Occurrence and development
   a. Average of 780 each year in the United States
   b. Most frequent from April through June
   c. Associated with severe thunderstorms
   d. Exact cause for tornadoes is not known
   e. Conditions for the formation of tornadoes
      1. Occur most often along a cold front
      2. During the spring months
      3. Associated with intense thunderstorms
4. Characteristics
   a. Diameter between 150 and 600 meters
   b. Speed across landscape is about 45 kilometers per hour
   c. Cut about a 10 km long path
   d. Most move toward the northeast
   e. Maximum winds are about 480 kilometers per hour
   f. Intensity measured by the Fujita intensity scale
5. Tornado forecasting
   a. Difficult to forecast because of their small size
   b. Tornado watch
      1. To alert people

2. Issued when the conditions are favorable
3. Covers 65,000 square kilometers

c. Tornado warning is issued when a funnel cloud is sighted or is indicated by radar

d. Use of Doppler radar helps increase the accuracy by detecting the air motion

C. Hurricane
1. Most violent storm on Earth
2. To be called a hurricane
   a. Wind speed in excess of 119 kilometers per hour
   b. Rotary circulation
3. Features
   a. Tropical cyclone
      1. Low pressure (lowest pressure system in the Western Hemisphere)
      2. Steep pressure gradient
      3. Rapid, inward-spiraling wind
   b. Parts of a hurricane
      1. Eyewall
         a. Near the center
         b. Rising air
         c. Intense convective activity
         d. Wall of cumulonimbus clouds
         e. Greatest wind speeds
         f. Heaviest rainfall
      2. Eye
         a. At the very center
         b. About 20 km diameter
         c. Precipitation ceases
         d. Wind subsides
         e. Air gradually descends and heats
         f. Warmest part of the storm
   c. Wind speeds reach 300 km/hr
   d. Generate 50 foot waves at sea
4. Cause great damage on land
5. Known by different names
   a. Typhoon in western Pacific
   b. Cyclone in Indian Ocean
6. Frequency
   a. Most (20 per year) occur in the North Pacific
   b. Fewer than 5 occur in the warm North Atlantic
7. Hurricane formation and decay
   a. Form in all tropical waters except the

1. South Atlantic and
2. Eastern South Pacific
   b. Energy comes from condensing water vapor
   c. Develop most often in late summer when warm water temperatures provide energy and moisture
   d. Initial stage is not well understood
      1. Tropical depression—winds do not exceed 61 kilometers per hour
      2. Tropical storm—winds 61 to 119 km/hr
   e. Diminish in intensity whenever
      1. They move over cooler ocean water
      2. They move onto land
      3. The flow aloft is unfavorable
8. Destruction from a hurricane
   a. Factors that affect amount of hurricane damage
      1. Strength of storm (the most important factor)
      2. Size and population density
      3. Shape of the ocean bottom near the shore
   b. Categories of hurricane damage
      1. Wind damage
      2. Storm surge
         a. Large dome of water
         b. 65 to 80 kilometers long
         c. Where eye makes landfall
      3. Inland freshwater flooding from torrential rains

— Key Terms —

Page numbers shown in ( ) refer to the textbook page where the term first appears.

air-mass weather (p. 438)   maritime (m) air mass (p. 439)   thunderstorm (p. 448)
cold front (p. 442)   mesocyclone (p. 454)   tornado (p. 452)
continental (c) air mass (p. 439) middle-latitude cyclone (p. 444)   tornado warning (p. 454)
Doppler radar (p. 454)   occluded front (p. 443)   tornado watch (p. 454)
eye (p. 456)   air mass (p. 438)   tropical (T) air mass (p. 439)
eye wall (p. 456)   polar (P) air mass (p. 439)   tropical depression (p. 457)
front (p. 441)   source region (p. 439)   tropical storm (p. 457)
hurricane (p. 451)   stationary front (p. 443)   warm front (p. 441)
lake-effect snow (p. 439)   storm surge (p. 460)

Vocabulary Review

Choosing from the list of key terms, furnish the most appropriate response for the following statements.

1. A(n) _____ is an immense body of air characterized by a similarity of temperature and moisture at any given altitude.

2. A tropical cyclonic storm having winds in excess of 119 kilometers (74 miles) per hour is called a(n) _____.

3. A(n) _____ is a boundary that separates different air masses, one warmer than the other and often higher in moisture content.

4. By international agreement, a(n) _____ is a tropical cyclone with winds between 61 and 119 kilometers (38 and 74 miles) per hour.

5. A(n) _____ is the area where an air mass acquires its characteristic properties of temperature and moisture.

6. A(n) _____ is a type of air mass that forms over land.

7. When the surface position of a front moves so that warm air occupies territory formerly covered by cooler air, it is called a(n) _____.

8. In the region between southern Florida and Alaska, the primary weather producer is the _____.

9. The _____ is the doughnut-shaped area of intense cumulonimbus development and very strong winds that surrounds the center of a hurricane.

10. An air mass that forms in low latitudes is called a(n) _____.

11. By international agreement, a(n) _____ is a tropical cyclone with maximum winds that do not exceed 61 kilometers (38 miles) per hour.

12. The _____ is a zone of scattered clouds and calm, averaging about 20 kilometers in diameter, at the center of a hurricane.

13. Fairly constant weather that may take several days to traverse an area often represents a weather situation called _____.

14. When cold air is actively advancing into a region occupied by warmer air, the boundary is called a(n) _____.

15. A(n) _____ is a small, very intense cyclonic storm with exceedingly high winds, most often produced along cold fronts in conjunction with severe thunderstorms.

16. A(n) _____ is a type of air mass that originates over water.

17. When conditions appear favorable for tornado formation, a _____ is issued for areas covering about 65,000 square kilometers (25,000 miles).

18. The new generation of radar that can detect motion directly, and hence greatly improve tornado and severe storm warnings, is called _____.

19. A(n) _____ is a storm of relatively short duration produced by a cumulonimbus cloud and accompanied by strong wind gusts, heavy rain, lightning, thunder, and sometimes hail.

20. A(n) _____ is a dome of water that sweeps across the coast near the point where the eye of the hurricane makes landfall.

21. A(n) _____ is issued when a tornado funnel cloud has actually been sighted or is indicated by radar.

22. An air mass that originates in high latitudes is called a(n) _____.

23. When an active cold front overtakes a warm front a new type of front, called a(n) _____, often forms.

24. During the winter, a highly localized storm occurring along the leeward shores of the Great Lakes often creates what is known as a(n) _____.

25. When the surface position of a front does not move, the front is referred to as a(n) _____.

Comprehensive Review

1. List and describe the two criteria used to classify air masses.

    1)

    2)

2. Using Figure 17.1, list the name and describe the temperature and moisture characteristics of the air mass identified on the figure by each of the following letters.

    A:

    B:

    F:

    G:

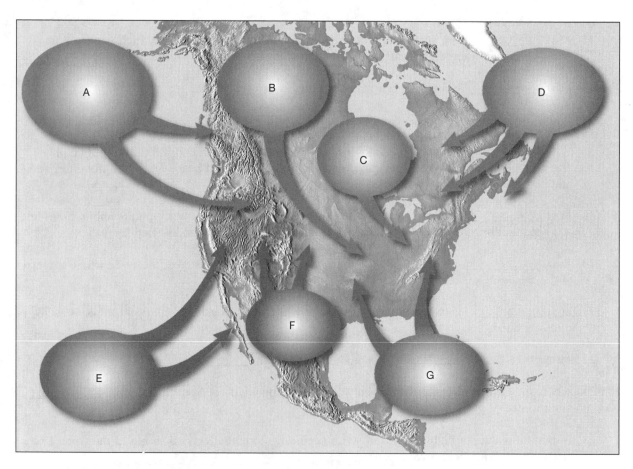

Figure 17.1

3. Which two air masses have the greatest influence on the weather of North America, especially east of the Rocky Mountains?

4. Sketch profiles (side-views) of a typical warm front and a typical cold front and briefly describe each.

   a) Warm front:

   b) Cold front:

5. Of the two diagrams illustrated in Figure 17.2, which represents the more advanced stage in the development of a middle-latitude cyclone? What is the reason for your choice?

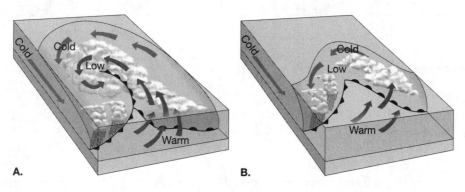

Figure 17.2

6. Using Figure 17.3, a diagram of a typical mature, middle-latitude cyclone, answer the following.

   a) Which type of front is shown at B?

   b) Which type of front is shown at D?

   c) Which air mass type would most likely be found at C?

   d) Which air mass type would most likely be found at G?

   e) The wind direction at C would most likely be from the _____.

   f) The wind direction at G would most likely be from the _____.

   g) As the center of the low moves to F, what would be the expected weather changes experienced by people living at C?

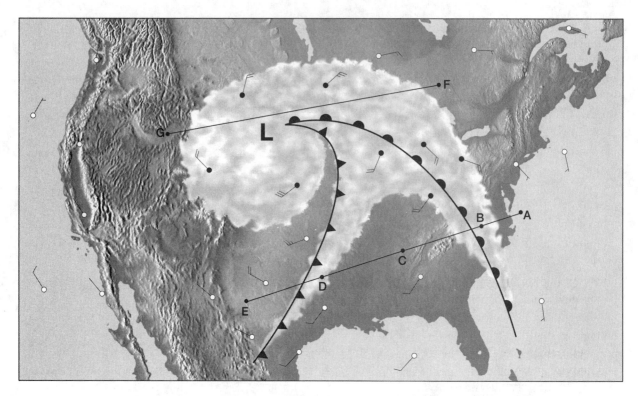

Figure 17.3

7. Describe the airflow at the surface and aloft for each of the following pressure systems.

    a) Cyclone:

    b) Anticyclone:

8. Compared to warm fronts, what are the two differences that largely account for the more violent nature of cold-front weather?

    1)

    2)

9. What are the atmospheric conditions associated with the formation of

    a) Thunderstorms?

    b) Tornadoes?

10. What are three conditions that can cause the intensity of hurricanes to diminish?

    1)

    2)

    3)

11. List the three categories of damage caused by hurricanes.

1)

2)

3)

Practice Test

Multiple choice. Choose the best answer for the following multiple choice questions.

1. An immense body of air characterized by a similarity of temperature and moisture at any given altitude is referred to as a(n) _____.
   a) cyclone   c) anticyclone   e) front
   b) air mass   d) air cell

2. Hurricanes are classified according to intensity using the _____.
   a) Richter scale   c) Fujita intensity scale   e) F-scale
   b) Saffir-Simpson scale   d) Doppler scale

3. Along which type of front is the intensity of precipitation greater, but the duration shorter?
   a) warm front
   b) cold front

4. The boundary that separates different air masses is called a(n) _____.
   a) front   c) storm   e) anticyclone
   b) cyclone   d) contact slope

5. Which air mass is most associated with lake-effect snows?
   a) mT   c) cT
   b) cP   d) mP

6. In the United States, thunderstorms typically form within _____ air masses.
   a) mT   c) cT
   b) cP   d) mP

7. The primary tornado warning system in use today involves both _____.
   a) observers and satellites   d) observers and radar
   b) barometers and wind vanes   e) satellites and radar
   c) observers and barometers

8. Which type of front has the steepest frontal surface?
   a) warm front
   b) cold front

9. The area in which an air mass acquires its characteristic properties of temperature and moisture is called its _____.
   a) area of origin   c) weather site   e) classification region
   b) location   d) source region

10. Surface airflow in a Northern Hemisphere middle-latitude cyclone is _____.
    a) divergent and clockwise   d) convergent and counterclockwise
    b) divergent and counterclockwise   e) both convergent and divergent
    c) convergent and clockwise

11. The greatest tornado frequency in the United States is during the period _____.
    a) January through March   c) July through September
    b) April through June   d) October through December

12. Along a front, which air is always forced aloft?
    a) cooler, denser air   d) the fastest moving air
    b) the driest air   e) warmer, less dense air
    c) the wettest air

13. Following the passage of a cold front, winds often come from the _____.
    a) south to southeast   c) north to northwest   e) south to southwest
    b) east to southeast   d) west to southwest

14. Which type of air mass originates in central Canada?
    a) mT   c) mP
    b) cP   d) cT

15. Pressure in a middle-latitude cyclone _____.
    a) decreases toward the center
    b) remains the same everywhere
    c) increases toward the center
    d) increases then decreases toward the center
    e) decreases then increases toward the center

16. Which one of the following is NOT an area where maritime tropical air masses that affect North America originate?
    a) Gulf of Mexico   c) Caribbean Sea   e) Pacific Ocean
    b) Hudson Bay   d) Atlantic Ocean

17. The first sign of the approach of a warm front is the appearance of _____ clouds overhead.
    a) stratus   c) cumulus   e) nimbostratus
    b) cirrus   d) cumulonimbus

18. The weather behind a cold front is dominated by a _____.
    a) subsiding, relatively cold air mass   d) rising, relatively cold air mass
    b) rising, relatively warm air mass   e) subsiding, relatively warm air mass
    c) mixed air mass

19. An mT air mass would be best described as _____.
    a) cold and dry   c) cold and wet   e) hot and dry
    b) warm and dry   d) warm and wet

20. Hurricanes form in tropical waters between the latitudes of _____.
    a) 0 and 5 degrees   c) 20 and 30 degrees
    b) 5 and 20 degrees   d) 30 and 40 degrees

21. The source of the heat necessary to maintain the upward development of a thunderstorm is _____.
    a) surface heating   c) latent heat   e) compressional heat
    b) friction   d) electrical heat

22. Which air mass is most associated with a "northeaster" in New England?
    a) mT   c) mP
    b) cP   d) cT

23. Following the passage of a warm front, winds often come from the _____.
    a) north to northeast   c) north to northwest   e) south to southwest
    b) east to southeast   d) west to northwest

24. Hurricanes develop most often in the _____.
    a) early winter   c) early summer   e) early spring
    b) late winter   d) late summer

25. The center of a tornado is best characterized by its _____.
    a) very high pressure   c) large calm area   e) subsidence
    b) eye wall   d) very low pressure

26. Along a(n) _____ front, the flow of air on both sides of the front is almost parallel.
    a) warm   c) cold   e) advancing
    b) stationary   d) occluded

27. During the life cycle of a middle-latitude cyclone, the process of _____ occurs when the cold front overtakes the warm front and the warm sector is displaced aloft.
    a) frontalysis   c) differentiation   e) occlusion
    b) cyclogenesis   d) cyclonics

28. In the United States, during the period from 1940 to 1992, most weather related deaths were caused by _____.
    a) lightning   c) flash floods
    b) tornadoes   d) hurricanes

*True/false. For the following true/false questions, if a statement is not completely true, mark it false. For each false statement, change the **italicized** word to correct the statement.*

1. ___ A *front* usually marks a change in weather.

2. ___ Air masses are typically *five* thousand miles or more across.

3. ___ Cyclones are typically the bearers of *stormy* weather.

4. ___ A cP air mass originates over land and is likely to be cold and *dry*.

5. ___ Cold fronts advance *less* rapidly than warm fronts.

6. ___ Guided by the westerlies aloft, middle-latitude cyclones generally move *westward* across the United States.

7. ___ *Continental* tropical air is the source of much, if not most, of the precipitation received in the eastern two-thirds of the United States.

8. ___ Most severe weather occurs along *cold* fronts.

9. ___ Doppler radar has the ability to detect *motion* directly.

10. ___ On a weather map, the surface position of a *warm* front is shown by a line with semicircles extending into the cooler air.

11. ___ After the passage of a cold front, air pressure will most likely *rise*.

12. ___ Most severe thunderstorms in the mid-latitudes form along or ahead of *warm* fronts.

13. ___ On the average, the slopes of cold fronts are about *half* as steep as warm front slopes.

14. ___ In a middle-latitude cyclone, *convergence* and forceful lifting initiate cloud development.

15. ___ Cyclones and anticyclones are typically found *adjacent* to one another.

16. ___ More tornadoes are generated in the *western* United States than any other area of the country.

17. ___ In the western Pacific, hurricanes are called *typhoons*.

18. ___ A tornado *warning* is issued when a funnel cloud has actually been sighted or is indicated by radar.

19. ___ In a middle-latitude cyclone, *divergence* is occurring in the air aloft.

20. ___ A tornado with an F4 rating on the Fujita intensity scale produces *devastating* damage.

21. ___ The lowest pressures ever recorded in the Western Hemisphere are associated with *hurricanes*.

22. ___ Airflow aloft plays an *important* role in maintaining cyclonic and anticyclonic circulation.

Word choice. Complete each of the following statements by selecting the most appropriate response.

1. The term [tornado/cyclone] simply refers to the circulation around any low pressure center, no matter how large or intense it is.

2. When a front advances, [cool/warm] air is forced aloft over the [cooler/warmer] air mass.

3. In 1992, hurricane [Betsy/Andrew] became the costliest natural disaster in U.S. history.

4. On a weather map, a cold front is shown by a line of [triangles/semicircles] extending into the warmer air.

5. The North [Atlantic/Pacific] Ocean produces the greatest number of hurricanes per year.

6. During the winter, a North American cP air mass would most likely be cold and [moist/dry].

7. The greatest number of thunderstorms occur in association with [nimbostratus/cumulonimbus] clouds.

8. After a cold front passes, the weather will usually be [clear/stormy].

9. The center of a middle-latitude cyclone is a [high/low] pressure center.

10. A(n) [occluded/stationary] front often forms when an active cold front overtakes a warm front.

11. In a Northern Hemisphere middle-latitude cyclone, the circulation of the air around the center is [clockwise/counterclockwise].

12. The air within a hurricane's eye gradually [rises/descends] and [heats/cools] by [expansion/compression].

13. Due to lake-effect snows, snowfall is greatest on the [windward/leeward] shores of the Great Lakes.

14. Tornadoes are most frequent during the [spring/fall] and are most often spawned along the [cold/warm] front of a middle-latitude cyclone.

15. Prior to the development of a middle-latitude cyclone, the air flow in the adjacent air masses is [parallel/perpendicular] to the front.

Written questions

1. Describe the temperature and moisture characteristics of 1) a continental polar (cP) air mass and 2) a maritime tropical (mT) air mass.

2. Describe the weather conditions that an observer would experience as a middle-latitude cyclone passes with its center to the north.

3. What is the difference between a tornado watch and a tornado warning?

Climate

18

Climate begins with a brief description of Earth's climate system, followed by an overview of world climates. The Köppen system of climate classification, used throughout the chapter, and its five principal groups are presented and discussed. Beginning with the humid tropical (A) climates, the location, characteristics, and subtypes of each of the climate groups is examined in detail.

Learning Objectives

After reading, studying, and discussing this chapter, you should be able to:

- Explain what is meant by Earth's climate system.
- Discuss the factors that give every location a distinctive climate.
- Describe the Köppen system of climate classification.
- List the five principal climate groups of the Köppen system and describe the criteria used to define each group.
- Describe the location and general characteristics of the principal climate groups.

Chapter Summary

- *Climate* is an aggregate of the weather conditions of a place or region over a long period of time. Earth's *climate system* involves the exchanges of energy and moisture that occur among the atmosphere, hydrosphere, solid Earth, biosphere, and *cryosphere* (the ice and snow that exist at Earth's surface). The varied nature of Earth's surface and the interactions that occur between atmospheric processes give every location a distinct climate.

- Climate classification brings order to large quantities of information, which aids comprehension and understanding, and facilitates analysis and explanation. *Temperature and precipitation are the most important elements in a climatic description.* Many climate-classification schemes have been devised, with the value of each determined by its intended use. The *Köppen classification*, which uses mean monthly and annual values of temperature and precipitation, has been the best-known and most used system for more than 70 years. The boundaries Köppen chose were largely based on the limits of certain plant associations. Five principal climate groups, each with subdivisions, were recognized. Each group is designated by a capital letter. Four of the climate groups (A, C, D, and E) are defined on the basis of temperature characteristics, and the fifth, the B group, has precipitation as its primary criterion.

- *Humid tropical (A) climates* are winterless with all months having a mean temperature above 18°C. *Wet tropical climates* (Af and Am), which lie near the equator, have constantly high temperatures, year-round rainfall, and the most luxuriant vegetation (tropical rain forest) found in any climatic realm. *Tropical wet and dry climates* (Aw) are found poleward of the wet tropics and equatorward of the tropical deserts where the rain forest gives way to the tropical grasslands and scattered drought-tolerant trees of the savanna. The most distinctive feature of this climate is the seasonal character of the rainfall.

• *Dry (B) climates*, in which the yearly precipitation is not as great as the potential loss of water by evaporation, are subdivided into two climatic types: *arid* or *desert* (BW) and *semiarid* or *steppe* (BS). Their differences are primarily a matter of degree, with semiarid being a marginal and more humid variant of arid. Low-latitude deserts and steppes coincide with the clear skies caused by subsiding air beneath the subtropical high-pressure belts. Middle-latitude deserts and steppes exist principally because of their position in the deep interiors of large landmasses far removed from the oceans. Because many middle-latitude deserts occupy sites on the leeward sides of mountains, they can also be classified as *rainshadow deserts*.

• *Middle-latitude climates with mild winters* (C climates) occur where the average temperature of the coldest month is below 18°C but above -3°C. Several C climate subgroups exist. *Humid subtropical climates* (Cfa) are located on the eastern sides of the continents, in the 25 to 40° latitude range. Summer weather is hot and sultry, and winters are mild. In North America, the *marine west coast climate* (Cfb, Cfc) extends from near the United States-Canadian border northward as a narrow belt into southern Alaska. The prevalence of maritime air masses means that mild winters and cool summers are the rule. *Dry-summer subtropical climates* (Csa, Csb) are typically located along the west sides of continents between latitudes 30 and 45°. In summer, the regions are dominated by stable, dry conditions associated with the oceanic subtropical highs. In winter they are within range of the cyclonic storms of the polar front.

• *Humid middle-latitude climates with severe winters* (D climates) are land-controlled climates that are absent in the Southern Hemisphere. D climates have severe winters. The average temperature of the coldest month is -3°C and the warmest monthly mean exceeds 10°C. *Humid continental climates* (Dfa, Dfb, Dwa, Dwb) are confined to the eastern portions of North America and Eurasia in the latitude range between approximately 40 and 50°N latitude. Both winter and summer temperatures may be characterized as relatively severe. Precipitation is generally greater in summer than in winter. *Subarctic climates* (Dfc, Dfd, Dwc, Dwd) are situated north of the humid continental climates and south of the polar tundras. The outstanding feature of subarctic climates is the dominance of winter. By contrast, summers in the subarctic are remarkably warm, despite their short duration. The highest annual temperature ranges on Earth occur here.

• *Polar (E) climates* are summerless with the average temperature of the warmest month below 10°C. Two types of polar climates are recognized. The *tundra climate* (ET) is a treeless climate found almost exclusively in the Northern Hemisphere. The *ice cap climate* (EF) does not have a single monthly mean above 0°C. As a consequence, the growth of vegetation is prohibited and the landscape is one of permanent ice and snow.

• Compared to nearby places of lower elevation, *highland climates* are cooler and usually wetter. Because atmospheric conditions fluctuate rapidly with changes in altitude and exposure, these climates are best described by their variety and changeability.

• Humans have had an impact on the global climate by their use of fire as well as the overgrazing of marginal lands by domesticated animals. The consumption of fossil fuels (coal, natural gas, and petroleum) has added great quantities of the gas carbon dioxide, an important heat absorber, to the atmosphere. If the use of fossil fuels continues to increase at projected rates, models predict an increase in the mean global surface temperature of between 1.5 and 4.5°C. The atmospheric buildup of trace gases such as methane, nitrous oxide, and certain chlorofluorocarbons will also contribute to a future global increase in temperature. Some possible consequences of a greenhouse warming of the atmosphere include 1) global mean surface warming (very probable), 2) global mean precipitation increase (very probable), 3) reduction of sea ice (very probable), 4) polar winter surface warming (very probable), 5) summer continental dryness/warming (likely in the long term), 6) high latitude precipitation increase (probable), and 7) a rise in global mean sea level (probable).

Chapter Outline

I.   The climate system
  A. Climate is an aggregate of weather
  B. Involves the exchanges of energy and moisture that occur among the
    1. Atmosphere
    2. Hydrosphere
    3. Solid Earth
    4. Biosphere, and
    5. Cryosphere (ice and snow)
II. World climates
  A. Every location has a distinctive climate
  B. The most important elements in a climatic description are
    1. Temperature, and
    2. Precipitation
III. Climate classification
  A. Brings order to large quantities of information
  B. Many climatic-classification schemes have been devised
  C. Köppen classification of climates
    1. Best known and most used system
    2. Uses mean monthly and annual values of temperature and precipitation
    3. Divides the world into climatic regions in a realistic way
    4. Boundaries Köppen chose were largely based on the limits of certain plant associations
    5. Five principal climate groups
      a. Humid tropical (A)
      b. Dry (B)
      c. Humid middle-latitude with mild winters (C)
      d. Humid middle-latitude with severe winters (D)
      e. Polar (E)
    6. A, C, D, and E climates are defined on the basis of temperature characteristics
    7. Precipitation is the primary criterion for the B group
IV. Köppen climates
  A. Humid tropical (A) climates
    1. Winterless climates, with all

months having a mean temperature above 18˚C
    2. Two main types
      a. Wet tropics
        1. High temperatures and year-round rain
        2. Luxuriant vegetation (tropical rain forest)
        3. Discontinuous belt astride the equator
        4. Influenced by the equatorial low pressures
      b. Tropical wet and dry
        1. Poleward of wet tropics and equatorward of the tropical deserts
        2. Tropical grassland (savanna)
        3. Seasonal rainfall
  B. Dry (B) climates
    1. Evaporation exceeds precipitation and a constant water deficiency
    2. Boundary determined by formulas involving the three variables
      a. Average annual precipitation
      b. Average annual temperature
      c. Seasonal distribution of precipitation
    3. Two climatic types
      a. Arid or desert (BW)
      b. Semiarid or steppe (BS)
        1. More humid than arid climate
        2. Surrounds desert
    4. Causes of deserts and steppes
      a. In the low latitudes
        1. e.g., North Africa to India, Mexico, southwestern U.S.
        2. Coincide with the dry, stable, subsiding air of the subtropical high-pressure belts
      b. Middle-latitude deserts and steppes
        1. Due to their position in the deep interiors of large land masses and/or the

presence of high
mountains

2. Most are found in the Northern
Hemisphere

C. Humid middle-latitude climates with
mild winters (C climates)

1. Average temperature of the coldest
month is below 18°C but above -3°C

2. Subgroups

a. Humid subtropics

1. Eastern sides of continents

2. 25 to 40° latitude range

3. Hot, sultry summers

4. Mild winters

5. Winter precipitation is generated along fronts

b. Marine west coast

1. Western (windward) side of
continents

2. 40 to 65° north and south
latitude

3. Onshore flow of ocean air

4. Mild winters and cool summers

c. Dry-summer subtropics

1. West sides of continents
between latitudes 30 and 45°

2. Strong winter rainfall maximum

3. Often called a Mediterranean
climate

D. Humid middle-latitude climates with
severe winters (D climates)

1. Average temperature of the coldest
month is below -3°C and the
warmest monthly mean exceeds 10°C

2. Land-controlled climates

3. Absent in the Southern Hemisphere

4. Subgroups

a. Humid continental

1. Confined to the central and
eastern portions of North
America and Eurasia between
40 and 50°N

2. Severe winter and summer
temperatures

3. High annual temperature
ranges

4. Precipitation is generally
greater in the summer than
in the winter

5. Snow remains on the ground
for extended periods

b. Subarctic

1. North of the humid continental
climate

2. Often referred to as the taiga
climate

3. Largest stretch of continuous
forests on Earth

4. Source regions of cP air masses

5. Frigid winters, remarkably
warm but short summers

E. Polar (E) climates

1. Mean temperature of the warmest
month is below 10°C

2. Enduring cold

3. Meager precipitation

4. Two types of polar climates

a. Tundra climate (ET)

1. Treeless climate

2. Almost exclusively in the
Northern Hemisphere

3. Severe winters, cool summers

4. High annual temperature range

b. Ice cap climate (EF)

1. No monthly mean above 0°C

2. Permanent ice and snow

F. Highland climates

1. Usually cooler and wetter than
adjacent lowlands

2. Great diversity of climatic conditions

V. Human impact on global climate

A. Humans have been modifying the
environment over extensive areas
for thousands of years

1. By using fire

2. By overgrazing of marginal lands

B. Most hypotheses of climatic change are
to some degree controversial

C. Global warming

1. Water vapor and carbon dioxide
absorb heat and are largely responsible for the greenhouse effect of
the atmosphere

2. Burning fossil fuels has added
great quantities of carbon dioxide
to the atmosphere

3. Models project a 1.5 to 4.°C increase
in the mean global surface temperature by the second half of the next
century

4. Trace gases that add to global
warming

a. Methane

b. Nitrous oxide

c. Certain chlorofluorocarbons
5. Polar regions will have a greater temperature response to global warming than the global average
6. Some consequences of global warming
   a. Global mean surface warming

b. Global mean precipitation will increase
c. Reduction in sea ice
d. Polar winter surface warming
e. Summer continental dryness
f. High-latitude precipitation will increase
g. Rise in global mean sea level

Key Terms

Page numbers shown in ( ) refer to the textbook page where the term first appears.

arid climate (p. 472)
climate system (p. 464)
desert climate (p. 472)
dry-summer subtropical climate (p. 474)
highland climate (p. 478)
humid continental climate (p. 474)

humid subtropical climate (p. 473)
ice cap climate (p. 478)
Köppen classification (p. 465)
marine west coast climate (p. 474)
rainshadow desert (p. 473)
semiarid climate (p. 472)
steppe climate (p. 472)

subarctic climate (p. 477)
tropical rain forest (p. 467)
tropical wet and dry climate (p. 471)
tundra climate (p. 477)

Vocabulary Review

Choosing from the list of key terms, furnish the most appropriate response for the following statements.

1. Earth's _____ consists of the atmosphere, hydrosphere, solid Earth, biosphere, and cryosphere and involves the exchanges of energy and moisture that occur among the five parts.

2. Located on the eastern sides of the continents, in the 25 to 40° latitude range, is a climatic region known as the _____, which dominates the southeastern United States, as well as other similarly situated areas around the world.

3. Nearly half of Australia is a(n) _____, and much of the remainder is a steppe climate.

4. One of the best-known and most-used systems of climate classification is the _____.

5. The _____ is a type of polar climate that does not have a single monthly mean above 0°C.

6. Situated north of the humid continental climate and south of the polar tundra is an extensive _____ region covering broad, uninterrupted expanses from western Alaska to Newfoundland in North America and from Norway to the Pacific coast of Russia in Eurasia.

7. The _____ is a climatic region situated on the western (windward) side of continents, from about 40 to 65° north and south latitude, that is dominated by the onshore flow of oceanic air.

8. The most luxuriant vegetation found in any climatic realm occurs in the _____.

9. A(n) _____ frequently occurs on the leeward side of a mountain where the conditions are often drier and more arid than the windward side.

10. The climatic region called the _____ is confined to the central and eastern portions of North America and Eurasia in the latitude range between approximately 40 and 50°N latitude.

11. The transitional climatic region located poleward of the wet tropics and equatorward of the tropical deserts is called the _____.

12. The _____ is a treeless climate found almost exclusively in the Northern Hemisphere.

13. A(n) _____ is a marginal and more humid variant of the arid climate and represents a transition zone that surrounds the desert and separates it from the bordering humid climates.

14. Typically located along the west sides of continents between latitudes 35° and 45° and situated between the marine west coast climate on the poleward side and the subtropical steppes on the equatorward side, the climatic region known as the _____ is best described as transitional in character.

15. Compared to nearby places at lower elevations, sites with a(n) _____ are cooler and usually wetter.

Comprehensive Review

1. What are the five parts of the climate system that constantly exchange energy and moisture?

2. What are the two most important elements in a climatic description? Why are these elements important?

3. What are the names, capital letter designations, and characteristics of the five principal Köppen climate groups?

   1)

   2)

   3)

   4)

   5)

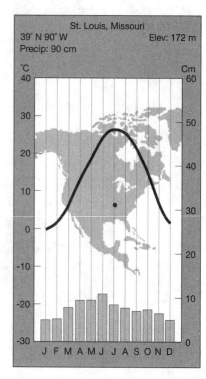

4. Using the climate diagram for St. Louis, Missouri, Figure 18.1, answer the following questions.

   a) During which month does the highest temperature occur? What is the highest temperature?

   b) During which month does the lowest temperature occur? What is the lowest temperature?

Figure 18.1

    c) What is the annual temperature range for St. Louis?

    d) Which month receives the greatest amount of precipitation?

    e) The precipitation is concentrated in which season?

    f) Using the Köppen system of climatic classification presented in Table 18.1 of the textbook and the data presented in the climate diagram, Figure 18.1, determine the Köppen climate classification for St. Louis, MO.

5. What are the three temperature/precipitation features that characterize the wet tropics? Where are the most continuous expanses of the wet tropics located?

6. What is the difference between the two climatic types: arid (desert) and semiarid (steppe)?

7. With what meteorological feature(s) do most low-latitude deserts and steppes coincide?

8. By referring to Figure 18.2 in the textbook, where in the United States is the location of the humid subtropical, warm summer, climate region?

9. Examine the locations of the humid middle-latitude climates with severe winters (D climates) in Figure 18.2 of the textbook. Why are D climates absent in the Southern Hemisphere?

10. Indicate by name the climate group best described by each of the following statements.

    a) Potential evaporation exceeds precipitation:

    b) Average temperature of the warmest month is below 10°C:

    c) Very little change in monthly precipitation and temperature throughout the year:

    d) Situated on the windward side of continents, from 40 to 65° latitude:

    e) Location coincides with the subtropical high-pressure belts:

    f) Vast areas of northern coniferous forest:

    g) Tropical rain forest:

11. Why are the summer temperatures at Death Valley, California, consistently among the highest in the Western Hemisphere?

12. Compared with a rural environment, what impact does a city have on the following elements?

    a) Winter temperature:

    b) Precipitation:

    c) Winter fog:

13. What are the two most important gases involved in the greenhouse effect of the atmosphere?

14. What are the two most prominent means by which humans contribute to the buildup of carbon dioxide in the atmosphere?

    1)

    2)

15. What are three potential weather changes that could occur as a result of the projected global warming?

    1)

    2)

    3)

Practice Test

Multiple choice. Choose the best answer for the following multiple choice questions.

1. The best-known and most used system for climate classification is the _____ classification.
    a) Smith   c) Klaus   e) Trewartha
    b) Köppen   d) Strahler

2. As a result of broad continents in the middle-latitudes, _____ climates are land-controlled climates.
    a) A   c) C   e) E
    b) B   d) D

3. The description of the aggregate weather conditions of a place or region is termed _____.
    a) weather   c) climate   e) averaging
    b) cyclogenesis   d) synthesis

4. Which one of the following is NOT a possible consequence of a greenhouse warming _____.
    a) more frequent and intense hurricanes
    b) reduction in secondary pollutants
    c) rising sea levels
    d) increase of heat waves and droughts
    e) shifts in the paths of large-scale cyclonic storms

5. A(n) _____ climate dominates the southeastern Unites States.
   a) marine west coast   d) tropical wet and dry
   b) humid subtropical   e) steppe
   c) humid continental

6. According to Köppen, the distribution of natural _____ was an excellent expression of the totality of climate.
   a) wind belts   c) cyclones   e) pressure belts
   b) air masses   d) vegetation

7. The _____ refers to the snow and ice that exist at Earth's surface.
   a) crystalsphere   c) hydrosphere   e) exosphere
   b) cryosphere   d) thermosphere

8. Many of the rainiest places in the world are located on _____ mountain slopes.
   a) windward   c) leeward   e) barren
   b) gradual   d) steep

9. A type of polar climate, called the _____ climate, is a treeless climate found almost exclusively in the Northern Hemisphere.
   a) taiga   c) plains   e) highland
   b) tundra   d) coastal

10. In the Köppen system, the _____ climate group has precipitation as its primary criterion.
    a) A   c) C   e) E
    b) B   d) D

11. Melting ice packs and reducing the albedo of a region, leading to increased warming, is an example of _____.
    a) a temperature inversion   d) deforestation
    b) a climatic feedback mechanism   e) a climatic control
    c) frost control

12. The highest temperature recorded in the Western Hemisphere (134°F) was set at _____.
    a) Celaya, Mexico   c) Death Valley, CA   e) Tucson, AZ
    b) Houston, TX   d) Las Vegas, NM

13. Af and Am climates form a discontinuous belt astride the _____.
    a) Antarctic Circle   d) Tropic of Capricorn
    b) Tropic of Cancer   e) Atlantic Ocean
    c) equator

14. The consensus in the scientific community is that altering atmospheric composition by the addition of carbon dioxide and trace gases will eventually lead to a _____ planet with a different distribution of climatic regimes.
    a) cooler   c) warmer   e) smaller
    b) wetter   d) drier

15. Poleward of the wet tropics and equatorward of the tropical deserts lies a transitional climatic region called the _____ climate.
    a) humid middle-latitude   d) tropical wet and dry
    b) polar   e) highland
    c) humid tropical

16. The two most important elements in a climatic description are temperature and _____.
    a) precipitation   c) wind speed   e) wind direction
    b) pressure   d) altitude

17. The lowest seasonal temperature variations occur in the _____ climates.
    a) A   c) C   e) highland
    b) B   d) D

18. The name _____ climate is often used as a synonym for the dry-summer subtropical climate.
    a) Canadian   c) African   e) Mediterranean
    b) Hudson Bay   d) European

19. Which climates are absent in the Southern Hemisphere?
    a) A   c) C   e) E
    b) B   d) D

20. The highest average annual rainfall in the world (some 486 inches) occurs in _____.
    a) California   c) New Zealand   e) Japan
    b) Brazil   d) Hawaii

21. The best known climatic effect of increased altitude is to cause lower _____.
    a) relative humidities   c) temperatures   e) rainfall amounts
    b) wind velocities   d) solar intensities

22. If carbon dioxide levels reach projected levels, models predict that by the year 2000 mean global surface temperatures will increase between _____.
    a) 0.5 and 1.0°C   c) 1.5 and 4.5°C
    b) 1.0 and 2.0°C   d) 4.0 and 7.5°C

23. The subarctic climate region is often referred to as the _____ climate.
    a) taiga   c) plains   e) polar
    b) tundra   d) coastal

24. The most studied and well-documented urban climate effect is the urban _____ island.
    a) pollution   c) pressure   e) heat
    b) wind   d) precipitation

25. Which one of the following is a climatic condition that is lowered by cities in comparison with the rural environment?
    a) relative humidity   c) thunderstorms   e) fog
    b) precipitation   d) temperature

*True/false. For the following true/false questions, if a statement is not completely true, mark it false. For each false statement, change the **italicized** word to correct the statement.*

1. ___ The Köppen climate classification recognizes *three* principal groups, each designated by a capital letter.

2. ___ The *leeward* sides of mountains are often wet.

3. ___ Wet tropical climates are strongly influenced by the equatorial *high* pressures.

4. ___ Temperature and *pressure* are the most important elements in a climatic description.

5. ___ In a humid tropical (A) climate, all months have a mean temperature *above* 18°C.

6. ___ Models indicate that the temperature response in *polar* regions due to global warming triggered by carbon dioxide and trace gases could be as much as two to three times greater than the global average.

7. ___ Climatologists define a dry climate as one in which the yearly precipitation is not as great as the potential loss of water by *infiltration*.

8. ___ Wet tropical climates are restricted to elevations *below* 1000 meters.

9. ___ Köppen believed that the distribution of natural *vegetation* was an excellent expression of the totality of climate.

10. ___ In a polar (E) climate, the average temperature of the warmest month is *above* 10°C.

11. ___ Much of California has a climate similar to that surrounding the *Mediterranean* Sea.

12. ___ A steppe climate is a marginal and *more* humid variant of a desert climate.

13. ___ Places with an Af or Am climate designation lie near the *equator*.

14. ___ Paralleling the rapid growth of *agriculture*, great quantities of carbon dioxide have been added to the atmosphere.

15. ___ Coinciding with the subtropical high-pressure belts are Earth's *low-latitude* deserts and steppes.

Word choice. Complete each of the following statements by selecting the most appropriate response.

1. The climate system involves the exchanges of energy and moisture among its [two/three/five] parts.

2. In the Köppen climate classification, precipitation is the primary criterion for the [B/C] group.

3. Greenhouse warming could cause a [rise/fall] in sea level.

4. The Köppen climate classification uses mean monthly and annual values of [pressure/temperature] and [precipitation/solar radiation].

5. The largest area of Cfb climate is found in [South America/Europe], for here there is no mountain barrier blocking the movement of cool maritime air.

6. The intense drought conditions of the tropical wet and dry climates are caused by the [equatorial low/subtropical high] pressure belts.

7. The taiga climate closely corresponds to the northern [deciduous/coniferous] forest region.

8. Total precipitation in the wet tropical climates often exceeds [200/800] centimeters per year.

9. Beneath the subtropical high-pressure belts, air is [rising/subsiding], compressed, and warmed.

10. Since 1880, when reliable temperature measurements began, the mean global temperature has risen approximately [0.5/5.0]°C.

11. Many of the rainiest places in the world are located on the [windward/leeward] slopes of mountains.

12. The maximum temperature record in North America is [114/134]°F, recorded at Death Valley, CA.

13. A climate diagram is a plot of the average monthly temperature and [precipitation/solar radiation] amounts that occur at a place throughout the year.

14. Along with carbon dioxide, [nitrogen/water vapor] is largely responsible for the greenhouse effect of the atmosphere.

15. Models indicate that global warming of the lower atmosphere triggered by carbon dioxide and trace gases [will/will not] be the same everywhere.

Written questions

1. Briefly describe the distribution of climates found in North America.

2. Why are wet tropical climates restricted to elevations below 1000 meters?

3. When compared to rural areas, what effects do cities have on climatic elements such as precipitation, temperature, wind speed, and solar radiation?

Earth's Place in the Universe

Earth's Place in the Universe opens by examining the astronomical observations and contributions made by early civilizations and the Greek philosophers, including Aristotle, Aristarchus, Hipparchus, as well as Ptolemy. An in-depth examination of the birth of modern astronomy centers on the contributions of Nicolaus Copernicus, Tycho Brahe, Johannes Kepler, Galileo Galilei, and Sir Isaac Newton. Following a brief overview of the constellations, a system for locating stars in the sky is presented. The primary motions of Earth are also described in detail. The chapter concludes with discussions of the phases of the moon, lunar motions, and eclipses.

Learning Objectives

After reading, studying, and discussing this chapter, you should be able to:

- Describe the geocentric theory of the universe held by many early Greeks.
- List the astronomical contributions of the ancient Greek philosophers Aristotle, Anaxagoras, Aristarchus, Eratosthenes, Hipparchus, and Ptolemy.
- Describe the Ptolemaic model of the universe.
- List the contributions to modern astronomy of Nicolaus Copernicus, Tycho Brahe, Johannes Kepler, Galileo Galilei, and Sir Isaac Newton.
- Describe the equatorial system for locating stars.
- List and describe the primary motions of Earth.
- Discuss the phases of the moon, lunar motions, and eclipses.

Chapter Summary

- Early Greeks held the *geocentric* ("Earth-centered") view of the universe, believing that Earth was a sphere that stayed motionless at the center of the universe. Orbiting Earth were the seven wanderers (*planetai* in Greek) which included the moon, sun, and the known planets—Mercury, Venus, Mars, Jupiter, and Saturn. To the early Greeks, the stars traveled daily around Earth on a transparent, hollow sphere called the *celestial sphere*. In A.D. 141, *Claudius Ptolemy* presented the geocentric outlook of the Greeks in its most sophisticated form in a model which became known as the *Ptolemaic system*. The Ptolemaic model had the planets moving in circular orbits around a motionless Earth. To explain the *retrograde motion* of planets (the apparent westward, or opposite motion planets exhibit for a period of time as Earth overtakes and passes them) Ptolemy proposed that the planets orbited in small circles (*epicycles*), revolving along large circles (*deferents*).

- In the fifth century B.C., the Greek *Anaxagoras* reasoned that the moon shines by reflected sunlight, and because it is a sphere, only half is illuminated at one time. *Aristotle* (384-322 B.C.) concluded that Earth is spherical. The first Greek to profess a sun-centered, or *heliocentric*, universe was *Aristarchus*

(312-230 B.C.). The first successful attempt to establish the size of Earth is credited to *Eratosthenes* (276-194 B.C.). The greatest of the early Greek astronomers was *Hipparchus* (second century B.C.), best known for his star catalog.

• Modern astronomy evolved through the work of many dedicated individuals during the 1500s and 1600s. *Nicolaus Copernicus* (1473-1543) reconstructed the solar system with the sun at the center and the planets orbiting around it, but erroneously continued to use circles to represent the orbits of planets. *Tycho Brahe's* (1546-1601) observations were far more precise than any made previously and are his legacy to astronomy. *Johannes Kepler* (1571-1630) ushered in the new astronomy with his three laws of planetary motion. After constructing his own telescope, *Galileo Galilei* (1564-1642) made many important discoveries that supported the Copernican view of a sun-centered solar system. *Sir Isaac Newton* (1643-1727) was the first to formulate and test the law of universal gravitation, develop the laws of motion, and prove that the force of *gravity*, combined with the tendency of an object to move in a straight line (*inertia*), results in the elliptical orbits discovered by Kepler.

• As early as 5000 years ago people began naming the configurations of stars, called *constellations*, in honor of mythological characters or great heroes. Today, 88 constellations are recognized that divide the sky into units, just as state boundaries divide the United States.

• One method for locating stars, called the *equatorial system*, divides the celestial sphere into a coordinate system similar to the latitude-longitude system used for locations on Earth's surface. *Declination*, like latitude, is the angular distance north or south of the *celestial equator*. *Right ascension* is the angular distance measured eastward from the position of the *vernal equinox* (the point in the sky where the sun crosses the celestial equator at the onset of spring).

• The two primary motions of Earth are *rotation* (the turning, or spinning, of a body on its axis) and *revolution* (the motion of a body, such as a planet or moon, along a path around some point in space). Another very slow motion of Earth is *precession* (the slow motion of Earth's axis which traces out a cone over a period of 26,000 years). Earth's rotation can be measured in two ways, making two kinds of days. The *mean solar day* is the time interval from one noon to the next, which averages about 24 hours. On the other hand, the *sidereal day* is the time it takes for Earth to make one complete rotation with respect to a star other than the sun, a period of 23 hours, 56 minutes, and 4 seconds. Earth revolves around the sun in an elliptical orbit at an average distance from the sun of 150 million kilometers (93 million miles). At *perihelion* (closest to the sun), which occurs in January, Earth is 147 million kilometers from the sun. At *aphelion* (farthest from the sun), which occurs in July, Earth is 152 million kilometers distant. The imaginary plane that connects Earth's orbit with the celestial sphere is called the *plane of the ecliptic*.

• One of the first astronomical phenomenon to be understood was the regular cycle of the phases of the moon. The cycle of the moon through its phases requires 29½ days, a time span called the *synodic month*. However, the true period of the moon's revolution around Earth takes 27⅓ days and is known as the *sidereal month*. The difference of nearly two days is due to the fact that as the moon orbits Earth, the Earth-moon system also moves in an orbit around the sun.

• In addition to understanding the moon's phases, the early Greeks also realized that eclipses are simply shadow effects. When the moon moves in a line directly between Earth and the sun, which can occur only during the new-moon phase, it casts a dark shadow on Earth, producing a *solar eclipse*. A *lunar eclipse* takes place when the moon moves within the shadow of Earth during the full-moon phase. Because the moon's orbit is inclined about 5 degrees to the plane that contains the Earth and sun (the plane of the ecliptic), during most new- and full-moon phases no eclipse occurs. Only if a new- or full-moon phase occurs as the moon crosses the plane of the ecliptic can an eclipse take place. The usual number of eclipses is four per year.

Chapter Outline

I. Early history of astronomy
  A. Ancient Greeks
    1. Used philosophical arguments to explain natural phenomena
    2. Also used some observational data
    3. Most ancient Greeks held a geocentric (Earth-centered) view of the universe
      a. "Earth-centered" view
        1. Earth was a motionless sphere at the center of the universe
        2. Stars were on the celestial sphere
          a. Transparent, hollow sphere
          b. Celestial sphere turns daily around Earth
      b. Seven heavenly bodies (planetai)
        1. Changed position in sky
        2. The seven wanderers included the
          a. Sun
          b. Moon
          c. Mercury through Saturn (excluding Earth)
    4. Aristarchus (312-230 B.C.) was the first Greek to profess a suncentered, or heliocentric, universe
    5. Planets exhibit an apparent westward drift
      a. Called retrograde motion
      b. Occurs as Earth, with its faster orbital speed, overtakes another planet
    6. Ptolemaic system
      a. A.D. 141
      b. Geocentric model
      c. To explain retrograde motion, Ptolemy used two motions for the planets
        1. Large orbital circles, called deferents, and
        2. Small circles, called epicycles
  B. Birth of modern astronomy
    1. 1500s and 1600s
    2. Five noted scientists
      a. Nicolaus Copernicus (1473-1543)
        1. Concluded Earth was a planet
        2. Constructed a model of the solar system that put the sun at the center, but he used circular orbits for the planets

        3. Ushered out old astronomy
      b. Tycho Brahe (1546-1601)
        1. Precise observer
        2. Tried to find stellar parallax—the apparent shift in a star's position due to the revolution of Earth
        3. Did not believe in the Copernican system because he was unable to observe stellar parallax
      c. Johannes Kepler (1571-1630)
        1. Ushered in new astronomy
        2. Planets revolve around the sun
        3. Three laws of planetary motion
          a. Orbits of the planets are elliptical
          b. Planets revolve around the sun at varying speed
          c. There is a proportional relation between a planet's orbital period and its distance to the sun (measured in astronomical units (AUs)—one AU averages about 150 million kilometers, or 93 million miles)
      d. Galileo Galilei (1564-1642)
        1. Supported Copernican theory
        2. Used experimental data
        3. Constructed an astronomical telescope in 1609
        4. Galileo's discoveries using the telescope
          a. Four large moons of Jupiter
          b. Planets appeared as disks
          c. Phases of Venus
          d. Features on the moon
          e. Sunspots
        5. Tried and convicted by the Inquisition
      e. Sir Isaac Newton (1643-1727)
        1. Law of universal gravitation
        2. Proved that the force of gravity, combined with the tendency of a planet to remain in straight-line motion, results in the elliptical orbits discovered by Kepler

II. Constellations
  A. Configuration of stars named in honor

of mythological characters or great heroes

B. Today 88 constellations are recognized

C. Constellations divide the sky into units, like state boundaries in the United States

D. The brightest stars in a constellation are identified in order of their brightness by the letters of the Greek alphabet—alpha, beta, and so on

III. Positions in the sky

A. Stars appear to be fixed on a spherical shell (the celestial sphere) that surrounds Earth

B. Equatorial system of location

1. A coordinate system that divides the celestial sphere

2. Similar to the latitude-longitude system that is used on Earth's surface

3. Two locational components

a. Declination—the angular distance north or south of the celestial equator

b. Right ascension—the angular distance measured eastward along the celestial equator from the position of the vernal equinox

IV. Earth motions

A. Two primary motions

1. Rotation

a. Turning, or spinning, of a body on its axis

b. Two measurements for rotation

1. Mean solar day—the time interval from one noon to the next, about 24 hours

2. Sidereal day—the time it takes for Earth to make one complete rotation (360°) with respect to a star other than the sun—23 hours, 56 minutes, 4 seconds

2. Revolution

a. The motion of a body, such as a planet or moon, along a path around some point in space

b. Earth's orbit is elliptical

1. Earth is closest to the sun (perihelion) in January

2. Earth is farthest from the sun (aphelion) in July

c. The plane of the ecliptic is an imaginary plane that connects Earth's orbit with the celestial sphere

B. Other Earth motions

1. Precession

a. Very slow Earth movement

b. Direction in which Earth's axis points continually changes

2. Movement with the solar system in the direction of the star Vega

3. Revolution with the sun around the galaxy

4. Movement with the galaxy within the universe

V. Motions of the Earth-moon system

A. Phases of the moon

1. When viewed from above the North Pole, the moon orbits Earth in a counterclockwise (eastward) direction

2. The relative positions of the sun, Earth, and moon constantly change

3. Lunar phases are a consequence of the motion of the moon and the sunlight that is reflected from its surface

B. Lunar motions

1. Earth-moon

a. Synodic month

1. Cycle of the phases

2. Takes 29½ days

b. Sidereal month

1. True period of the moon's revolution around Earth

2. Takes 27⅓ days

c. The difference of two days between the synodic and sidereal cycles is due to the Earth-moon system also moving in an orbit around the sun

2. Moon's period of rotation about its axis and its revolution around Earth are the same, 27⅓ days

a. Causes the same lunar hemisphere to always face Earth

b. Causes high surface temperature on the day side of the moon

C. Eclipses

1. Simply shadow effects that were first understood by the early Greeks

2. Two types of eclipses

a. Solar eclipse

1. Moon moves in a line directly between Earth and the sun
2. Can only occur during the new-moon phase

b. Lunar eclipse
1. Moon moves within the shadow of Earth
2. Only occurs during the full-moon phase

3. For any eclipse to take place, the moon must be in the plane of the ecliptic at the time of new- or full-moon

4. Because the moon's orbit is inclined about 5 degrees to the plane of the ecliptic, during most of the times of new- and full-moon the moon is above or below the plane, and no eclipse can occur

5. The usual number of eclipses is four per year

Key Terms

Page numbers shown in ( ) refer to the textbook page where the term first appears.

aphelion (p. 505)
astronomical unit (AU) (p. 497)
celestial sphere (p. 490)
constellation (p. 502)
declination (p. 505)
ecliptic (p. 505)
equatorial system (p. 502)
geocentric (p. 490)
heliocentric (p. 491)

lunar eclipse (p. 511)
mean solar day (p. 505)
perihelion (p. 505)
perturbation (p. 501)
phases of the moon (p. 508)
plane of the ecliptic (p. 506)
precession (p. 505)
Ptolemaic system (p. 491)
retrograde motion (p. 492)

revolution (p. 505)
right ascension (p. 505)
rotation (p. 503)
sidereal day (p. 505)
sidereal month (p. 510)
solar eclipse (p. 510)
synodic month (p. 510)

Vocabulary Review

Choosing from the list of key terms, furnish the most appropriate response for the following statements.

1. The apparent westward drift of the planets with respect to the stars is called _____.

2. The _____, the time interval it takes for Earth to make one complete rotation (360°) with respect to a star other than the sun, has a period of 23 hours, 56 minutes, and 4 seconds.

3. Aristarchus (312-230 B.C.) was the first Greek to profess a sun-centered, or _____, universe.

4. The true period of the moon's revolution around Earth, 27⅓ days, is known as the _____.

5. Early Greeks held the _____ view of the universe, believing that Earth was a sphere that stayed motionless at its center.

6. _____ is the turning, or spinning, of a body on its axis.

7. The average distance from Earth to the sun, about 150 million kilometers (93 million miles), is a unit of distance called the _____.

8. When the moon moves in a line directly between Earth and the sun it casts a dark shadow on Earth, producing a _____.

9. The angular distance north or south of the celestial equator denoting the position of a stellar body is referred to as _____.

10. Any variance in the orbit of a body from its predicted path is called _____.

11. The point in the orbit of a planet where it is closest to the sun is called _____.

12. The apparent annual path of the sun against the backdrop of the celestial sphere is called the _____.

13. The _____ was an imaginary, transparent, hollow sphere on which the ancients believed the stars traveled daily around Earth.

14. A(n) _____ is an apparent group of stars originally named in honor of a mythological character or great hero.

15. The imaginary plane that connects Earth's orbit with the celestial sphere is called the _____.

16. The _____ is an Earth-centered model of the universe proposed in A.D. 141 which uses epicycles and deferents to describe a planet's motion.

17. _____ is the motion of a body, such as a planet or moon, along a path around some point in space.

18. The _____ refers to the progression of changes in the moon's appearance during the month.

19. The angular distance of a stellar object measured eastward along the celestial equator from the position of the vernal equinox is called _____.

20. The cycle of the moon through its phases requires 29½ days, a time span called the _____.

21. During the full-moon phase, when the moon moves within the shadow of Earth, a _____ occurs.

22. The _____ is a method used to locate stellar objects that is very similar to the latitude-longitude system used on Earth's surface.

23. The point in the orbit of a planet where it is farthest from the sun is referred to as _____.

24. The very slight movement of Earth's axis over a period of 26,000 years is known as _____.

25. The _____, the time interval from one noon to the next, averages about 24 hours.

Comprehensive Review

1. Describe the geocentric model of the universe held by the early Greeks.

2. What is retrograde motion? In Figure 19.1, a diagram of the orbit of Mars as observed from Earth over many weeks, between which two points is retrograde motion occurring? How did the geocentric model of Ptolemy explain retrograde motion?

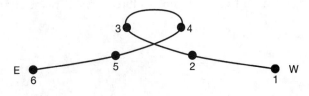

Figure 19.1

3. Why did the early Greeks reject the idea of a rotating Earth?

4. For each of the following statements, list the name of the scientist credited with the contribution.

   a) First to use the telescope for astronomy:

   b) Proposed circular orbits of the planets around the sun in 1500s:

   c) Systematically measured the locations of heavenly bodies:

   d) Proposed three laws of planetary motion:

   e) Monumental work, *De Revolutionibus:*

   f) Proposed elliptical orbits for planets around the sun:

   g) Discovered four satellites, or moons, orbiting Jupiter:

   h) Wrote *Dialogue of the Great World Systems:*

   i) Law of universal gravitation:

   j) The first Greek to profess a heliocentric universe:

   k) An early Greek astronomer best known for his star catalog:

   l) Credited with the first successful attempt to establish the size of Earth:

5. Describe stellar parallax.

6. List each of Kepler's three laws of planetary motion.

   1)

   2)

   3)

7. List four of Galileo's astronomical discoveries.

   1)

   2)

   3)

   4)

8. Use Figure 19.2, a diagram of the orbit of a hypothetical planet, to answer the following questions.

   a) During which months is the planet moving fastest in its orbit?

   b) During which months is the planet moving slowest in its orbit?

   c) In addition to inertia, what force is keeping the planet in orbit?

   d) If the planet's mean distance from the sun is 4 AUs, what would be its orbital period?

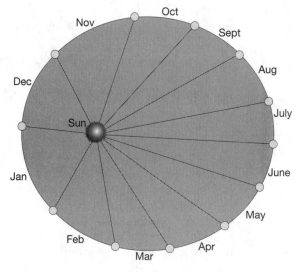

Figure 19.2

9. Briefly explain the equatorial system used for locating objects on the celestial sphere.

10. Using Figure 19.3, select the letter that identifies each of the following components of the celestial sphere.

    a) Celestial equator:

    b) North celestial pole:

    c) Declination of star G:

    d) Right ascension of star G:

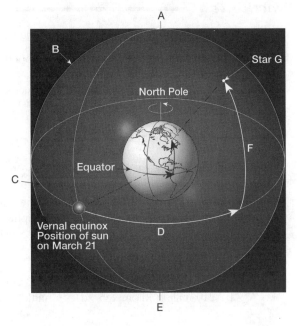

Figure 19.3

11. List six motions that Earth is continuously experiencing.

12. Describe the difference between the moon's synodic and sidereal cycles.

13. Using Figure 19.4, select the number that illustrates the moon's position in its orbit for each of the following phases.

    a) Full:

    b) Third quarter:

    c) Crescent (waxing):

    d) New:

    e) Gibbous (waning):

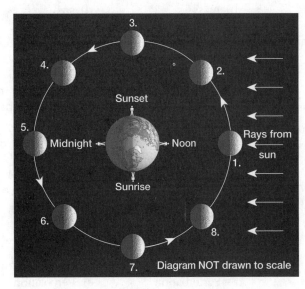

Figure 19.4

Practice Test

Multiple choice. Choose the best answer for the following multiple choice questions.

1. Who was the ancient Greek that developed a geocentric model of the universe explaining the observable motions of the planets?
   a) Aristotle   c) Copernicus   e) Newton
   b) Ptolemy   d) Kepler

2. During the period that the moon's phases are changing from new to full, it is _____.
   a) waning   c) waxing   e) exhibiting retrograde motion
   b) approaching Earth d) receding from Earth

3. The apparent path of the sun against the backdrop of the celestial sphere is called _____.
   a) declination   c) the ecliptic   e) the precession
   b) perturbation   d) right ascension

4. One astronomical unit averages about _____.
   a) 93 million miles   c) 39 million miles   e) 150 million miles
   b) 210 million kilometers d) 93 million kilometers

5. At its closest position to the sun, called _____, Earth is 147 million kilometers (91 million miles) distant.
   a) aphelion   c) perihelion   e) ascension
   b) quadrature   d) declination

6. Ancient astronomers believed that _____.
   a) Earth revolved around the celestial sphere   d) Earth was a "wanderer"
   b) the sun was the center of the universe   e) Earth was the center of the universe
   c) the sun was on the celestial sphere

7. The _____ of an object is a measure of the total amount of matter it contains.
   a) density   c) volume   e) velocity
   b) mass   d) size

8. It takes the moon about _____ weeks to go from full-moon phase to new-moon phase.
   a) one   c) three   e) five
   b) two   d) four

9. The apparent westward movement of a planet against the background of stars is called _____.
   a) retrograde motion c) westward drift   e) deferent motion
   b) reverse motion   d) Ptolemaic motion

10. The most visible stars in a constellation are generally named in order of their _____ by the letters of the Greek alphabet.
    a) color   c) size   e) brightness
    b) temperature   d) mass

11. _____ refers to fact that the direction in which Earth's axis points continuously changes.
    a) Perturbation   c) Percolation   e) Progression
    b) Precession   d) Progradation

12. Which one of the following planets was unknown to the ancient Greeks?
    a) Earth   c) Uranus   e) Venus
    b) Mars   d) Mercury

13. Earth moves forward in its orbit about _____ kilometers each second.
    a) 11   c) 25   e) 43
    b) 16   d) 30

14. Using Tycho Brahe's data, which scientist proposed three laws of planetary motion?
    a) Newton   c) Kepler   e) Ptolemy
    b) Galileo   d) Copernicus

15. A(n) _____ eclipse occurs when the moon casts its shadow on Earth.
    a) lunar   c) sidereal   e) synodic
    b) umbral   d) solar

16. The first accurate measurement of the size of Earth was made by _____.
    a) Aristotle   c) Aristarchus   e) Newton
    b) Eratosthenes   d) Ptolemy

17. Another name for the North Star is _____.
    a) Vega   c) Andromeda   e) Polaris
    b) Altair   d) Antares

18. The first modern astronomer to propose a sun-centered solar system was _____.
    a) Galileo   c) Ptolemy   e) Brahe
    b) Newton   d) Copernicus

19. The variance in the orbit of a planet from its predicted path is called _____.
    a) perturbation   c) percolation   e) progression
    b) precession   d) progradation

20. Which scientist discovered that Venus has phases, just like the moon?
    a) Ptolemy   c) Galileo   e) Newton
    b) Copernicus   d) Kepler

21. Thirty degrees of right ascension is equivalent to _____ minutes.
    a) 15   c) 45   e) 120
    b) 30   d) 60

22. The apparent shift in the position of a star caused by Earth's motion is called _____.
    a) stellar shift   c) an epicycle   e) revolution
    b) retrograde motion   d) stellar parallax

23. The force that gravity exerts on an object is referred to as _____.
    a) mass   c) perturbation   e) density
    b) pressure   d) weight

24. The ancient Greek _____ reasoned that the moon shines by reflected sunlight.
    a) Aristotle   c) Aristarchus   e) Almagest
    b) Anaxagoras   d) Artemis

25. The shape of a planet's orbit is _____.
    a) circular   c) constantly changing   e) parabolic
    b) irregular   d) elliptical

26. According to the ancients, the stars traveled around Earth on the transparent, hollow, _____.
    a) deferent   c) equatorial sphere   e) stellar globe
    b) celestial sphere   d) solar orb

27. Which of the following is correctly paired with its analogous measurement on Earth?
    a) right ascension-declination   d) declination-right ascension
    b) right ascension-latitude   e) declination-longitude
    c) declination-latitude

28. A day measured by using the stars rather than the sun is called a(n) _____ day.
    a) sidereal   c) synodic   e) average
    b) mean solar   d) ecliptic

29. According to Kepler's third law, the period of revolution of a planet is related to the planet's _____.
    a) distance to the sun   c) mass   e) gravitational attraction
    b) size   d) acceleration curve

30. The period of revolution of the moon around Earth is 27⅓ days; while the period of rotation of the moon about its axis is _____ days.
    a) 27⅓   c) 29½   e) 33½
    b) 28   d) 30⅓

*True/false. For the following true/false questions, if a statement is not completely true, mark it false. For each false statement, change the **italicized** word to correct the statement.*

1. ___ The orbits of the planets are *circular*.

2. ___ The early Greeks believed that there were *nine* wandering heavenly bodies.

3. ___ Probably the greatest of the early Greek astronomers was *Tycho Brahe,* best known for his star catalog.

4. ___ The farther away a star is, the *greater* will be its parallax.

5. ___ The cycle of the moon through its phases as seen from Earth is called the *sidereal* month.

6. ___ The force of gravity is *inversely* proportional to the square of the distance between two objects.

7. ___ The mean solar day is *shorter* than the sidereal day.

8. ___ Total solar eclipses are only observed by people in the dark part of the moon's shadow, called the *umbra*.

9. ___ A *geocentric* model holds that Earth is the center of the universe.

10. ___ Any variance in the orbit of a body from its predicted path is called *precession*.

11. ___ Gravity is directly proportional to *mass*.

12. ___ The orbital periods of the planets and their *distances* to the sun are proportional.

13. ___ A *lunar* eclipse occurs when Earth's shadow is cast on the moon.

14. ___ Our present calendar is accurate to within 1 *second* in 3000 years.

15. ___ Lunar eclipses occur during *full-moon* phase.

16. ___ Earth is the center of the universe in a *heliocentric* model.

17. ___ During a new moon, the lunar hemisphere that faces Earth is the *opposite* hemisphere that faces Earth during full moon.

18. ___ In his model, Copernicus used *circles* to represent the orbits of the planets.

19. ___ The declination of a star is usually expressed in *hours*.

20. ___ Beginning in 1609, *Galileo Galilei* published his three laws of planetary motion.

Word choice. Complete each of the following statements by selecting the most appropriate response.

1. The actual time it takes for the moon to complete one 360° revolution around Earth is called the [sidereal/synodic] month.

2. Most of the time planets move [eastward/westward] among the stars.

3. A [lunar/solar] eclipse occurs when the moon casts its shadow on Earth.

4. The [nautical/astronomical] unit is equal to Earth's mean distance to the sun, about 150 million kilometers (93 million miles).

5. The mass of an object is [constant/variable] but its weight [remains constant/varies] when gravitational forces change.

6. The synodic cycle of the moon is [shorter/longer] than the sidereal cycle.

7. The calendar that we currently use is called the [Julian/Gregorian] calendar.

8. The belief that Earth is the center of the universe is a [geocentric/heliocentric] view.

9. Early Greeks believed that the moon was illuminated by [reflected sunlight/internal fires].

10. The concept that once an object is in motion it will continue traveling at a uniform speed in a straight line is called [inertia/momentum].

11. A total [solar/lunar] eclipse is visible to anyone on the side of Earth facing the moon.

12. Earth is closest to the sun at [aphelion/perihelion].

13. The Ptolemaic model of the universe states that the planets move in [circular/elliptical] orbits around [Earth, the sun].

14. [Aristotle/Aristarchus] believed that the universe was centered on the sun.

15. The moon's period of rotation is [more than/equal to/less than] its period of revolution around Earth.

16. All stars in a constellation [are/are not] located at the same distance from Earth.

17. The plane of the ecliptic is tilted relative to the celestial equator due to the inclination of Earth's [axis/orbit].

18. Solar eclipses occur during the [new-moon/full-moon] phase.

19. In the 1630s [Kepler/Galileo/Newton] was tried by the Inquisition and convicted of proclaiming doctrines contrary to religious beliefs.

20. A [quadrant/pendulum] can be used to prove that Earth rotates.

Written questions

1. Explain why planets appear to exhibit retrograde motion.

2. Briefly explain why eclipses don't occur every month.

3. List and describe four motions that Earth continuously experiences.

4. Describe the model of the universe proposed by Ptolemy in A.D. 141.

Touring Our Solar System

20

Touring Our Solar System opens with an investigation of the differences between the terrestrial and Jovian planets, followed by a discussion of the evolution of the planets. Included in the chapter is a detailed study of the moon's physical characteristics and history. An inventory of the solar system presents the prominent features and peculiarities of each planet (excluding Earth). The chapter closes with a discussion of the minor members of the solar system—asteroids, comets, and meteoroids.

Learning Objectives

After reading, studying, and discussing this chapter, you should be able to:

- Describe the general characteristics of the two groups of planets in the solar system.
- Discuss the evolution of the planets.
- Describe the major features of the lunar surface and discuss the moon's history.
- List the distinguishing features of each planet in the solar system.
- List and describe the minor members of the solar system.

Chapter Summary

- The planets can be arranged into two groups: the *terrestrial* (Earthlike) *planets* (Mercury, Venus, Earth, and Mars) and the *Jovian* (Jupiterlike) *planets* (Jupiter, Saturn, Uranus, and Neptune). Pluto is not included in either group. When compared to the Jovian planets, the terrestrial planets are smaller, more dense, contain proportionally more rocky material, have slower rates of rotation, and lower *escape velocities*.

- The *nebular hypothesis* (discussed in Chapter 11) describes the formation of the solar system. Due to *chemical differentiation*, as the terrestrial planets formed, the denser metallic elements (iron and nickel) sank toward their centers, whereas the lighter substances (silicate minerals, oxygen, hydrogen) migrated toward their surfaces. Due to their surface gravities, Earth and Venus were able to retain the heavier gases, like nitrogen, oxygen, and carbon dioxide. Because of the very cold temperatures existing far from the sun, the fragments from which the Jovian planets formed contained a high percentage of ices—water, carbon dioxide, ammonia, and methane.

- The lunar surface exhibits several types of features. Most *craters* were produced by the impact of rapidly moving interplanetary debris (*meteoroids*). Bright, densely cratered *highlands* make up most of the lunar surface. The dark, fairly smooth lowlands are called *maria* (singular, *mare*). Maria basins are enormous impact craters that have been flooded with layer-upon-layer of very fluid basaltic lava. All lunar terrains are mantled with a soil-like layer of gray, unconsolidated debris, called *lunar regolith*, which has been derived from a few billion years of meteoric bombardment. Much is still unknown about the moon's origin. One hypothesis suggests that a giant asteroid collided with Earth to produce the moon. Scientists conclude that *the moon evolved in three phases* 1) *the original crust (highlands)*, 2) *maria basins*, and 3) *youthful rayed craters*.

• *Mercury* is a small, dense planet that has no atmosphere and exhibits the greatest temperature extremes of any planet. *Venus*, the brightest planet in the sky, has a thick, heavy atmosphere composed of 97 percent carbon dioxide, a surface of relatively subdued plains and inactive volcanic features, a surface atmospheric pressure 90 times that of Earth's, and surface temperatures of 475°C (900°F). *Mars*, the "Red Planet," has a carbon dioxide atmosphere only 1 percent as dense as Earth's, extensive dust storms, numerous inactive volcanoes, many large canyons, and several valleys of debatable origin exhibiting drainage patterns similar to stream valleys on Earth. *Jupiter*, the largest planet, rotates rapidly, has a banded appearance caused by huge convection currents driven by the planet's interior heat, a *Great Red Spot* that varies in size, a thin ring system, and at least sixteen moons (one of the moons, *Io*, is a volcanically active body). *Saturn* is best known for its system of rings. It also has a dynamic atmosphere with winds up to 930 miles per hour and "storms" similar to Jupiter's Great Red Spot. *Uranus* and *Neptune* are often called "the twins" because of similar structure and composition. A unique feature of Uranus is the fact that it rotates "on its side." Neptune has white, cirruslike clouds above its main cloud deck and an Earth-sized *Great Dark Spot*, assumed to be a large rotating storm similar to Jupiter's Great Red Spot. *Pluto*, a small frozen world with one moon (Charon), may have once been a satellite of Neptune. Pluto's noticeably elongated orbit causes it to occasionally travel inside the orbit of Neptune, but with no chance of collision.

• The minor members of the solar system include the *asteroids, comets,* and *meteoroids*. Most asteroids lie between the orbits of Mars and Jupiter. No conclusive evidence has been found to explain their origin. Comets are made of frozen gases (water, ammonia, methane, carbon dioxide, and carbon monoxide) with small pieces of rocky and metallic material. Many travel in very elongated orbits that carry them beyond Pluto and little is known about their origin. Meteoroids, small solid particles that travel through interplanetary space, become *meteors* when they enter Earth's atmosphere and vaporize with a flash of light. *Meteor showers* occur when Earth encounters a swarm of meteoroids, probably material lost by a comet. *Meteorites* are the remains of meteoroids found on Earth. The *three types of meteorites* (classified by their composition) are 1) *irons*, 2) *stony*, and 3) *stony-irons*. One rare kind of meteorite, called a *carbonaceous chondrite*, was found to contain amino acids and other organic compounds.

Chapter Outline

I. Overview of the solar system
    A. Solar system includes
        1. Sun
        2. Nine planets and their satellites
        3. Asteroids
        4. Comets
        5. Meteoroids
    B. A planet's orbit lies in an orbital plane
        1. Similar to a flat sheet of paper
        2. The orbital planes of the planets are inclined
            a. Planes of seven planets lie within 3° of the sun's equator
            b. Mercury's is inclined 7°
            c. Pluto's is inclined 17°

C. Two groups of planets occur in the solar system
    1. Terrestrial (Earthlike) planets
        a. Mercury through Mars
        b. Small, dense, rocky
        c. Low escape velocities
    2. Jovian (Jupiterlike) planets
        a. Jupiter through Neptune
        b. Large, low density, gaseous
        c. Massive
        d. Thick atmospheres composed of
            1. Hydrogen
            2. Helium
            3. Methane
            4. Ammonia

e. High escape velocities
3. Pluto not included in either group
D. Planets are composed of
1. Gases
   a. Hydrogen
   b. Helium
2. Rocks
   a. Silicate minerals
   b. Metallic iron
3. Ices
   a. Ammonia (NH_3)
   b. Methane (CH_4)
   c. Carbon dioxide (CO_2)
   d. Water (H_2O)

II. Evolution of the planets
A. Nebular hypothesis (see Chapter 11)
   1. Planets formed about 5 billion years ago
   2. Solar system condensed from a gaseous nebula
B. As the planets formed, the materials that compose them separated
   1. Dense metallic elements (iron and nickel) sank toward their centers
   2. Lighter elements (silicate minerals, oxygen, hydrogen) migrated toward their surfaces
   3. Process called chemical differentiation
C. Due to their surface gravities, Venus and Earth retained atmospheric gases
D. Due to frigid temperatures, the Jovian planets contain a high percentage of ices

III. Earth's moon
A. General characteristics
   1. Diameter of 2150 miles is unusually large compared to its parent planet
   2. Density
      a. 3.3 times that of water
      b. Comparable to Earth's crustal rocks
      c. Perhaps the moon has a small iron core
   3. Gravitational attraction is one-sixth of Earth's
   4. No atmosphere
   5. Tectonics no longer active
   6. Surface is bombarded by micro-meteorites from space which gradually makes the landscape smooth
B. Lunar surface
   1. Two types of terrain

a. Maria (singular, mare), Latin for "sea"
   1. Dark regions
   2. Fairly smooth lowlands
   3. Originated from asteroid impacts and lava flooding the surface
b. Highlands
   1. Bright, densely cratered regions
   2. Make up most of the moon
   3. Make up all of the "back" side of the moon
   4. Older than maria
2. Craters
   a. Most obvious features of the lunar surface
   b. Most are produced by an impact from a meteoroid which produces
      1. Ejecta
      2. Occasional rays (associated with younger craters)
3. Lunar regolith
   a. Covers all lunar terrains
   b. Gray, unconsolidated debris
   c. Composed of
      1. Igneous rocks
      2. Breccia
      3. Glass beads
      4. Fine lunar dust
   d. "Soil-like" layer
   e. Produced by meteoric bombardment
C. Lunar History
   1. New hypothesis suggests that a giant asteroid collided with Earth to produce the moon
   2. One method used to work out lunar history is to observe crater density
      a. Older areas have a higher density
      b. Younger areas are still smooth
   3. Moon evolved in three phases
      a. Original crust (highlands)
         1. As moon formed, its outer shell melted, cooled, solidified, and became the highlands
         2. About 4.5 billion years old
      b. Formation of maria basins
         1. Younger than highlands
         2. Between 3.2 and 3.8 billion years old
      c. Formation of rayed craters

1. Material ejected from craters is still visible
2. e.g., Copernicus (a rayed crater)

IV. Planets
 A. Mercury
  1. Innermost planet
  2. Smallest
  3. No atmosphere
  4. Cratered highlands
  5. Vast, smooth terrains
  6. Very dense
  7. Revolves quickly
  8. Rotates slowly
   a. Cold nights (-280°F)
   b. Hot days (800°F)
 B. Venus
  1. Second to the moon in brilliance
  2. Similar to Earth in
   a. Size
   b. Density
   c. Location in the solar system
  3. Shrouded in thick clouds
   a. Impenetrable by visible light
   b. Atmosphere is 97 percent carbon dioxide
   c. Surface atmospheric pressure is 90 times that of Earth's
  4. Surface
   a. Mapped by radar
   b. Features
    1. 80 percent of surface is sub-dued plains that are mantled by volcanic flows
    2. Low density of impact craters
    3. Tectonic deformation must have been active during the recent geologic past
    4. Thousands of volcanic structures
 C. Mars
  1. Called the "Red Planet"
  2. Atmosphere
   a. 1 percent as dense as Earth's
   b. Primarily carbon dioxide
   c. Cold polar temperatures (-193°F)
   d. Polar caps of water ice, covered by a thin layer of frozen carbon dioxide
   e. Extensive dust storms with winds up to 170 miles per hour
  3. Surface
   a. Numerous large volcanoes—largest is Mons Olympus

   b. Less-abundant impact craters
   c. Tectonically dead
   d. Several canyons
    1. Some larger than Grand Canyon
    2. Valles Marineras—the largest canyon
     a. Almost 5000 km long
     b. Formed from huge faults
   e. "Stream drainage" patterns
    1. Found in some valleys
    2. No surface water on the planet
    3. Possible origins
     a. Past rainfall
     b. Surface material collapses as the subsurface ice melts
   4. Moons
    a. Two moons
     1. Phobos
     2. Deimos
    b. Captured asteroids

 D. Jupiter
  1. Largest planet
  2. Very massive
   a. 2.5 more massive than combined mass of the planets, satellites, and asteroids
   b. If it had been ten times larger, it would have been a small star
  3. Rapid rotation
   a. Slightly less than 10 hours
   b. Slightly bulged equatorial region
  4. Banded appearance
   a. Multicolored
   b. Bands are aligned parallel to Jupiter's equator
   c. Generated by wind systems
  5. Great Red Spot
   a. In planet's southern hemisphere
   b. Counterclockwise rotating cyclonic storm
  6. Structure
   a. Surface thought to be a gigantic ocean of liquid hydrogen
   b. Halfway into the interior, pressure causes liquid metallic hydrogen
   c. Perhaps rocky and metallic material exists in a central core
  7. Moons
   a. At least 16 moons
   b. Four largest moons
    1. Discovered by Galileo
    2. Called Galilean satellites

3. Each has its own character
   a. Callisto
      1. Outermost Galilean moon
      2. Densely cratered
   b. Europa
      1. Smallest Galilean moon
      2. Icy surface
      3. Many linear surface features
   c. Ganymede
      1. Largest Jovian satellite
      2. Diverse terrains
      3. Surface has numerous parallel grooves
   d. Io
      1. Innermost Galilean moon
      2. Volcanically active (heat source could be from tidal energy)
      3. Sulfurous
8. Ring system
E. Saturn
   1. Similar to Jupiter in its
      a. Atmosphere
      b. Composition
      c. Internal structure
   2. Rings
      a. Most prominent feature
      b. Discovered by Galileo in 1610
      c. Complex
      d. Composed of small particles (moonlets) that orbit the planet
   3. Other features
      a. Dynamic atmosphere
      b. Large cyclonic storms similar to Jupiter's Great Red Spot
      c. At least 21 moons
      d. Titan—the largest Saturnian moon
         1. Second largest moon (after Jupiter's Ganymede) in the solar system
         2. Has a substantial atmosphere
            a. 80 percent nitrogen
            b. 6 percent methane
            c. Atmospheric pressure is 1.5 times Earth's
         3. May have polar ice caps
F. Uranus
   1. Uranus and Neptune are considered twins
   2. Rotates "on its side"
   3. Rings

4. Large moons have varied terrains
G. Neptune
   1. Dynamic atmosphere
      a. One of the windiest places in the solar system
      b. Great Dark Spot
      c. White cirruslike clouds above the main cloud deck
   2. Eight satellites
   3. Triton—largest Neptune moon
      a. Orbit is opposite the direction that all the planets travel
      b. Lowest surface temperature in the solar system (-391°F)
      c. Thin atmosphere
      d. Volcaniclike activity
      e. Composed largely of water ice, covered with layers of solid nitrogen and methane
H. Pluto
   1. Not visible with the unaided eye
   2. Discovered in 1930
   3. Highly elongated orbit causes it to occasionally travel inside the orbit of Neptune, where it currently is
   4. Moon (Charon) discovered in 1978
   5. Average temperature is -210°C
   6. May have once been a moon of Neptune
V. Minor members of the solar system
A. Asteroids
   1. Most lie between Mars and Jupiter
   2. Small bodies—largest (Ceres) is about 620 miles in diameter
   3. Some have very eccentric orbits
   4. Many of the recent impacts on the moon and Earth were collisions with asteroids
   5. Irregular shapes
   6. Origin is uncertain
B. Comets
   1. Often compared to large, "dirty snowballs"
   2. Composition
      a. Frozen gases
      b. Rocky and metallic materials
   3. Frozen gases vaporize when near the sun
      a. Produces a glowing head called the coma
      b. Some may develop a tail that points away from sun due to

1. Radiation pressure and
2. Solar wind
4. Origin
   a. Not well known
   b. Form at great distance from
      the sun
5. Most famous short-period comet
   is Halley's comet
   a. 76-year orbital period
   b. Potato-shaped nucleus (16 km
      by 8 km)
C. Meteoroids
   1. Called meteors when they enter
      Earth's atmosphere
   2. A meteor shower occurs when Earth
      encounters a swarm of meteoroids
      associated with a comet's path
   3. Meteoroids are referred to as mete-
      orites when they are found on Earth

a. Types of meteorites classified by
   their composition
   1. Irons
      a. Mostly iron
      b. 5-20 percent nickel
   2. Stony
      a. Silicate minerals
      b. Inclusions of other
         minerals
   3. Stony-irons—mixtures
   4. Carbonaceous chondrites
      a. Rare
      b. Composition
         1. Simple amino acids
         2. Other organic material
b. May give an idea as to the composition of Earth's core
c. Give an idea as to the age of
   the solar system

— Key Terms —

Page numbers shown in ( ) refer to the textbook page where the term first appears.

asteroid (p. 537)   Jovian planet (p. 516)   meteoroid (p. 540)
coma (p. 537)   lunar regolith (p. 521)   meteor shower (p. 540)
comet (p. 537)   maria (p. 519)   stony-iron meteorite (p. 541)
escape velocity (p. 518)   meteor (p. 540)   stony meteorite (p. 541)
iron meteorite (p. 541)   meteorite (p. 541)   terrestrial planet (p. 516)

Vocabulary Review

Choosing from the list of key terms, furnish the most appropriate response for the following statements.

1. The glowing head of a comet is called the _____.

2. A(n) _____ is a planet that has physical characteristics similar to those of Earth.

3. A(n) _____ is a small, planetlike body whose orbit lies mainly between Mars and Jupiter.

4. The dark regions on the moon that resemble "seas" on Earth are _____.

5. A(n) _____ is a small solid particle that travels through interplanetary space.

6. The physical characteristics of a(n) _____ are like those of Jupiter.

7. _____ is the soil-like layer of gray, unconsolidated debris that mantles all lunar terrains.

8. A(n) _____ is a luminous phenomenon observed when a small solid particle enters Earth's atmosphere and burns up.

9. The initial speed that an object needs before it can leave a planet and go into space is known as the _____.

10. A solid fragment from space that strikes Earth's surface and contains mostly silicate minerals (with inclusions of other minerals) is called a(n) _____.

11. A small body made of frozen gases and small pieces of rocky and metallic materials which generally revolves about the sun in an elongated orbit is a(n) _____.

12. The remains of a meteoroid, when found on Earth, is referred to as a(n) _____.

13. A spectacular display of numerous meteor sightings that occurs when Earth encounters a swarm of meteoroids is called a(n) _____.

14. A meteorite that contains mostly iron with 5-20 percent nickel is called a(n) _____.

15. A meteorite that is a mixture of iron, nickel, and silicate minerals is called a(n) _____.

Comprehensive Review

1. Which celestial objects constitute the solar system?

2. What are the gas, rock, and ice substances that compose the majority of the solar system?

   a) Gases:

   b) Rocks:

   c) Ices:

3. Describe the following two groups of planets and list the planets that are included in each group.

   a) Terrestrial planets:

   b) Jovian planets:

4. For each of the following characteristics, write a brief statement that compares the terrestrial planets to the Jovian planets.

   a) Size:

   b) Density:

   c) Period of rotation:

   d) Mass:

   e) Number of known satellites:

   f) Period of revolution:

   g) General composition:

5. What is the process of chemical differentiation? What was the effect of the process on Earth?

6. Using the photograph of the moon, Figure 20.1, select the letter that identifies each of the following lunar features.

   a) Mare: ____

   b) Crater: ____

   c) Ray(s): ____

   d) Highland: ____

   e) Oldest feature: ____

7. Why would a 100-pound person weigh only 17 pounds on the moon?

8. Describe the origin of lunar maria.

Figure 20.1

9. What are two hypotheses that have been proposed for the origin of the moon?

   1)

   2)

10. By placing a number in front of the event, arrange the following lunar events in correct order from oldest (1) to most recent (5).

    a) ____ Crust solidifies

    b) ____ Debris gathers in space

    c) ____ Formation of rayed craters

    d) ____ Maria formation

    e) ____ Bombardment melts the outer shell, and possibly the interior

11. List the name of the planet described by each of the following statements.

    a) Red planet:

    b) Thick clouds, dense carbon dioxide atmosphere, volcanic surface:

    c) Greatest diameter:

    d) Has only satellite (Titan) with substantial atmosphere:

    e) Hot surface temperature and no atmosphere:

    f) Cassini's division:

    g) Second only to the moon in brilliance in the night sky:

    h) Axis of rotation lies near the plane of its orbit:

    i) Great Red Spot:

    j) Moon (Triton) has lowest surface temperature in solar system:

    k) Most prominent system of rings:

    l) Carbon dioxide atmosphere 1 percent as dense as Earth's, volcanic surface:

    m) Moons include Io, Callisto, Ganymede, and Europa:

    n) Earth's "twin":

    o) Great Dark Spot:

    p) Revolves quickly but rotates slowly:

    q) Longest period of revolution:

12. What is the most probable origin of the Martian moons?

13. Write the name of the planet to which each of the following moons belong.

    a) Callisto: e) Miranda:

    b) Deimos: f) Charon:

    c) Io: g) Titan:

    d) Triton:

14. What is one theory for the origin of Pluto and its moon Charon?

15. What are the two hypotheses for the origin of the asteroids?

    1)

    2)

16. Using Figure 20.2, list the letter that identifies each of the following comet parts.

    a) Ionized gas tail: ____

    b) Coma: ____

    c) Nucleus: ____

    d) Dust tail: ____

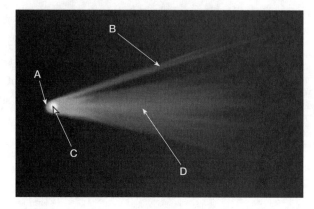

Figure 20.2

17. What are the two forces that contribute to the formation of a comet's tail?

    1)

    2)

18. What is the difference between a meteor and a meteorite?

19. List and briefly describe the three most common types of meteorites based upon composition.

    1)

    2)

    3)

Practice Test

Multiple choice. Choose the best answer for the following multiple choice questions.

1. Which one of the following is proportionally more abundant on terrestrial planets than on Jovian planets?
   a) hydrogen
   b) silicate minerals
   c) methane
   d) ammonia ice
   e) helium

2. Which one of the following is NOT a terrestrial planet?
   a) Mercury
   b) Earth
   c) Mars
   d) Venus
   e) Jupiter

3. The fact that Mercury is very dense implies that it contains a(n) _____.
   a) asthenosphere
   b) liquid interior
   c) magma reservoir
   d) iron core
   e) highly volatile surface

4. The densely cratered lunar highlands have been estimated to be about _____.
   a) the same age as Earth
   b) 200 million years old
   c) much older than Earth
   d) 2.6 billion years old
   e) 360 million years old

5. The large dark regions on the moon are _____.
   a) highlands
   b) craters
   c) mountains
   d) maria
   e) rays

6. The most important agents currently modifying the moon's surface are _____.
   a) moonquakes
   b) volcanoes
   c) faults
   d) wind storms
   e) micrometeorites

7. Which planet has a mass greater than the combined mass of all the remaining planets, satellites, and asteroids?
   a) Venus
   b) Mars
   c) Jupiter
   d) Uranus
   e) Saturn

8. The oldest lunar features are _____.
   a) highlands
   b) rayed craters
   c) ejecta
   d) maria
   e) regolith deposits

9. In 1995 and 1996, _____ planets were detected outside our solar system.
   a) three
   b) five
   c) seven
   d) nine
   e) eleven

10. Which one of the following is NOT a material commonly found in lunar regolith?
    a) glass beads
    b) clay
    c) lunar dust
    d) igneous rocks

11. During the formation of the solar system, the inner planets were unable to accumulate much hydrogen, ammonia, methane, and water because of their _____.
    a) high temperatures and weak magnetism
    b) high temperatures and weak gravities
    c) high magnetism and slow velocities
    d) low temperatures and high velocities
    e) low temperatures and strong gravities

12. Which planet in the solar system is experiencing a "runaway" greenhouse effect?
    a) Earth   c) Venus   e) Uranus
    b) Mars   d) Mercury

13. Which one of Jupiter's moons is volcanically active?
    a) Callisto   c) Ganymede
    b) Europa   d) Io

14. Which features on Earth offer clear evidence that comets and asteroids have struck its surface?
    a) rayed craters   c) astroblemes   e) giant faults
    b) terra   d) volcanoes

15. Which chemical is responsible for the greenish-blue color of Uranus and Neptune?
    a) hydrogen sulfide c) ammonia   e) carbon dioxide
    b) methane   d) hydrogen chloride

16. Which planet has a density less than that of water?
    a) Mercury   c) Jupiter   e) Neptune
    b) Venus   d) Saturn

17. Which planet might best be described as a large, dirty iceball?
    a) Mercury   c) Mars   e) Pluto
    b) Venus   d) Saturn

18. The Martian polar caps are made of _____, covered by a thin layer of frozen carbon dioxide.
    a) silicate minerals c) volcanic rocks   e) frozen methane
    b) ammonia ice   d) water ice

19. Although this planet is shrouded in thick clouds, radar mapping has revealed a varied topography consisting of subdued plains, highlands, and thousands of volcanic structures.
    a) Mercury   c) Mars   e) Uranus
    b) Venus   d) Jupiter

20. Which one of the following is currently responsible for shaping the surface of Mars?
    a) marsquakes   c) wind   e) volcanoes
    b) water   d) tectonics

21. Which moon is the only satellite in the solar system with a substantial atmosphere?
    a) Io   c) Miranda   e) Europa
    b) Triton   d) Titan

22. The lowest surface temperature in the solar system (-200°C) occurs on _____.
    a) Titan   c) Neptune   e) Callisto
    b) Pluto   d) Triton

23. The moon's density is comparable to that of Earth's _____.
    a) core   c) crustal rocks   e) lower mantle
    b) atmosphere   d) asthenosphere

24. Which planet's atmosphere contains the Great Dark Spot?
    a) Venus   c) Saturn   e) Neptune
    b) Jupiter   d) Uranus

25. Which feature(s) on Mars have raised the question about the possibility of liquid water on the planet?
    a) mountain ranges with faults
    b) impact craters with sharp rims
    c) volcanic cones with craters
    d) deep, long canyons
    e) valleys with tributaries

26. The Jovian planets contain a large percentage of the gases _____.
    a) nitrogen and argon
    b) hydrogen and oxygen
    c) oxygen and nitrogen
    d) helium and oxygen
    e) hydrogen and helium

27. Which one of the following is NOT included in the solar system?
    a) comets
    b) galaxies
    c) sun
    d) planets
    e) asteroids

28. Second only to the moon in brilliance in the night sky is _____.
    a) Mercury
    b) Venus
    c) Mars
    d) Jupiter
    e) Saturn

29. Which planet has the most eccentric orbit?
    a) Mercury
    b) Pluto
    c) Earth
    d) Uranus
    e) Mars

30. Although the atmosphere of this planet is very thin, extensive dust storms with wind speeds in excess of 150 miles per hour do occur.
    a) Mercury
    b) Venus
    c) Mars
    d) Saturn
    e) Uranus

31. Which planet's axis lies only 8 degrees from the plane of its orbit?
    a) Mercury
    b) Mars
    c) Jupiter
    d) Saturn
    e) Uranus

32. Which planet completes one rotation in slightly less than 10 hours?
    a) Mercury
    b) Venus
    c) Mars
    d) Jupiter
    e) Saturn

33. Relatively young lunar craters often exhibit _____.
    a) rays
    b) maria
    c) eroded rims
    d) highlands
    e) active volcanoes

34. Which one of the following planets does NOT have rings?
    a) Mars
    b) Jupiter
    c) Saturn
    d) Uranus
    e) Neptune

35. Through the telescope, this planet appears as a reddish ball interrupted by some permanent dark regions that change intensity.
    a) Mercury
    b) Venus
    c) Mars
    d) Saturn
    e) Uranus

36. Most asteroids lie between the orbits of _____.
    a) Mercury and Venus
    b) Venus and Earth
    c) Earth and Mars
    d) Mars and Jupiter
    e) Jupiter and Saturn

37. Most meteor showers are associated with the orbits of _____.
    a) satellites   c) planets   e) meteorites
    b) comets   d) asteroids

*True/false. For the following true/false questions, if a statement is not completely true, mark it false. For each false statement, change the **italicized** word to correct the statement.*

1. ___ The orbital planes of seven planets lie within *three* degrees of the plane of the sun's equator.

2. ___ The most prominent feature of *Pluto* is its ring system.

3. ___ Compared to other planet-satellite systems, Earth's moon is unusually *large*.

4. ___ Most of the volcanoes on Venus are small *shield* cones.

5. ___ *Earth* has the greatest temperature extremes of any planet.

6. ___ Most asteroids orbit the sun between the orbits of Mars and *Earth*.

7. ___ The Jovian planets have *high* densities and large masses.

8. ___ Bombardment of the moon by *micrometeorites* gradually smooths the landscape.

9. ___ Maria basins are filled with very fluid basaltic *ejecta*.

10. ___ The Martian atmosphere is *fifty* percent as dense as that of Earth.

11. ___ Most of the lunar surface is made up of *maria*.

12. ___ Comets typically *gain* material with every pass by the sun.

13. ___ Over 90 percent of the mass of the solar system is contained within the *planets*.

14. ___ The most striking feature on the face of *Mars* is the Great Red Spot.

15. ___ The escape velocity for Earth is *seven* miles per second.

16. ___ The formation of rings is believed be related to the *gravitational* force of a planet.

17. ___ The *surface* of Venus reaches temperatures of 475°C (900°F).

18. ___ The *retrograde* motion of Triton suggests that it formed independently of Neptune.

19. ___ In 1978, the moon Charon was discovered orbiting *Pluto*.

20. ___ The *Voyager 2* spacecraft has surveyed the greatest number of planets, including Neptune in 1989.

Word choice. Complete each of the following statements by selecting the most appropriate response.

1. The tail of a comet always points [toward/away from] the sun.

2. The moon's density is considerably [more/less] than Earth's average density and its gravity is approximately one-[quarter/sixth/eighth] of Earth's.

3. The processes that formed the very large valleys on Mars would be comparable to those which created [glacial/rift] valleys on Earth.

4. Because of its high temperature and small size, the planet [Mercury/Venus/Mars] was unable to retain an atmosphere during its formation.

5. Terrestrial bodies with no atmosphere [reflect/absorb] most of the sunlight that strikes them.

6. The process of plate tectonics [does/does not] appear to have contributed to the present Venusian topography.

7. Jupiter is covered with alternating bands of multicolored clouds that are [parallel/perpendicular] to its equator.

8. The planets orbit the sun in the [same/different] direction(s).

9. A bright streak of light caused by a solid particle entering Earth's atmosphere is called a [meteor/meteorite].

10. The most prominent features of Mars visible through a telescope are the planet's [canals/polar caps].

11. The gradual smoothing of the lunar landscape is caused by [tectonic forces/micrometeorite impacts].

12. Most asteroids are about [1/10/100] kilometer(s) across.

13. The thickness of Saturn's ring system is approximately a few [hundred/thousand] meters.

14. On the moon, greater crater density indicates that an area is [older/younger] than an area with lower crater density.

15. Compared to features on Earth, most Martian features are [old/young].

16. Earth and [Mercury/Venus/Mars] are often referred to as "twins."

17. Most comets are believed to orbit the sun [within/beyond] the orbit of Pluto.

18. The most prominent feature of Venus visible through a telescope is the planet's [barren surface/cloud cover].

19. Currently Pluto is [closer to/further from] the sun than Neptune.

20. Mercury rotates [slowly/rapidly] and revolves [slowly/quickly].

Written questions

1. Compare the physical characteristics of the terrestrial planets to those of the Jovian planets.

2. Distinguish between a meteoroid, meteor, and meteorite.

3. Briefly compare the two different types of lunar terrain.

Light, Astronomical Observations, and the Sun | 21

Light, Astronomical Observations, and the Sun begins with an examination of the nature of electromagnetic radiation and how it is used to gather information concerning the state of matter, composition, temperature, and motion of stars and other celestial objects. After investigating the nature of light, the focus shifts to astronomical tools and how they are used to intercept and study the energy emitted by distant objects in the universe. The chapter concludes with descriptions of the structure of the sun, some features that occur on the active sun, and the source of the sun's energy.

Learning Objectives

After reading, studying, and discussing this chapter, you should be able to:

- Describe electromagnetic radiation and the two models used to explain its properties.
- List and describe the three types of light spectra.
- Explain how light (electromagnetic radiation) can be used to investigate the properties of a star.
- Describe the two types of optical telescopes and list their component parts.
- List and describe the three properties of optical telescopes that aid astronomers in their work.
- Describe radio telescopes and list some of their advantages over optical telescopes.
- List and describe the four parts of the sun.
- Describe several features found on the active sun.
- Describe the source of the sun's energy.

Chapter Summary

- Visible light constitutes only a small part of an array of energy generally referred to as *electromagnetic radiation*. Light, a type of electromagnetic radiation, can be described in two ways 1) as waves and 2) as a stream of particles, called *photons*. The wavelengths of electromagnetic radiation vary from several kilometers for *radio waves* to less than a billionth of a centimeter for *gamma rays*. The shorter wavelengths correspond to more energetic photons.

- *Spectroscopy* is the study of the properties of light that depend on wavelength. When a prism is used to disperse visible light into its component parts (wavelengths), one of three possible types of *spectra* (a *spectrum*, the singular form of spectra, is the light pattern produced by passing light through a prism) is produced. The three types of spectra are 1) *continuous spectrum*, 2) *dark-line (absorption) spectrum*, and 3) *bright-line (emission) spectrum*. The spectra of most stars are of the dark-line type. Spectroscopy can be used to determine 1) the state of matter of an object (solid, liquid, high or low pressure gas), 2) the composition of gaseous objects, 3) the temperature of a radiating body, and 4) the motion of an object. Motion (direction toward or away and velocity) is determined using the *Doppler effect*—the apparent change in the wavelength of radiation emitted by an object caused by the relative motions of the source and the observer.

- There are two types of optical telescopes; 1) the *refracting telescope*, which uses a *lens* as its *objective* to bend or refract light so that it converges at an area called the *focus*, and 2) the *reflecting telescope*, which uses a *concave mirror* to focus (gather) the light. When examining an image directly, both types of telescopes require a second lens, called an *eyepiece*, which magnifies the image produced by the objective.

- Telescopes have three properties that aid astronomers: 1) *light-gathering power*, which is a function of the size of the objective—large objectives gather more light and therefore "see" farther into space, 2) *resolving power*, which allows for sharper images and finer detail, is the ability of a telescope to separate objects that are close together, e.g. Pluto and its moon Charon, and 3) *magnifying power*, the ability to make an object larger. Most modern telescopes have supplemental devices that enhance the image.

- Invisible radio wave radiation is detected by "big dishes" called *radio telescopes*. A parabolic shaped dish, often consisting of a wire mesh, operates in the same manner as the mirror of a reflecting telescope. Radio telescopes have poor resolution, making it difficult to pinpoint a radio source. To reduce this problem, several can be wired together into a network called a *radio interferometer*. The advantages of radio telescopes over optical telescopes are that radio telescopes are less affected by the weather, they are less expensive to construct, "viewing" is possible 24 hours a day, they can detect material in the universe too cool to emit visible radiation, and they can "see" through interstellar dust clouds.

- The *sun* is one of the 200 billion stars that make up the Milky Way galaxy. The sun can be divided into four parts 1) the *solar interior*, 2) the *photosphere* (visible surface), and the two layers of its atmosphere, 3) the *chromosphere* and 4) *corona*. The photosphere radiates most of the light we see. Unlike most surfaces, it consists of a layer of incandescent gas 300 kilometers (200 miles) thick with a grainy texture consisting of numerous, relatively small, bright markings called *granules*. Just above the photosphere lies the chromosphere, a relatively thin layer of hot, incandescent gases a few thousand kilometers thick. At the edge of the uppermost portion of the solar atmosphere, called the corona, ionized gases escape the gravitational pull of the sun and stream toward Earth at high speeds producing the *solar wind*.

- Numerous features have been identified on the active sun. *Sunspots* are dark blemishes with a black center, the *umbra*, which is rimmed by a lighter region, the *penumbra*. The number of sunspots observable on the solar disk varies in an 11-year cycle. *Plages* are large "clouds" that appear as bright centers of solar activity often directly above sunspot clusters. *Prominences*, huge cloudlike structures best observed when they are on the edge, or limb, of the sun, are apparently condensations of coronal material gracefully "sliding down" lines of magnetic force back to the chromosphere. The most explosive events associated with sunspots are *solar flares*. Flares are brief outbursts that release enormous quantities of energy that appear as a sudden brightening of the region above sunspot clusters. During the event, radiation and fast-moving atomic particles are ejected, causing the solar wind to intensify. When the ejected particles reach Earth and disturb the ionosphere, radio communication is disrupted and the *auroras*, also called the northern and southern lights, occur.

- The source of the sun's energy is *nuclear fusion*. Deep in the solar interior, at a temperature of 15 million K, a nuclear reaction called the *proton-proton chain* converts four hydrogen nuclei (protons) into the nucleus of a helium atom. During the reaction some of the matter is converted to the energy of the sun. A star the size of the sun can exist in its present stable state for 10 billion years. Since the sun is already 5 billion years old, it is a "middle-aged" star.

Chapter Outline

I. The study of light
  A. Electromagnetic radiation
    1. Visible light is only one small part of an array of energy
    2. Electromagnetic radiation includes
      a. Gamma rays
      b. X-rays
      c. Ultraviolet light
      d. Visible light
      e. Infrared light
      f. Radio waves
    3. All forms of radiation travel at 300,000 kilometers (186,000 miles) per second
  B. Light (electromagnetic radiation) can be described in two ways
    1. As waves
      a. Wavelengths of radiation vary
        1. Radio waves measure up to several kilometers long
        2. Gamma ray waves are less than a billionth of a centimeter long
      b. White light consists of several wavelengths corresponding to the colors of the rainbow
    2. As particles
      a. Called photons
      b. Exert a pressure, called radiation pressure, on matter
      c. Shorter wavelengths correspond to more energetic photons
  C. Spectroscopy
    1. The study of the properties of light that depend on wavelength
    2. The light pattern produced by passing light through a prism, which spreads out the various wavelengths, is called a spectrum (plural: spectra)
    3. Types of spectra
      a. Continuous spectrum
        1. Produced by an incandescent solid, liquid, or high pressure gas
        2. Uninterrupted band of color
      b. Dark-line (absorption) spectrum
        1. Produced when white light is passed through a comparatively cool, low pressure gas
        2. Appears as a continuous spectrum but with dark lines running through it
      c. Bright-line (emission) spectrum
        1. Produced by a hot (incandescent) gas under low pressure
        2. Appears as a series of bright lines of particular wavelengths depending on the gas that produced them
      4. Most stars have a dark-line spectrum
      5. Instrument used to spread out the light is called a spectroscope
  D. Doppler effect
    1. The apparent change in wavelength of radiation caused by the relative motions of the source and observer
    2. Used to determine
      a. Direction of motion
        1. Increasing distance—wavelength is longer ("stretches")
        2. Decreasing distance—makes wavelength shorter ("compresses")
      b. Velocity—larger Doppler shifts indicate higher velocities

II. Astronomical tools
  A. Optical (visible light) telescopes
    1. Two basic types
      a. Refracting telescope
        1. Uses a lens (called the objective) to bend (refract) the light to produce an image
        2. Light converges at an area called the focus
        3. Distance between the lens and the focus is called the focal length
        4. The eyepiece is a second lens used to examine the image directly
        5. Have an optical defect called chromatic aberration (color distortion)

b. Reflecting telescope
1. Uses a concave mirror to gather the light
2. No color distortion
3. Nearly all large telescopes are of this type
2. Properties of optical telescopes
a. Light-gathering power
1. Larger lens (or mirror) intercepts more light
2. Determines the brightness
b. Resolving power
1. The ability to separate close objects
2. Allows for a sharper image and finer detail
c. Magnifying power
1. The ability to make an image larger
2. Calculated by dividing the focal length of the objective by the focal length of the eyepiece
3. Can be changed by changing the eyepiece
4. Limited by atmospheric conditions and the resolving power of the telescope
5. Even with the largest telescopes, stars (other than the sun) appear only as points of light
B. Detecting invisible radiation
1. Photographic films are used to detect ultraviolet and infrared wavelengths
2. Most invisible wavelengths do not penetrate Earth's atmosphere, so balloons, rockets, and satellites are used
3. Radio radiation
a. Reaches Earth's surface
b. Gathered by "big dishes" called radio telescopes
1. Large because radio waves are about 100,000 longer than visible radiation
2. Often made of a wire mesh
3. Have rather poor resolution
4. Can be wired together into a network called a radio interferometer
5. Advantages over optical telescopes
a. Less affected by weather

b. Less expensive
c. Can be used 24 hours a day
d. Detects material that does not emit visible radiation
e. Can "see" through interstellar dust clouds
6. A disadvantage is that they are hindered by man-made radio interference
III. Sun
A. One of 200 billion stars that make up the Milky Way galaxy
B. Only star close enough to allow the surface features to be studied
C. An average star
D. Structure can be divided into four parts
1. Solar interior
2. Photosphere
a. "Sphere of light"
b. Sun's "surface"—actually a layer of incandescent gas 300 kilometers (200 miles) thick
c. Grainy texture made up of many small, bright markings, called granules, produced by convection
d. Most of the elements found on Earth also occur on the sun
e. Temperature averages approximately 6000 K (10,000°F)
3. Chromosphere
a. Just above photosphere
b. Lowermost atmosphere
c. Relatively thin, hot layer of incandescent gases a few thousand kilometers thick
d. Top contains numerous spicules—narrow jets of rising material
4. Corona
a. Outermost portion of the solar atmosphere
b. Very tenuous
c. Ionized gases escape from the outer fringe and produce the solar wind
d. Temperature at the top exceeds 1 million K
E. Solar features
1. Sunspots
a. On the solar surface
b. Dark center, the umbra, surrounded by a lighter region, the penumbra
c. Dark color is due to a cooler temperature (1500 K less than the solar surface)

d. Follow an 11-year cycle

e. Large spots are strongly magnetized

f. Pairs have opposite magnetic poles

2. Plages

a. Bright centers of solar activity

b. Occur above sunspot clusters

3. Prominences

a. Huge arching cloudlike structures that extend into the corona

b. Condensations of material in the corona

4. Flares

a. Explosive events that normally last an hour or so

b. Sudden brightening above a sunspot cluster

c. Release enormous quantities of energy

d. Eject particles that reach Earth in about one day and interact with the atmosphere to cause the auroras (the northern and southern lights)

IV. Solar interior

A. Cannot be observed directly

B. Nuclear fusion occurs here

1. Source of the sun's energy

2. Occurs in the deep interior

3. Nuclear reaction that produces the sun's energy is called the proton-proton reaction

a. Four hydrogen nuclei are converted into a helium nuclei

b. Matter is converted to energy

c. 600 million tons of hydrogen is consumed each second

4. Sun has enough fuel to last another five billion years

Key Terms

Page numbers shown in ( ) refer to the textbook page where the term first appears.

aurora (p. 562)
bright-line (emission) spectrum (p. 547)
chromatic aberration (p. 550)
chromosphere (p. 558)
continuous spectrum (p. 547)
corona (p. 559)
dark-line (absorption) spectrum (p. 547)
Doppler effect (p. 548)
electromagnetic radiation (p. 546)

eyepiece (p. 549)
focal length (p. 549)
focus (p. 549)
granules (p. 558)
nuclear fusion (p. 562)
objective lens (p. 547)
photon (p. 547)
photosphere (p. 558)
plage (p. 561)
prominence (p. 561)
proton-proton chain (p. 562)
radiation pressure (p. 547)

radio interferometer (p. 555)
radio telescope (p. 554)
reflecting telescope (p. 550)
refracting telescope (p. 549)
solar flare (p. 562)
solar wind (p. 559)
spectroscope (p. 548)
spectroscopy (p. 547)
spicule (p. 559)
sunspot (p. 560)

Vocabulary Review

Choosing from the list of key terms, furnish the most appropriate response for the following statements.

1. The array of energy that consists of radiations of electromagnetic waves is referred to as _____.

2. The very tenuous, outermost portion of the solar atmosphere is called the _____.

3. A(n) _____ uses a lens to bend and concentrate the light from distant objects.

4. The _____ is an instrument for directly viewing the spectrum of a light source.

5. The visible surface region of the sun that radiates energy to space is called the _____.

6. The source of the sun's energy is _____.

7. A(n) _____ is the type of spectrum produced by an incandescent solid, liquid, or gas under high pressure.

8. The _____ concentrates light from distant objects by using a concave mirror.

9. A(n) _____ is a discrete amount (quantum) of electromagnetic energy that moves as a unit with the velocity of light.

10. A(n) _____ is designed to make observations using the radio wavelengths of electromagnetic radiation.

11. The apparent change in wavelength of radiation caused by the relative motions of the source and the observer is called the _____.

12. The streams of protons and electrons that "boil" from the sun's corona constitute the _____.

13. The most important lens in a refracting telescope is the _____.

14. Deep in the sun's interior, a nuclear reaction called the _____ converts four hydrogen nuclei (protons) into the nucleus of a helium atom.

15. A sudden brightening of an area on the sun that normally lasts an hour or so but releases enormous quantities of energy is called a(n) _____.

16. A(n) _____ is the type of spectrum produced by a hot (incandescent) gas under low pressure.

17. The sun's _____ is a relatively thin layer of hot, incandescent gases a few thousand kilometers thick that lies just above the photosphere.

18. A solar _____ is a huge cloudlike concentration of material that appears as a bright arch-like structure above the solar surface.

19. _____ is the study of the properties of light that depend on wavelength.

20. A conspicuous dark blemish on the surface of the sun is called a(n) _____.

21. The _____ is the area where the light rays from a lens or mirror converge.

22. A(n) _____ is the type of spectrum produced when white light is passed through a comparatively cool gas under low pressure.

23. _____ is the property of a lens whereby light of different colors is focused at different places.

24. The pressure ("push") on matter exerted by photons is called _____.

25. The _____ of a telescope is the distance between the lens (or mirror) and the telescope's focus.

26. The numerous, relatively small, bright markings on the sun's surface are called _____.

27. A(n) _____ is a bright region in the solar atmosphere located above a sunspot.

28. The short-focal-length lens used to magnify the image in a telescope is called the _____.

29. When two or more radio telescopes are wired together, the resulting network is called a(n) _____.

30. A narrow jet of rising material which extends upward from the sun's chromosphere into the corona is called a(n) _____.

31. The _____ is a bright display of ever-changing light caused by solar radiation interacting with the upper atmosphere of Earth in the region of the poles.

Comprehensive Review

1. List and describe the two models used to explain the properties of light.

    1)

    2)

2. Other than the colors of visible light, list three types of the electromagnetic radiation. In what way(s) are they similar? In what way(s) are they different?

3. Briefly describe what happens to white light as it passes through a prism and emerges as the "colors of the rainbow."

4. List and describe the three types of spectra.

    1)

    2)

    3)

5. How do astronomers identify the elements that are present in the sun and other stars?

6. What two aspects of stellar motion can be detected from a star's Doppler shift?

7. List and describe the two basic types of optical telescopes.

    1)

    2)

8. Using Figure 21.1, select the appropriate letter that identifies each of the following.

    a) Objective lens: ____

    b) Focus: ____

    c) Focal length of the objective: ____

    d) Eyepiece: ____

    e) Focal length of the eyepiece: ____

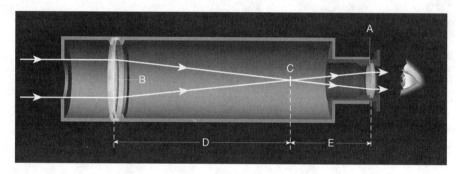

Figure 21.1

9. List and briefly describe the three properties of optical telescopes that astronomers are most interested in.

    1)

    2)

    3)

10. What are some advantages that a reflecting telescope has over a refracting telescope?

11. If a telescope has an objective with a focal length of 50 cm and an eyepiece with a focal length of 25 mm, what will be the magnification?

12. How do astronomers detect and study invisible infrared and ultraviolet radiation?

13. List two advantages that radio telescopes have over visible light telescopes.

    1)

    2)

14. Using Figure 21.2, select the letter that identifies each of the following solar features.

    a) Corona: ____

    b) Core: ____

    c) Prominence: ____

    d) Spicules: ____

    e) Chromosphere: ____

    f) Sunspots: ____

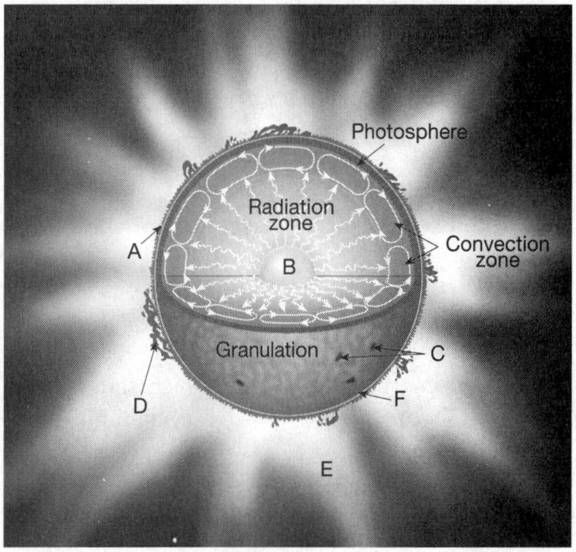

15. What is the effect on Earth's atmosphere of a strong solar flare?

Figure 21.2

16. Briefly describe how the sun (as well as most other stars) produce their energy.

Practice Test

Multiple choice. Choose the best answer for the following multiple choice questions.

1. Gamma rays, x-rays, visible light, and radio waves are all types of _____.
   a) electromagnetic radiation
   b) nuclear energy
   c) gravitational energy
   d) ultraviolet radiation
   e) chromatic aberration

2. The part of a telescope that gathers the light from a distant object is called the _____.
   a) eyepiece
   b) objective
   c) interferometer
   d) focus
   e) aberration

3. The fact that light can exert a pressure ("push") on matter suggests it is made of particles called _____.
   a) electrons
   b) protons
   c) photons
   d) nuclei
   e) neutrons

4. The solar wind consists of streams of _____ that travel outward from the sun at very high speeds.
   a) helium atoms
   b) photons
   c) protons and electrons
   d) radio waves
   e) protons and neutrons

5. The source of the sun's energy is _____.
   a) magnetism   c) radiation pressure   e) combustion
   b) nuclear fusion   d) nuclear fission

6. The world's largest conventional, single mirror telescope, located in the Caucasus Mountains in Russia, has a diameter of _____ meters.
   a) 2   c) 6   e) 10
   b) 4   d) 8

7. When several radio telescopes are wired together, the resulting network is called a radio _____.
   a) receiver   c) refractor   e) exoresolver
   b) interferometer   d) tuner

8. The primary material from which stars are made is _____.
   a) hydrogen   c) helium   e) nitrogen
   b) silicates   d) oxygen

9. The sun should be able to exist in its present state for approximately _____ billion more years.
   a) 1   c) 5   e) 9
   b) 3   d) 7

10. If the temperature of a radiating body is increased, the total amount of energy emitted _____.
    a) remains the same   c) decreases
    b) increases

11. The numerous, relatively small, bright markings on the sun's photosphere are referred to as _____.
    a) spicules   c) sunspots   e) prominences
    b) plages   d) granules

12. The light-gathering power of a telescope is directly related to the _____.
    a) diameter of the eyepiece   d) chromatic aberration of the eyepiece
    b) focal length of the eyepiece   e) diameter of the objective
    c) focal length of the objective

13. Which one of the following stellar properties can NOT be determined from the analysis of the star's light?
    a) rate of movement   d) direction of movement
    b) composition   e) surface features
    c) temperature

14. Which layer of the sun is considered its "surface"?
    a) corona   c) photosphere   e) chromosphere
    b) spicule   d) auroral

15. The distance between the objective lens and the focus of a telescope is called the _____.
    a) magnification   c) image diameter   e) refractive distance
    b) aberration length   d) focal length

16. The energy of a photon is related to its _____.
    a) size   c) density   e) wavelength
    b) mass   d) speed

17. The largest earthbound telescopes cannot photograph lunar features less than _____ kilo-meter in size.
    a) 0.1   c) 0.3   e) 0.5
    b) 0.2   d) 0.4

18. As the temperature of an object increases, a larger proportion of its energy is radiated at _____ wavelengths.
    a) longer   c) shorter   e) slower
    b) the same   d) faster

19. The magnification of a telescope is changed by changing the _____.
    a) focal length of the objective   d) eyepiece
    b) chromatic aberration   e) wavelength
    c) interferometer

20. Which color of light has the most energetic photons?
    a) red   c) yellow   e) violet
    b) orange   d) blue

21. The spectra of most stars are _____.
    a) bright-line   c) dark-line   e) invisible
    b) continuous   d) emission

22. The thin red rim seen around the sun during a total solar eclipse is the _____.
    a) aurora   c) corona   e) photosphere
    b) chromosphere   d) solar wind

23. As light passes through a lens, which color will undergo the greatest amount of bending?
    a) red   c) yellow   e) violet
    b) orange   d) blue

24. The apparent change in the wavelength as a light-emitting object moves rapidly away from an observer is called _____.
    a) diffraction   c) relativity   e) refraction
    b) the Doppler effect d) chromatic aberration

25. _____ are huge cloudlike structures that appear as great arches when they are on the limb of the sun.
    a) Spicules   c) Plages   e) Sunspots
    b) Prominences   d) Flares

*True/false. For the following true/false questions, if a statement is not completely true, mark it false. For each false statement, change the **italicized** word to correct the statement.*

1. ___ The outermost portion of the sun's atmosphere is the *photosphere*.

2. ___ Compared to other stars, the sun is best described as *middle-aged*.

3. ___ In a *reflecting* telescope, the light is focused in front of the objective.

4. ___ The energy producing reaction in the sun converts *hydrogen* nuclei into the nuclei of helium atoms.

5. ___ The ability of a telescope to separate objects is referred to as *resolving* power.

6. ___ The internal temperature of the sun is estimated to be approximately *fifteen* million K.

7. ___ The average density of the sun is nearly the density of *lead*.

8. ___ The study of the properties of light that depend on wavelength is called *optimography*.

9. ___ *Convection* is the process believed to be responsible for the transfer of energy in the upper-most part of the sun's interior.

10. ___ On a night when the stars twinkle, seeing is *poor*.

11. ___ The pressure ("push") on matter exerted by photons is called *particle* pressure.

12. ___ Neutral hydrogen (hydrogen atoms that lack an electrical charge) emit *radio* radiation with a 21-centimeter (8 inch) wavelength.

13. ___ The diameter of the sun is equal to *one hundred nine* Earth diameters.

14. ___ To record *ultraviolet* and infrared radiation from a star, balloons, rockets, and satellites must transport cameras "above" the atmosphere.

15. ___ The larger the diameter of the telescope's objective, the *greater* the resolving power.

16. ___ *Refracting* telescopes have the ability to "see" through interstellar dust clouds.

17. ___ All radiant energy travels through the vacuum of space in a straight line at the rate of 186,000 miles per *hour*.

18. ___ The variation in sunspot activity has a(n) *twenty*-year cycle.

19. ___ During nuclear fusion, some of the matter is converted to *energy*.

20. ___ Magnification is calculated by dividing the focal length of the objective by the *diameter* of the eyepiece.

21. ___ Upward from the photosphere, the temperature of the sun *decreases*.

Word choice. Complete each of the following statements by selecting the most appropriate response.

1. Short wavelengths of radiation have [more/less] energetic photons than radiation with longer wavelengths.

2. The lens that magnifies an image in a telescope is the [objective/eyepiece].

3. Different wavelengths of light will appear as different [intensities/colors] to the human eye.

4. Radio wavelengths are [longer/shorter] than visible wavelengths.

5. The [chromosphere/photosphere] is the layer of the sun that radiates most of the sunlight we see.

6. If a light source is moving very rapidly away from an observer, its light will appear [bluer/redder] than it actually is.

7. Increasing the magnification of a telescope [increases/decreases] the brightness of an object.

8. Sunspots appear dark because they are [warmer/cooler] than the surrounding solar surface.

9. X-rays are a type of [nuclear/electromagnetic] radiation.

10. The absorption spectrum of the sun's photosphere reveals that 90 percent of the surface atoms are [hydrogen/helium].

11. The largest optical telescopes are [refracting/reflecting] telescopes.

12. The greater the Doppler shift, the [greater/slower] the velocity of the object.

13. The sun is a(n) [very large/average/very small] star.

14. All radiation travels through the vacuum of space in a [curved/straight] line at the rate of [300,000/186,000] kilometers per second.

15. The factor that determines both the light-gathering power and the resolving power of a telescope is the [thickness/diameter] of the objective lens or mirror.

16. [Red/Violet] is the color of visible light with the shortest wavelength; while [gamma/radio] radiation is the shortest wavelength of electromagnetic radiation.

17. Radio telescopes can detect [hotter/cooler] objects than those that are detectable with optical telescopes.

18. On a night when the stars shine steadily, seeing is [good/poor].

19. The [chromosphere/corona] is the lower portion of the solar atmosphere.

20. When passing through a lens, the longer wavelengths of light are refracted at a [larger/smaller] angle than the shorter wavelengths.

Written questions

1. Describe the source of the sun's energy.

2. What are three properties of a star that can be determined by spectroscopic analysis of its light?

3. List and briefly describe each of the four parts of the sun.

4. Compared to a refracting telescope, what are some advantages of a reflecting telescope?

Beyond Our Solar System

22

Beyond Our Solar System begins with an examination of the intrinsic properties of stars—distance, brightness, color, and temperature. Binary star systems, stellar mass, and the Hertzsprung-Russell diagram are discussed in detail. Also investigated are the various types of nebulae. Stellar evolution, from birth through protostar, main-sequence, red giant, burnout and death, is presented. Following stellar evolution are descriptions of the various stellar remnants—dwarf stars, neutron stars, and black holes. The chapter continues with a detailed discussion of the Milky Way galaxy and a general description of types of galaxies, galactic clusters, and red shifts. The chapter closes with a commentary on the origin of the universe and the Big Bang theory.

Learning Objectives

After reading, studying, and discussing this chapter, you should be able to:

- Discuss the principle of parallax and explain how it is used to measure the distance to a star.
- List and describe the major intrinsic properties of stars.
- Describe the different types of nebulae.
- Describe the most plausible model for stellar evolution and list the stages in the life cycle of a star.
- Describe the possible final states that a star may assume after it consumes its nuclear fuel and collapses.
- List and describe the major types of galaxies.
- Describe the Big Bang theory of the origin of the universe.

Chapter Summary

- One method for determining the distance to a star is to use a measurement called *stellar parallax*, the extremely slight back-and-forth shifting in a nearby star's position due to the orbital motion of Earth. *The farther away a star is, the less its parallax.* A unit used to express stellar distance is the *light-year*, which is the distance light travels in one Earth year—about 9.5 trillion kilometers (5.8 trillion miles).

- The intrinsic properties of stars include *brightness, color, temperature, mass,* and *size*. Three factors control the brightness of a star as seen from Earth: how big it is, how hot it is, and how far away it is. *Magnitude* is the measure of a star's brightness. *Apparent magnitude* is how bright a star appears when viewed from Earth. *Absolute magnitude* is the "true" brightness if a star were at a standard distance of about 32.6 light-years. The difference between the two magnitudes is directly related to a star's distance. Color is a manifestation of a star's temperature. Very hot stars (surface temperatures above 30,000 K) appear blue; red stars are much cooler (surface temperatures generally less than 3000 K). Stars with surface temperatures between 5000 and 6000 K appear yellow, like the sun. The center of mass of orbiting *binary stars* (two stars revolving around a common center of mass under their mutual gravitational attraction) is used to determine the mass of the individual stars in a binary system.

• A *Hertzsprung-Russell diagram* is constructed by plotting the absolute magnitudes and temperatures of stars on a graph. A great deal about the sizes of stars can be learned from H-R diagrams. Stars located in the upper-right position of an H-R diagram are called *giants*, luminous stars of large radius. *Supergiants* are very large. Very small *white dwarf* stars are located in the lower-central portion of an H-R diagram. Ninety percent of all stars, called *main-sequence stars*, are in a band that runs from the upper-left corner to the lower-right corner of an H-R diagram.

• *Variable stars* fluctuate in brightness. Some, called *pulsating variables*, fluctuate regularly in brightness by expanding and contracting in size. When a star explosively brightens, it is called a *nova*. During the outburst, the outer layer of the star is ejected at high speed. After reaching maximum brightness in a few days, the nova slowly returns in a year or so to its original brightness.

• New stars are born out of enormous accumulations of dust and gases, called *nebula*, that are scattered between existing stars. A *bright nebula* glows because the matter is close to a very hot (blue) star. The two main types of bright nebulae are *emission nebulae* (which derive their visible light from the fluorescence of the ultraviolet light from a star in or near the nebula) and *reflection nebulae* (relatively dense dust clouds in interstellar space that are illuminated by reflecting the light of nearby stars). When a nebula is not close enough to a bright star to be illuminated, it is referred to as a *dark nebula*.

• Stars are born when their nuclear furnaces are ignited by the unimaginable pressures and temperatures in collapsing nebulae. New stars not yet hot enough for nuclear fusion are called *protostars*. When collapse causes the core of a protostar to reach a temperature of at least 10 million K, the fusion of hydrogen nuclei into helium nuclei begins in a process called *hydrogen burning*. The opposing forces acting on a star are *gravity* trying to contract it and *gas pressure* (*thermal nuclear energy*) trying to expand it. When the two forces are balanced, the star becomes a stable *main-sequence star*. When the hydrogen in a star's core is consumed, its outer envelope expands enormously and a *red giant* star, hundreds-to-thousands of times larger than its main-sequence size, forms. When all the usable nuclear fuel in these giants is exhausted and gravity takes over, the stellar remnant collapses into a small dense body.

• The *final fate of a star is determined by its mass*. Stars with less than one-half the mass of the sun collapse into hot, dense *white dwarf* stars. Medium-mass stars (between 0.5 and 3.0 times the mass of the sun) become red giants, collapse, and end up as white dwarf stars, often surrounded by expanding spherical clouds of glowing gas called *planetary nebulae*. Stars more than three times the mass of the sun terminate in a brilliant explosion called a *supernova*. Supernovae events can produce small, extremely dense *neutron stars*, composed entirely of subatomic particles called neutrons; or even smaller and more dense *black holes*, objects that have such immense gravity that light cannot escape their surface.

• The *Milky Way galaxy* is a large, disk-shaped, *spiral galaxy* about 100,000 light-years wide and about 10,000 light-years thick at the center. There are three distinct *spiral arms* of stars, with some showing splintering. The sun is positioned in one of these arms about two-thirds of the way from the galactic center, at a distance of about 30,000 light-years. Surrounding the galactic disk is a nearly spherical halo made of very tenuous gas and numerous *globular clusters* (nearly spherically shaped groups of densely packed stars).

• The various types of galaxies include 1) *spiral galaxies*, which are typically disk-shaped with a somewhat greater concentration of stars near their centers, often containing arms of stars extending from their central nucleus, 2) *barred spiral galaxies*, a type of spiral galaxy that has the stars arranged in the shape of a bar, which rotates as a rigid system, 3) *elliptical galaxies*, the most abundant type, which have an ellipsoidal shape that ranges to nearly spherical, and lack spiral arms, and 4) *irregular galaxies*, which lack symmetry and account for only 10 percent of the known galaxies.

• Galaxies are not randomly distributed throughout the universe. They are grouped in *galactic clusters*, some containing thousands of galaxies. Our own, called the *Local Group*, contains at least 28 galaxies.

• By applying the *Doppler effect* (the apparent change in wavelength of radiation caused by the motions of the source and the observer) to the light of galaxies, galactic motion can be determined. Most galaxies have Doppler shifts toward the red end of the spectrum, indicating increasing distance. The amount of Doppler shift is dependent on the velocity at which the object is moving. Because the most distant galaxies have the greatest red shifts, Edwin Hubble concluded in the early 1900s that they were retreating from us with greater recessional velocities than more nearby galaxies. It was soon realized that an *expanding universe* can adequately account for the observed red shifts.

• The belief in the expanding universe led to the widely accepted *Big Bang theory* of the origin of the universe. According to this theory, the entire universe was at one time confined in a dense, hot, supermassive concentration. About 20 billion years ago, a cataclysmic explosion hurled this material in all directions, creating all matter and space. Eventually the ejected masses of gas cooled and condensed, forming the stellar systems we now observe fleeing from their place of origin.

Chapter Outline

I. Properties of stars
   A. Distance
      1. Measuring a star's distance can be very difficult
      2. Stellar parallax
         a. Used for measuring distance to a star
         b. Apparent shift in a star's position due to the orbital motion of Earth
         c. Measured as an angle
         d. Near stars have the largest parallax
         e. Largest parallax is less than one second of arc
      3. Distances to the stars are very large
      4. Units of measurement
         a. Kilometers or astronomical units are too cumbersome to use
         b. Light-year is used most often
            1. Distance that light travels in 1 year
            2. One light-year is 9.5 trillion km (5.8 trillion miles)
      5. Other methods for measuring distance are also used
   B. Stellar brightness
      1. Controlled by three factors
         a. Size
         b. Temperature
         c. Distance
      2. Magnitude

      a. Measure of a star's brightness
      b. Two types of measurement
         1. Apparent magnitude
            a. Brightness when a star is viewed from Earth
            b. Decreases with distance
            c. Numbers are used to designate magnitudes
               1. Dim stars have large numbers
                  a. First magnitude appear brighter
                  b. Sixth magnitude are the faintest visible to the eye
               2. Negative numbers are also used
         2. Absolute magnitude
            a. "True" or intrinsic brightness of a star
            b. Brightness at a standard distance of 32.6 light-years
            c. Most stars' absolute magnitudes are between -5 and +15
   C. Color and temperature
      1. Hot star
         a. Temperature above 30,000 K
         b. Emits short-wavelength light
         c. Appears blue
      2. Cool star
         a. Temperature less than 3000 K

b. Emits longer-wavelength light

c. Appears red

3. Between 5000 and 6000 K

a. Stars appear yellow

b. e.g., Sun

D. Binary stars and stellar mass

1. Binary stars

a. Two stars orbiting one another

1. Stars are held together by mutual gravitation

2. Both orbit around a common center of mass

b. Visual binaries are resolved telescopically

c. More than 50% of the stars in the universe are binary stars

d. Used to determine stellar mass

2. Stellar mass

a. Determined using binary stars— the center of mass is closest to the most massive star

b. Mass of most stars is between one-tenth and fifty times the mass of the sun

II. Hertzsprung-Russell diagram

A. Shows the relation between stellar

1. Brightness (absolute magnitude) and

2. Temperature

B. Diagram is made by plotting (graphing) each star's

1. Luminosity (brightness) and

2. Temperature

C. Parts of an H-R diagram

1. Main-sequence stars

a. 90% of all stars

b. Band through the center of the H-R diagram

c. Sun is in the main-sequence

2. Giants (or red giants)

a. Very luminous

b. Large

c. Upper-right on the H-R diagram

d. Very large giants are called supergiants

e. Only a few percent of all stars

3. White dwarfs

a. Fainter than main-sequence stars

b. Small (approximate the size of Earth)

c. Lower-central area on the H-R diagram

d. Not all are white in color

e. Perhaps 10% of all stars

D. Used to study stellar evolution

III. Variable stars

A. Stars that fluctuate in brightness

B. Types of variable stars

1. Pulsating variables

a. Fluctuate regularly in brightness

b. Expand and contract in size

2. Eruptive variables

a. Explosive event

b. Sudden brightening

c. Called a nova

IV. Interstellar matter

A. Between the stars is "the vacuum of space"

B. Nebula

1. Cloud of dust and gases

2. Two major types of nebulae

a. Bright nebula

1. Glows if close to a very hot star

2. Two types of bright nebulae

a. Emission nebula

1. Largely hydrogen

2. Absorb ultraviolet radiation

3. Emit visible light

b. Reflection nebula

1. Reflect the light of nearby stars

2. Composed of interstellar dust

b. Dark nebula

1. Not close to any bright star

2. Appear dark

3. Contains the material that forms stars and planets

V. Stellar evolution

A. Stars exist because of gravity

B. Two opposing forces in a star are

1. Gravity—contracts

2. Thermal nuclear energy—expands

C. Stages

1. Birth

a. In dark, cool, interstellar clouds

b. Gravity contracts the cloud

c. Temperature rises

d. Radiates long-wavelength (red) light

e. Becomes a protostar

2. Protostar

a. Gravitational contraction of gaseous cloud continues

b. Core reaches 10 million K

c. Hydrogen nuclei fuse

1. Become helium nuclei
2. Process is called hydrogen burning
   d. Energy is released
   e. Outward pressure increases
   f. Outward pressure balanced by gravity pulling in
   g. Star becomes a stable main-sequence star
3. Main-sequence stage
   a. Stars age at different rates
      1. Massive stars use fuel faster and exist for only a few million year
      2. Small stars use fuel slowly and exist for perhaps hundreds of billions of years
   b. 90% of a star's life is in the main-sequence
4. Red giant stage
   a. Hydrogen burning migrates outward
   b. Star's outer envelope expands
      1. Surface cools
      2. Surface becomes red
   c. Core is collapsing as helium is converted to carbon
   d. Eventually all nuclear fuel is used
   e. Gravity squeezes the star
5. Burnout and death
   a. Final stage depends on mass
   b. Possibilities
      1. Low-mass star
         a. 0.5 solar mass
         b. Red giant collapses
         c. Becomes a white dwarf
      2. Medium-mass star
         a. Between 0.5 and 3 solar masses
         b. Red giant collapses
         c. Planetary nebula forms
            1. Cloud of gas
            2. Outer layer of the star
         d. Becomes a white dwarf
      3. Massive star
         a. Over 3 solar masses
         b. Short life span
         c. Terminates in a brilliant explosion called a super-nova
         d. Interior condenses
         e. May produce a hot, dense object that is either a

1. Neutron star (from a massive star) or a
2. Black hole (from a very massive star)
   D. H-R diagrams are used to study stellar evolution
VI. Stellar remnants
   A. White dwarf
      1. Small (some no larger than Earth)
      2. Dense
         a. Can be more massive than the sun
         b. Spoonful weighs several tons
         c. Atoms take up less space
            1. Electrons displaced inward
            2. Called degenerate matter
            3. Hot surface
            4. Cools to become a black dwarf
   B. Neutron star
      1. Forms from a more massive star
         a. Star has more gravity
         b. Squeezes itself smaller
      2. Remnant of a supernova
      3. Gravitational force collapses atoms
         a. Electrons combine with protons to produce neutrons
         b. Small size
      4. Pea size sample
         a. Weighs 100 million tons
         b. Same density as an atomic nucleus
      5. Strong magnetic field
      6. First one discovered in early 1970s
         a. Pulsar (pulsating radio source)
         b. Found in the Crab nebula (rem-nant of an A.D. 1054 supernova)
   C. Black hole
      1. More dense than a neutron star
      2. Intense surface gravity lets no light escape
      3. As matter is pulled into it
         a. Becomes very hot
         b. Emits x-rays
      4. Likely candidate is Cygnus X-1, a strong x-ray source
VII. Galaxies
   A. Milky Way galaxy
      1. Structure
         a. Determined by using radio telescopes
         b. Large spiral galaxy
            1. About 100,000 light-years wide
            2. Thickness at the galactic

nucleus is about 10,000
light-years
c. Three spiral arms of stars
d. Sun is 30,000 light-years from
the center
2. Rotation
a. Around the galactic nucleus
b. Outermost stars move the slowest
c. Sun rotates around the galactic
nucleus once about every 200
million years
3. Halo surrounds the galactic disk
a. Spherical
b. Very tenuous gas
c. Numerous globular clusters
B. Other galaxies
1. Existence was first proposed in mid-
1700s by Immanuel Kant
2. Four basic types of galaxies
a. Spiral galaxy
1. Arms extending from nucleus
2. About 30% of all galaxies
3. Large diameter of 20,000 to
125,000 light years
4. Contains both young and
old stars
5. e.g., Milky Way
b. Barred spiral galaxy
1. Stars arranged in the shape
of a bar
2. Generally quite large
3. About 10% of all galaxies
c. Elliptical galaxy
1. Ellipsoidal shape
2. About 60% of all galaxies
3. Most are smaller than spiral
galaxies; however, they are
also the largest known galaxies
d. Irregular galaxy
1. Lacks symmetry
2. About 10% of all galaxies
3. Contains mostly young stars
4. e.g., Magellanic Clouds
C. Galactic cluster
1. Group of galaxies
2. Some contain thousands of galaxies
3. Local Group
a. Our own group of galaxies
b. Contains at least 28 galaxies
4. Supercluster
a. Huge swarm of galaxies

b. May be the largest entity in
the universe
VIII. Red shifts
A. Doppler effect
1. Change in the wavelength of
light emitted by an object due to
its motion
a. Movement away stretches
the wavelength
1. Longer wavelength
2. Light appears redder
b. Movement toward squeezes
the wavelength
1. Shorter wavelength
2. Light shifted toward the blue
2. Amount of the Doppler shift indicates the rate of movement
a. Large Doppler shift indicates
a high velocity
b. Small Doppler shift indicates
a lower velocity
B. Expanding universe
1. Most galaxies exhibit a red
Doppler shift
a. Moving away
b. Far galaxies
1. Exhibit the greatest shift
2. Greater velocity
2. Discovered in 1929 by
Edwin Hubble
3. Hubble's Law—the recessional
speed of galaxies is proportional to
their distance
4. Accounts for red shifts
IX. Big Bang theory
A. Accounts for galaxies moving away
from us
B. Universe was once confined to a "ball"
that was
1. Supermassive
2. Dense
3. Hot
C. Big Bang marks the inception of
the universe
1. Occurred about 20 billion years ago
2. All matter and space was created
D. Matter is moving outward
E. Outward expansion could stop if the
concentration of matter is great enough
F. Gravitational contraction could perhaps
happen 20 billion years in the future

Key Terms

Page numbers shown in ( ) refer to the textbook page where the term first appears.

absolute magnitude (p. 570)

apparent magnitude (p. 569)

Big Bang (p. 588)

black hole (p. 582)

bright nebula (p. 573)

dark nebula (p. 574)

degenerate matter (p. 580)

elliptical galaxy (p. 585)

emission nebula (p. 574)

eruptive variables (p. 573)

galactic cluster (p. 586)

Hertzsprung-Russell (H-R) diagram (p. 571)

Hubble's law (p. 588)

hydrogen burning (p. 575)

interstellar dust (p. 574)

irregular galaxy (p. 585)

light-year (p. 569)

Local Group (p. 586)

magnitude (p. 569)

main-sequence stars (p. 571)

nebula (p. 573)

neutron star (p. 580)

nova (p. 573)

planetary nebula (p. 577)

protostar (p. 575)

pulsar (p. 581)

pulsating variables (p. 573)

red giant (p. 571)

reflection nebula (p. 574)

spiral galaxy (p. 585)

stellar parallax (p. 568)

supergiant (p. 572)

supernova (p. 578)

white dwarf (p. 572)

Vocabulary Review

Choosing from the list of key terms, furnish the most appropriate response for the following statements.

1. The measurement of a star's brightness is called its _____.

2. A(n) _____ is a type of galaxy that lacks symmetry.

3. A(n) _____ is a small, dense star that was once a low-mass or medium-mass star whose internal heat energy was able to keep these gaseous bodies from collapsing under their own weight.

4. A graph showing the relation between the true brightness and temperature of stars is called a(n) _____.

5. The fusion of groups of four hydrogen nuclei into a single helium nuclei is called _____.

6. _____ is a type of very dense matter that forms when electrons are displaced inward from their regular orbits around an atom's nucleus.

7. _____ is the extremely slight back-and-forth shifting in the position of nearby star due to the orbital motion of Earth.

8. An interstellar cloud consisting of dust and gases is referred to as a(n) _____.

9. A(n) _____ is an exceptionally large star, such as Betelgeuse, with a radius about 800 times that of the sun and high luminosity.

10. A(n) _____ is a gaseous interstellar mass that absorbs ultraviolet light from an embedded or nearby hot star and reradiates, or emits, this energy as visible light.

11. A collapsing cloud of gases and dust not yet hot enough to engage in nuclear fusion but destined to become a star is referred to as a(n) _____.

12. A star consisting of matter formed from the combination of electrons with protons that is smaller and more massive than a white dwarf is called a(n) _____.

13. A(n) _____ is a very large, cool, reddish colored star of high luminosity.

14. Stars exceeding three solar masses have relatively short life spans and terminate in a brilliant explosion called a(n) _____.

15. When a dense cloud of interstellar material is not close enough to a bright star to be illuminated, it is referred to as a(n) _____.

16. Discovered in the early 1970s, a(n) _____ is a neutron star that radiates short bursts of radio energy.

17. _____ states that galaxies are receding from us at a speed that is proportional to their distance.

18. A star's brightness as it appears when viewed from Earth is referred to as its _____.

19. A type of galaxy that has an ellipsoidal shape that ranges to nearly spherical and lacks spiral arms is called a(n) _____.

20. A(n) _____ is a unit used to express stellar distance equal to about 9.5 trillion kilometers (5.8 trillion miles).

21. A(n) _____ is an interstellar cloud that glows because of its proximity to a very hot (blue) star.

22. An object even smaller and denser than a neutron star, with a surface gravity so immense that light cannot escape, is called a(n) _____.

23. An interstellar cloud of gases and dust that merely reflects the light of nearby stars is called a(n) _____.

24. The Milky Way is a type of galaxy called a(n) _____.

25. The brightness of a star, if it were viewed at a distance of 32.6 light-years, is called its _____.

26. A reflection nebula is thought to be composed of a rather dense cloud of large particles called _____.

27. On an H-R diagram, the "ordinary" stars that are located along a band that extends from the upper-left corner to the lower-right corner are called _____.

28. An expanding spherical cloud of gas, called a(n) _____, forms when a medium-mass star collapses from a red giant to a white dwarf star.

29. The theory which proposes that the universe originated as a single mass that subsequently exploded is referred to as the _____ theory.

30. Our own group of galaxies, called the _____, contains at least 28 galaxies.

31. Stars that fluctuate regularly in brightness by expanding and contracting in size are called _____.

32. A system of galaxies containing from several to thousands of member galaxies is called a(n) _____.

33. A(n) _____ is a star that explosively increases in brightness.

Comprehensive Review

1. Briefly describe the relation between a star's distance and its parallax.

2. What causes the difference between a star's apparent magnitude and its absolute magnitude?

3. List the three factors that control the apparent brightness of a star as seen from Earth.

    1)

    2)

    3)

4. What is the approximate surface temperature of a star with the following color?

    a) Red:

    b) Yellow:

    c) Blue:

5. Describe how binary stars are used to determine stellar mass.

6. List the two properties of stars that are used to construct an H-R diagram.

    1)

    2)

7. Using Figure 22.1, an idealized Hertzsprung-Russell diagram, select the letter that indicates the location of each of the following types of stars.

        a) White dwarfs: ____

        b) Blue stars: ____

        c) Red stars (low mass): ____

        d) Giants: ____

8. In Figure 22.1, which letter is most representative of our sun? ____

9. In Figure 22.1, which stars are on the main sequence?

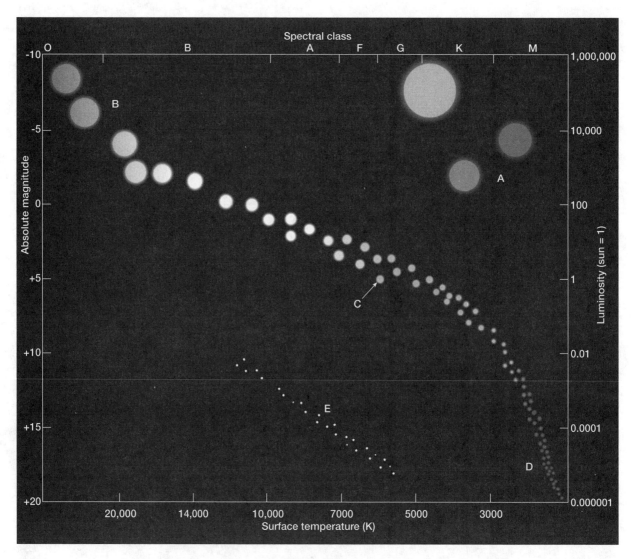

Figure 22.1

10. List and briefly describe the two main types of bright nebulae.

    1)

    2)

11. What circumstance produces a dark nebula?

12. Describe the process that astronomers refer to as hydrogen burning.

13. What will be the most likely final stage in the burnout and death of stars in each of the following mass categories?

    a) Low-mass stars:

    b) Medium-mass (sunlike) stars:

    c) Massive stars:

14. On Figure 22.2, beginning with dust and gases, draw a sequence of arrows that show in correct order the stages that a star about as massive as the sun will follow in its evolution.

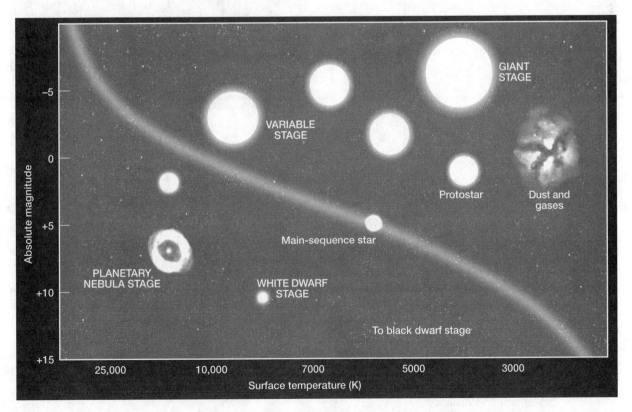

Figure 22.2

15. Why are radio telescopes rather than optical telescopes used to determine the gross structure of the Milky Way galaxy?

16. Describe the size and structure of the Milky Way.

17. List and describe the four basic types of galaxies.

   1)

   2)

   3)

   4)

18. The larger galaxy shown in Figure 22.3 is of which type?

19. What is Hubble's law?

Figure 22.3

20. Briefly describe the Big Bang theory.

Practice Test

Multiple choice. Choose the best answer for the following multiple choice questions.

1. The two properties of a star that are plotted on an H-R diagram are _____.
   a) size and distance
   b) brightness and temperature
   c) speed and distance
   d) size and temperature
   e) brightness and distance

2. Stars with surface temperatures between 5000 and 6000 K appear _____.
   a) red
   b) green
   c) violet
   d) yellow
   e) blue

3. A star with which one of the following magnitudes would appear the brightest?
   a) 15
   b) 10
   c) 5
   d) 1
   e) -5

4. Which force is most responsible for the formation of a star?
   a) magnetic
   b) gravity
   c) nuclear
   d) interstellar
   e) light pressure

5. The difference in the brightness of two stars having the same surface temperature is attributable to their relative _____.
   a) densities
   b) colors
   c) compositions
   d) ages
   e) sizes

6. In the cores of incredibly hot red giant stars, nuclear reactions convert helium to _____.
   a) carbon
   b) hydrogen
   c) oxygen
   d) argon
   e) nitrogen

7. Stellar parallax is used to determine a star's _____.
   a) motion
   b) brightness
   c) distance
   d) mass
   e) temperature

8. During the process referred to as hydrogen burning, hydrogen nuclei in the core of a star are fused together and become _____ nuclei.
   a) oxygen
   b) nitrogen
   c) helium
   d) carbon
   e) argon

9. Binary stars can be used to determine stellar _____.
   a) mass
   b) temperature
   c) brightness
   d) distance
   e) composition

10. One light-year is about _____.
    a) 6 trillion kilometers
    b) 9.5 trillion miles
    c) 8 trillion kilometers
    d) 5.8 trillion miles
    e) 11 trillion kilometers

11. Which one of the following stellar remnants has the greatest density?
    a) neutron star
    b) black hole
    c) white dwarf
    d) black dwarf

12. An interstellar accumulation of dust and gases is referred to as a(n) _____.
    a) red giant
    b) nebula
    c) white dwarf
    d) galaxy
    e) neutron star

13. Which one of the following is NOT a type of galaxy?
    a) nebular
    b) irregular
    c) spiral
    d) elliptical
    e) barred spiral

14. The most abundant gas in dark, cool, interstellar clouds in the neighborhood of the Milky Way is _____.
    a) hydrogen
    b) argon
    c) helium
    d) oxygen
    e) nitrogen

15. Very massive stars terminate in a brilliant explosion called a(n) _____.
    - a) red giant
    - b) supernova
    - c) protostar
    - d) neutron star
    - e) planetary nebula

16. According to the Big Bang theory, a cataclysmic explosion created the matter in the universe about _____ years ago.
    - a) 600 million
    - b) 2 billion
    - c) 4.6 billion
    - d) 12 billion
    - e) 20 billion

17. Which one of the following is NOT a type of nebula?
    - a) reflection
    - b) emission
    - c) spiral
    - d) dark

18. The apparent change in the position of a celestial object due to the orbital motion of Earth is called _____.
    - a) parallax
    - b) recessional velocity
    - c) magnitude
    - d) brightness
    - e) orbital shifting

19. Matter that is pulled into a black hole should become very hot and emit _____ before being engulfed.
    - a) x-rays
    - b) hydrogen nuclei
    - c) atoms
    - d) degenerate matter
    - e) infrared radiation

20. Which one of the following stars would appear brightest in the night sky?
    - a) Sirius
    - b) Deneb
    - c) Alpha Centauri
    - d) Arcturus
    - e) Betelgeuse

21. The Milky Way is a(n) _____.
    - a) constellation
    - b) red giant
    - c) spiral galaxy
    - d) dark nebula
    - e) galactic cluster

22. On an H-R diagram, main-sequence blue stars are massive and _____.
    - a) cool
    - b) white dwarfs
    - c) hot
    - d) novae
    - e) supergiants

23. The measurement of a star's brightness is called _____.
    - a) temperature
    - b) magnitude
    - c) parallax
    - d) period
    - e) pulsation

24. A very young, large, red object that is soon to become a star, but is not yet hot enough for nuclear fusion, is termed a(n) _____.
    - a) black hole
    - b) supernova
    - c) red giant
    - d) protostar
    - e) white dwarf

25. The apparent change in wavelength of radiation caused by the relative motions of the source and the observer is referred to as _____.
    - a) Hubble's law
    - b) the acceleration factor
    - c) Newton's law of radiation
    - d) relativity
    - e) the Doppler effect

26. The density of a white dwarf is such that a spoonful of its matter would weigh _____.
    - a) a kilogram
    - b) 500 pounds
    - c) 40 grams
    - d) a ton
    - e) several tons

27. The sun is positioned about _____ of the way from the center of the galaxy.
    a) one-fourth   c) one-half   e) three-fourths
    b) one-third   d) two-thirds

28. The brightness of a star, if it were at a distance of 32.6 light-years, is its _____.
    a) apparent magnitude   c) apparent brightness   e) mass-distance ratio
    b) parallax   d) absolute magnitude

29. Which type of galaxy is composed mostly of young stars?
    a) barred spiral   c) irregular   e) nebular
    b) elliptical   d) spiral

30. The red shifts exhibited by galaxies can be explained by _____.
    a) calculation errors   c) Kepler's first law   e) a collapsing universe
    b) an expanding universe   d) expansion of the Milky Way

31. On an H-R diagram, about 90 percent of the stars are _____.
    a) giants   c) white dwarfs   e) novae
    b) main-sequence stars d) super giants

32. The difference between a star's apparent and absolute magnitudes is related to _____.
    a) surface temperature   c) core temperature   e) distance
    b) color   d) time

*True/false. For the following true/false questions, if a statement is not completely true, mark it false. For each false statement, change the **italicized** word to correct the statement.*

1. ___ A hot, massive blue star will age more *slowly* than a small (red) main-sequence star.

2. ___ The parallax of a far star is *less* than that of a near star.

3. ___ A stable main-sequence star is balanced between the force of gravity and internal *gas* pressure.

4. ___ Galaxies are receding from us at a speed that is proportional to their *size*.

5. ___ *Absolute* magnitude is the brightness of a star when viewed from Earth.

6. ___ Stars with equal surface *temperatures* radiate the same amount of energy per unit area.

7. ___ On an H-R diagram, the sun is a *main-sequence* star.

8. ___ The larger the magnitude number, the *dimmer* the star.

9. ___ *Low-mass* stars evolve to become white dwarfs.

10. ___ On an H-R diagram, the hottest main-sequence stars are intrinsically the *brightest*.

11. ___ A star's color is primarily a manifestation of its *temperature*.

12. ___ Giant stars are *cooler* than white dwarfs.

13. ___ *Dark* nebulae glow because they are close to very hot stars.

14. ___ The light-year is a unit used to measure the *age* of a star.

15. ___ All stars eventually exhaust their nuclear fuel and collapse in response to *gravity*.

16. ___ A star with a surface temperature less than 3000 K will generally appear *red*.

17. ___ A rapidly rotating neutron star that radiates short pulses of radio energy is called a *pulsar*.

18. ___ The Milky Way is about 100,000 *kilometers* wide.

19. ___ *Spiral* galaxies are the most abundant type of galaxies.

20. ___ A first-magnitude star is about 100 times *dimmer* than a sixth-magnitude star.

21. ___ Large Doppler shifts indicate *low* velocities.

22. ___ Stars that orbit one another are called *neutron* stars.

23. ___ If there is enough *matter* in the universe, the outward expansion of the galaxies could eventually stop.

Word choice. Complete each of the following statements by selecting the most appropriate response.

1. There [are/are no] stars with negative magnitudes.

2. A [white dwarf/protostar] is the first stage in the evolution of a star.

3. As matter is pulled into a black hole, it emits a steady stream of [radio waves/x-rays].

4. On an H-R diagram, among main-sequence stars, hot stars are [more/less] massive than cooler stars.

5. The first naked-eye supernova in 383 years occurred in the southern sky in [1965/1973/1987].

6. Most of the stars in the Milky Way galaxy are distributed in a [disk/sphere] with spiral arms.

7. Stellar parallax [is/is not] detectable with the unaided eye.

8. Stars with surface temperatures less than 3000 K appear [blue/yellow/red].

9. Neutron stars are believed to be produced by [nuclear fission/supernovae/degenerate matter].

10. The apparent magnitude of the sun is [18.7/5.0/-26.7]; while its absolute magnitude is [18.7/5.0/-26.7].

11. By determining the light period of a cepheid variable, its [color/size/absolute magnitude] can be determined.

12. The atoms that compose our sun, as well as the rest of the solar system, formed during a [supernova/galactic collision] event trillions of kilometers away.

13. Planetary nebulae are the cast-off outer [planets/atmospheres] of collapsing red giants that formed from medium-mass stars.

14. The core of a red giant star is [expanding/contracting].

15. A star that is 5 units of magnitude brighter than another is [2.5/5.0/100] times brighter.

16. Concentrations of interstellar dust and gases are called [nebulae/novas].

17. If the sizes of the orbits of binary stars can be measured, a determination of their individual [masses/temperatures] can be made.

18. The term "Big Bang" describes the [beginning/end] of the [solar system/galaxy/universe].

19. A star's brightness as viewed from Earth is called its [absolute/apparent] brightness.

20. Neutron stars rotate very [slowly/quickly].

Written questions

1. Describe the movement through the H-R diagram that an average star like the sun follows during its lifetime.

2. What is the final state for 1) a low-mass (red) main-sequence star, 2) a medium-mass (sunlike) star, and 3) a very massive star?

3. Describe the Big Bang theory.

ANSWER KEY

INTRODUCTION

Vocabulary Review

1. core; mantle; crust
2. hypothesis
3. geology
4. Meteorology
5. theory
6. hydrosphere
7. oceanography
8. renewable resource
9. biosphere
10. nonrenewable resource
11. Astronomy
12. atmosphere
13. law
14. lithosphere

Comprehensive Review

1. The sciences traditionally included in Earth science are geology, oceanography, meteorology, and astronomy.

2. Renewable resources, such as forest products and wind energy, can be replenished over relatively short time spans. By contrast, nonrenewable resources, such as oil and copper, form so slowly that significant deposits often take millions of years to accumulate. Therefore, practically speaking, the Earth contains fixed quantities of these materials.

3. 1) hydrosphere: a dynamic mass of water that is continually on the move from the oceans to the atmosphere, precipitating back to the land, and returning back to the ocean to begin the cycle again; 2) atmosphere: Earth's blanket of air; 3) solid Earth: which includes the core, mantle, and crust; 4) biosphere: which includes all life on Earth

4. Earth science is the science which collectively seeks to understand Earth and its neighbors in space. It includes geology, oceanography, meteorology, and astronomy.

5. A scientific theory is a well-tested and widely accepted view that best explains certain observable facts; while a scientific law is a generalization from which there has been no known deviation.

6. Within about one second after the Big Bang protons and neutrons appear and quickly become atoms of hydrogen and helium. During the first billion years, galaxies, including the Milky Way, form. Massive stars in the Milky Way die violently and produce complex atoms such as oxygen, carbon, and iron. It was from this debris, scattered during the death of a star, or stars, that our solar system formed.

7. 1) collection of facts (data) through observation and measurement; 2) development of a working hypothesis to explain the facts; 3) construction of experiments to validate the hypothesis; 4) acceptance, modification, or rejection of the hypothesis on the basis of extensive testing

8. A (inner core); B (outer core); C (mantle); D (crust)

9. a) Physical geology studies the materials that compose Earth and seeks to understand the many processes that operate beneath and upon its surface. b) Historical geology seeks to understand the origin of Earth and the development of the planet through its 4.6-billion-year history.

10. On Earth, the hydrosphere, atmosphere, biosphere, and solid Earth act as a group of interrelated, interacting, or interdependent parts that form a complex whole that we call the Earth system.

Practice Test ━━━━━━━━━━━━━━━━━━━━━━━━━━━━━

Multiple choice

1. d	5. d	9. d	13. c	17. d
2. d	6. b	10. a	14. b	18. c
3. c	7. b	11. d	15. a	19. d
4. a	8. a	12. d	16. e	

True/false

1. T	5. F (predictable)	9. F (crust)
2. F (hydrosphere)	6. T	10. T
3. F (nonrenewable)	7. F (environment)	11. T
4. F (historical)	8. T	12. T

Word choice

1. dynamic	3. shelf	5. thinnest
2. 20	4. 6; 30	6. trenches

Written questions

1. The four steps that scientists often use to conduct experiments and gain scientific knowledge are 1) collection of facts (data) through observation and measurement, 2) development of a working hypothesis to explain the facts, 3) construction of experiments to validate the hypothesis, and 4) acceptance, modification, or rejection of the hypothesis on the basis of extensive testing.

2. Earth's four "spheres" include the 1) hydrosphere, a dynamic mass of water that is continually on the move from the oceans to the atmosphere, precipitating back to the land, and returning back to the ocean to begin the cycle again, 2) atmosphere, Earth's blanket of air, 3) solid Earth, which includes the core, mantle, and crust, and 4) biosphere, which includes all life on Earth.

3. Earth science is the science which collectively seeks to understand Earth and its neighbors in space. It includes geology, oceanography, meteorology, and astronomy.

CHAPTER ONE

Vocabulary Review

1. neutron
2. atom
3. ore
4. mineral
5. ion
6. cleavage
7. proton
8. compound
9. silicate
10. reserve

11. electron
12. rock
13. nucleus
14. atomic number
15. element
16. mass number
17. energy level
18. isotope
19. crystal form
20. radioactivity

21. luster
22. silicon-oxygen tetrahedron
23. mineral resource
24. streak
25. hardness
26. fracture
27. specific gravity
28. Mohs hardness scale
29. color

Comprehensive Review

1. 1) It must be naturally occurring. 2) It must be inorganic. 3) It must be solid. 4) It must possess a definite chemical structure.

2. 1) protons: particles with positive electrical charges that are found in an atom's nucleus (Letter: C); 2) neutrons: particles with neutral electrical charges that are found in an atom's nucleus (Letter: B); 3) electrons: negative electrical charges surrounding the atom's nucleus in energy levels (Letter: A)

3. Minerals are the same throughout, while most rocks are mixtures, or aggregates, of minerals where each mineral retains its distinctive properties.

4. a) 6; b) 14

5. a) 17; b) 17; c) 18

6. The sodium atom loses one electron and becomes a positive ion. The chlorine atom gains one electron and becomes a negative ion. These oppositely charged ions attract one another and the bond produces the compound sodium chloride.

7. Yes, the atom is an ion because it contains two more electrons (10) than protons (8). a) 13; b) 8

8. No, the atom contains an equal number of protons (17) and electrons (17) and is neutral.

9. A mineral is a naturally occurring inorganic solid that possesses a definite chemical structure.

10. An isotope will have a different mass number due to a different number of neutrons.

11. a) the appearance of light reflected from the surface of a mineral; b) the external expression of a mineral's orderly arrangement of atoms; c) the color of a mineral in its powdered form; d) a measure of the resistance of a mineral to abrasion or scratching; e) the tendency of a mineral to break along planes of weak bonding; f) the uneven, or irregular, breaking of a mineral; minerals that do not exhibit cleavage when broken are said to fracture; g) the weight of a mineral compared to the weight of an equal volume of water

12. Diagram A: three planes of cleavage that meet at 90-degree angles; Diagram B: three planes of cleavage that meet at approximately 70- and 120-degree angles

13. oxygen (symbol: O, 46.6%) and silicon (symbol: Si, 27.7%)

14. letter A (oxygen); letter B (silicon)

15. a) feldspars, quartz; b) micas (biotite, muscovite); c) amphibole group; d) pyroxene group; e) olivine

16. a) feldspar, quartz, and muscovite; b) calcite

17. A rock is an aggregate of minerals.

18. Most silicates crystallize from molten rock as it cools.

19. a) cinnabar; b) galena; c) sphalerite; d) hematite (see textbook Table 1.3)

20. As an isotope decays through a process called radioactivity, it actively radiates energy and particles. The decay often produces a different isotope of the same element, but with a different mass number.

21. No. Before aluminum can be extracted profitably it must be concentrated to four times its average crustal percentage of 8.13 percent. The sample, with only 10 percent aluminum, lacks the necessary concentration.

22. Diagram A: amphibole group; Diagram B: pyroxene group

23. iron (Fe), magnesium (Mg), potassium (K), sodium (Na), and calcium (Ca)

Practice Test

Multiple choice

1. c	6. a	11. d	16. c	21. b
2. b	7. c	12. b	17. a	22. e
3. e	8. c	13. a	18. b	23. a
4. c	9. c	14. e	19. e	24. a
5. a	10. b	15. b	20. d	25. c

True/false

1. F (electrons)	10. T	19. F (compound)
2. T	11. F (one-fourth)	20. T
3. F (four-thousand)	12. F (carbonate)	21. F (electrons)
4. T	13. F (Gypsum)	22. T
5. F (oxygen)	14. T	23. F (atom)
6. T	15. F (Crystal form)	24. T
7. T	16. T	25. F (harder)
8. T	17. F (nonmetallic)	
9. F (between)	18. T	

Word choice

1. atomic number; mass number
2. atom
3. mass; atomic; similar
4. luster
5. fracture

6. silicon; tetrahedron
7. cleavage
8. calcite
9. halite; calcite
10. hematite; galena

Written questions

1. Protons and neutrons are found in the atom's nucleus and represent practically all of an atom's mass. Protons are positively charged, while neutrons are neutral. Electrons orbit the nucleus in regions called energy levels (or shells) and carry a negative charge.

2. Most mineral samples don't demonstrate visibly their crystal form because, most of the time, crystal growth is severely constrained because of space restrictions where and when the mineral forms.

3. The basic building block of all silicate minerals is the silicon-oxygen tetrahedron. It consists of four oxygen atoms surrounding a much smaller silicon atom.

4. For any material to be considered a mineral it must be 1) naturally occurring, 2) inorganic, 3) solid, and 4) possess a definite chemical structure.

5. A mineral resource is any useful mineral that can be recovered for use. Resources include already identified deposits from which minerals can be extracted profitably, called reserves.

ANSWER KEY

CHAPTER TWO

Vocabulary Review

1. rock cycle
2. vein deposit
3. Lithification
4. metamorphic rock
5. Magma
6. crystallization
7. regional metamorphism
8. sedimentary rock
9. lava

10. texture
11. foliated texture
12. extrusive (volcanic)
13. sediment
14. igneous rock
15. nonfoliated texture
16. fine-grained texture
17. hydrothermal solution
18. chemical sedimentary rock

19. porphyritic texture
20. strata (beds)
21. evaporite deposit
22. contact metamorphism
23. glassy texture
24. detrital sedimentary rock
25. coarse-grained texture
26. disseminated deposit

Comprehensive Review

1. A (Crystallization); B (Igneous); C (Weathering, transportation, and deposition); D (Sediment); E (Lithification); F (Sedimentary); G (Metamorphism); H (Metamorphic); I (Melting)

2. James Hutton; 1700s

3. weathering, transportation, and deposition

4. running water, wind, waves, and glacial ice

5. a) Igneous rocks form as magma cools and solidifies either beneath or at Earth's surface. b) Sedimentary rocks are the lithified products of weathering. c) Metamorphic rocks can form from igneous, sedimentary, or other metamorphic rocks that are subjected to heat, pressure, and chemically active fluids.

6. The rate of cooling most influences the size of mineral crystals in igneous rock. Slow cooling results in the formation of large crystals; while rapid cooling often results in a mass of very small intergrown crystals.

7. Magma is molten rock beneath the surface. Lava, molten rock on the surface, is similar to magma except that most of the gaseous component has escaped.

8. 1) texture: the overall appearance of an igneous rock based on the size and arrangement of its interlocking crystals; 2) mineral composition: the mineral makeup of an igneous rock which depends on the composition of the magma from which it originated

9. a) rapid cooling at the surface or as small masses within the upper crust (letter A.); b) very rapid cooling, perhaps when molten rock is ejected into the atmosphere (letter D.); c) slow cooling of large masses of magma far below the surface (letter B.); d) two different rates of cooling, resulting in two different crystal sizes (letter C.)

10. N.L. Bowen discovered that as magma cools in the laboratory, certain minerals crystallize first at very high temperatures. At successively lower temperatures, other minerals crystallize.

11. olivine; quartz

12. Yes, providing that they have different textures resulting from different rates of cooling.

13. Basaltic group: pyroxene, amphibole, plagioclase feldspar Granitic group: potassium feldspar, muscovite mica, quartz

14. If the earlier formed minerals in a cooling magma are denser (heavier) than the liquid portion of the magma, they will settle to the bottom of the magma chamber. The process is called crystal settling.

15. compaction and cementation

16. a) Detrital sedimentary rocks are made of a wide variety of mineral and rock fragments, clay minerals and quartz dominate. (example: sandstone) b) Chemical sedimentary rocks are derived from material that is carried in solution to lakes and seas and, when conditions are right, precipitates to form chemical sediments. (example: limestone)

17. particle size

18. Angular fragments indicate that the particles were not transported very far from their source prior to deposition.

19. mineral composition

20. Limestone is the most abundant chemical sedimentary rock. Ninety percent of limestone is biochemical sediment. The rest precipitates directly from seawater.

21. swamp environment, peat, lignite, bituminous, anthracite

22. a) detrital; b) breccia; c) C

23. Fossils are traces or remains of prehistoric life. They are useful for interpreting past environments, as time indicators, and for matching up rocks from different places that are the same age.

24. a) Regional metamorphism occurs during mountain building when great quantities of rock are subjected to intense stress and deformation. b) Contact metamorphism occurs when rock is in contact with, or near, a mass of magma.

25. The three agents of metamorphism include heat, pressure, and chemically active fluids.

26. a) A foliated texture results from the alignment of mineral crystals with a preferred orientation, giving a metamorphic rock a layered or banded appearance. (example: slate) b) The minerals in a nonfoliated metamorphic rock are generally equidimensional and not aligned. (example: marble)

27. foliated (rock name: gneiss)

28. The ore deposits in both areas were generated from hot, metal-rich fluids, called hydrothermal solutions, that were associated with cooling magma bodies.

Practice Test

Multiple choice

1. b	6. b	11. b	16. d	21. c
2. b	7. d	12. b	17. b	22. b

3. e	8. a	13. a	18. a	23. d
4. d	9. d	14. d	19. d	24. a
5. a	10. c	15. d	20. b	25. a

True/false

1. T	10. F (less)	19. F (sedimentary)
2. F (basalt)	11. T	20. T
3. T	12. T	21. T
4. T	13. F (crystallization)	22. T
5. T	14. T	23. F (gypsum)
6. T	15. F (Metamorphism)	24. T
7. F (Sedimentary)	16. F (lava)	25. F (large)
8. F (metamorphic)	17. T	
9. T	18. T	

Word choice

1. igneous	5. deficient; not	9. biochemical
2. less; toward	6. slate	10. calcite
3. silicate	7. calcite	11. gneiss
4. size	8. heat	12. can

Written questions

1. Magma, the material of igneous rocks, is produced when any rock is melted. Sedimentary rocks are formed from the weathered products of any pre-existing rock—igneous, sedimentary, or metamorphic. Metamorphic rocks are created when any rock type undergoes metamorphism.

2. Bowen's reaction series illustrates the sequence in which minerals crystallize from magma. It can be used to help explain the mineral makeup of an igneous rock, in that minerals that crystallize at about the same time (temperature) are most often found together in the same igneous rock.

3. Quartz and clay minerals are the chief constituents of detrital sedimentary rocks. Clay minerals are the most abundant product of chemical weathering. Quartz is an abundant sediment because it is very resistant to chemical weathering.

4. Metamorphic processes cause many changes in existing rocks, including increased density, growth of larger crystals, foliation, and the transformation of low-temperature minerals into high-temperature minerals. Furthermore, the introduction of ions generates new minerals, some of which are economically important.

CHAPTER THREE

Vocabulary Review

1. erosion
2. frost wedging
3. regolith
4. Leaching
5. exfoliation dome
6. pedalfer
7. slump
8. weathering
9. permafrost
10. Secondary enrichment

11. Soil
12. humus
13. mechanical weathering
14. laterite
15. rockslide
16. chemical weathering
17. Eluviation
18. lahar
19. mass wasting
20. Differential weathering

21. soil texture
22. pedocal
23. sheeting
24. horizon
25. Creep
26. parent material
27. talus slope
28. solum
29. angle of repose
30. soil profile

Comprehensive Review

1. The three processes that are continually removing material from higher elevations and transporting them to lower elevations are 1) weathering, the disintegration and decomposition of rock at or near Earth's surface, 2) mass wasting, the transfer of rock material downslope under the influence of gravity, and 3) erosion, the incorporation and transportation of material by a mobile agent, usually water, wind, or ice.

2. 1) Mechanical weathering: the breaking of rock into smaller pieces, each retaining the characteristics of the original material. 2) Chemical weathering: those processes that alter the internal structure of a mineral by removing and/or adding elements.

3. The four natural processes that break rocks into smaller fragments (mechanically weather) include frost wedging, unloading, thermal expansion, and biological activity.

4. Water becomes mildly acidic when carbon dioxide (CO_2) dissolves in rain water and produces the weak acid, carbonic acid (H_2CO_3). Decaying organisms also produce acids.

5. The three factors that influence the rate of weathering are 1) the degree of mechanical weathering, which determines the amount of surface area exposed to chemical weathering, 2) mineral makeup —minerals that crystallize first from magma (Bowen's reaction series) are least resistant to chemical weathering, and 3) climate—the optimum environment for chemical weathering is a combination of warm temperatures and abundant moisture.

6. Originally angular blocks formed along cracks called joints. Chemical weathering by the water that entered the joints caused the corners and edges of blocks to become more rounded. The corners are attacked by the acidic water more readily because of their greater surface area, as compared to the edges and faces of the blocks. This process, called spheroidal weathering, gives the weathered rock a more rounded or spherical shape.

7. clay minerals, soluble salt (potassium bicarbonate), silica in solution

8. Regolith is the layer of rock and mineral fragments produced by weathering. Soil, on the other hand, is a combination of mineral and organic matter, water, and air—that portion of the regolith that supports the growth of plants.

9. The three particle size classes for determining soil texture are 1) clay, 2) silt, and 3) sand.

10. Clay: 20 percent, Silt: 40 percent, Sand: 40 percent

11. silty loam

12. The four basic soil structures are 1) platy, 2) prismatic, 3) blocky, and 3) spheroidal.

13. The five controls of soil formation include 1) parent material, 2) time, 3) climate, 4) plants and animals, and 5) slope.

14. From top to bottom the horizons are: O, A, E, B, and C.

15. a) at location B; b) location E; c) location C

16. a) pedocal; b) laterite; c) pedalfer

17. Strong winds and the expansion of agriculture brought about by mechanization were two other factors that contributed to the "Dust Bowl" of the 1930s.

18. Secondary enrichment takes place in one of two ways: 1) chemical weathering coupled with the downward-percolating water in soil removes undesired materials from decomposing rock, leaving the desired elements enriched in the upper zones, or 2) desirable elements near the surface are removed and carried to lower soil zones where they are redeposited and become more concentrated.

19. The combined effects of mass wasting and running water produce stream valleys.

20. Among the factors are 1) saturation of the material with water, 2) oversteepening of the slopes, 3) removal of anchoring vegetation, and 4) ground vibrations from earthquakes.

21. Diagram A: slump; Diagram B: rockslide; Diagram C: mudflow; Diagram D: earthflow

Practice Test

Multiple choice

1. b	7. b	13. c	19. c	25. c
2. b	8. d	14. a	20. d	26. d
3. e	9. b	15. d	21. b	27. a
4. b	10. a	16. d	22. d	28. a
5. d	11. b	17. d	23. b	29. a
6. a	12. a	18. d	24. e	30. b

True/false

1. T	9. T	17. T
2. F (subsoil)	10. F (thinner)	18. T
3. T	11. F (pedalfers)	19. T
4. F (first)	12. T	20. F (flowing water)

5. T	13. T	21. T
6. F (profile)	14. F (horizon)	22. F (pedocal)
7. T	15. T	
8. T	16. F (flow)	

Word choice

1. more	11. weak; acid
2. immature	12. greater
3. microorganisms	13. will not
4. middle	14. leaching
5. topsoil	15. falls
6. poor	16. expands
7. bauxite	17. calcium carbonate; pedocal
8. oxidized	18. curved
9. increase	19. reduction; removed
10. hardpan	20. creep

Written questions

1. The two types of weathering are mechanical weathering and chemical weathering. Mechanical weathering involves the breaking of rock into smaller pieces, each retaining the characteristics of the original material. Chemical weathering refers to those processes that alter the internal structure of a mineral by removing and/or adding elements. Advanced mechanical weathering aids chemical weathering by increasing the amount of surface area available for chemical attack.

2. Soil is a combination of mineral and organic matter, water, and air—that portion of the regolith that supports the growth of plants. Since the soil forming processes operate from the surface downward, variations in composition, texture, structure, and color evolve at varying depths. These vertical differences divide the soil into zones or layers that soil scientists call horizons. Such a vertical section through all the soil horizons constitutes the soil profile.

3. Gravity is the controlling force of mass wasting. Other important factors include 1) saturation of the material with water, 2) oversteepening of the slopes, 3) removal of anchoring vegetation, and 4) ground vibrations from earthquakes.

ANSWER KEY

CHAPTER FOUR

Vocabulary Review

1. Infiltration
2. drainage basin
3. zone of saturation
4. porosity
5. Gradient
6. alluvium
7. capacity
8. groundwater
9. yazoo tributary

10. transpiration
11. meander
12. water table
13. discharge
14. artesian well
15. natural levee
16. permeability
17. hydrologic cycle
18. floodplain

19. competence
20. divide
21. aquiclude
22. Base level
23. spring
24. aquifer
25. suspended load
26. karst topography
27. runoff

Comprehensive Review

1. mass wasting and running water

2. a) C; b) B; c) A; d) D

3. a) the distance that water travels in a unit of time; b) the slope of a stream channel expressed as the vertical drop of a stream over a specified distance; c) the volume of water flowing past a certain point in a given unit of time

4. 4 meters/kilometer (40 meters divided by 10 kilometers)

5. a) decreases; b) increases; c) decreases; d) increases; e) increase

6. a) Ultimate base level is sea level, the lowest level to which stream erosion could lower the land. b) Temporary base levels include lakes and main streams, which act as base level for the streams and tributaries entering them.

7. 1) as dissolved load; brought to a stream by groundwater; 2) as suspended load; sand, silt, and clay carried in the water; 3) as bed load; coarser particles moving along the bottom of a stream

8. During a flood, the increase in discharge of a stream results in a greater capacity and the increase in velocity results in greater competence.

9. a) C; b) B; c) A; d) D

10. 1) A narrow valley is typically V-shaped because the primary work of the stream has been downcutting toward base level. The features of narrow valleys include rapids and waterfalls. 2) In a wide valley the stream's energy is directed from side-to-side, and downward erosion becomes less dominant. The features of wide valleys include meanders, floodplains, bars, cutoffs, and oxbow lakes.

11. radial pattern; where streams diverge from a central area like the spokes of a wheel

12. a) B; b) C; c) A

13. a) downcutting, rapids, waterfalls, V-shaped, no floodplain, straight course; b) lateral erosion, meanders, floodplain beginning, low gradient; c) large floodplain, widespread meanders, natural levees, back swamps, yazoo tributaries

14. a) D; b) B; c) A; d) C; e) E

15. The shape of the water table is usually a subdued replica of the surface, reaching its highest elevations beneath hills and decreasing in height toward valleys.

16. The source of heat for most hot springs and geysers is in cooling igneous rock beneath the surface.

17. A rock layer or sediment that transmits water freely because of its high porosity and high permeability is called an aquifer.

18. 1) land subsidence caused by groundwater withdrawal; 2) groundwater contamination

19. 1) slowly, over many years by dissolving the limestone below the soil; 2) suddenly, when the roof of a cavern collapses under its own weight

20. Diagram A: dendritic pattern; Diagram B: rectangular pattern

Practice Test

Multiple choice

1. c	6. b	11. e	16. c	21. c
2. c	7. a	12. a	17. e	22. a
3. d	8. d	13. b	18. e	23. b
4. c	9. b	14. c	19. d	24. d
5. b	10. d	15. b	20. b	25. a

True/false

1. T	11. T	21. F (saturation)
2. F (Temporary)	12. F (Impermeable)	22. T
3. T	13. T	23. F (transpiration)
4. T	14. T	24. T
5. T	15. T	25. T
6. F (wide)	16. F (suspension)	26. T
7. T	17. T	27. T
8. F (decreases)	18. T	28. T
9. F (rectangular)	19. T	29. T
10. T	20. T	30. F (Running water)

Word choice

1. center	6. aquifer	11. raise
2. trellis	7. decreases; increases	12. outside
3. distributaries	8. limestone; wet	13. carbonic
4. water table	9. gentle; light	14. increases
5. decreases	10. discharge	15. porosity

Written questions

1. Evaporation, primarily from the ocean, transport via the atmosphere, and eventually precipitation back to the surface. If the water falls on the continents, much of it will find its way back to the atmosphere by evaporation and transpiration. Some water, however, will be absorbed by the surface (infiltration) and some will run off back to the ocean.

2. To learn about the development of a valley, it is helpful to divide its development into three stages: youth, maturity, and old age. A youthful valley is characterized by a stream that is downcutting, waterfalls, rapids, and a V-shape. In a mature valley, the downward erosion of the stream diminishes and lateral erosion dominates. The stream is meandering and a floodplain is beginning. In old age, the floodplain is large, meandering is widespread, and the river is primarily reworking unconsolidated floodplain deposits. Natural levees, back swamps, and yazoo tributaries are also common features.

3. Porosity is the percentage of the total volume of rock or sediment that consists of pore spaces. Permeability is a measure of a material's ability to transmit water through interconnected pore spaces. It is possible for a material to have a high porosity but a low permeability if the pore spaces lack interconnections.

4. Generally, for karst topography to develop the area must be humid and underlain by soluble rock, usually limestone.

CHAPTER FIVE

Vocabulary Review

1. ice sheet
2. ephemeral stream
3. drift
4. drumlin
5. Pleistocene epoch
6. glacial striations
7. interior drainage
8. medial moraine

9. zone of wastage
10. glacial trough
11. cirque
12. barchan dune
13. zone of accumulation
14. alluvial fan
15. longitudinal dune
16. loess

17. till
18. outwash plain
19. blowout
20. cross beds
21. glacier
22. piedmont glacier

Comprehensive Review

1. 1) Valley, or alpine, glaciers are relatively small streams of ice that exist in mountain areas, where they usually follow valleys originally occupied by streams. 2) Ice sheets exist on a large scale and flow out in all directions from one or more centers, completely obscuring all but the very high areas of underlying terrain.

2. This upper part of the glacier consists of brittle ice that is subjected to tension when the ice flows over irregular terrain, with cracks called crevasses resulting.

3. a) Plucking occurs when a glacier flows over a fractured bedrock surface and the ice incorporates loose blocks of rock and carries them off. b) Abrasion occurs as the ice with its load of rock fragments moves along and acts as a giant rasp or file, grinding the surface below as well as the rocks within the ice.

4. a) D; b) E; c) A; d) C; e) B

5. Till is unsorted material deposited directly by the glacier. Stratified drift is characteristically sorted sediment laid down by glacial meltwater.

6. Two of the many hypotheses for the cause of glacial ages involve 1) plate tectonics and 2) variations in Earth's orbit.

7. B; a) C; b) D

8. a) forms at the terminus of a glacier; b) forms along the side of a valley glacier; c) a layer of till that is laid down as the glacier recedes

9. a) A; b) G; c) C; d) B

10. Periodic slides of sand down the slip face of a dune cause the slow migration of the dune in the direction of air movement.

11. The indirect effects of Ice Age glaciers include plant and animal migrations, changes in stream and river courses, adjustment of the crust by rebounding, climate changes, and the worldwide change in sea level that accompanied each advance and retreat of the ice sheets.

12. moraines, glacial striations, glacial troughs, hanging valleys, cirques, horns, **aretes**, fiords, glacial drift, glacial erratics

13. The existence of several layers of glacial drift with well-developed zones of chemical weathering and soil formation, as well as the remains of plants that require warm temperatures, indicate several glacial advances separated by periods of warmer climates.

14. Wind is not capable of picking up and transporting coarse materials and, because it is not confined to channels, it can spread over large areas as well as high into the atmosphere.

15. Diagram A: barchan dunes; Diagram B: transverse dunes; Diagram C: barchanoid dunes; Diagram D: longitudinal dunes

Practice Test

Multiple choice

1. c	6. a	11. b	16. b	21. a
2. c	7. d	12. d	17. a	22. b
3. b	8. b	13. a	18. b	23. b
4. d	9. b	14. c	19. e	24. d
5. a	10. b	15. e	20. d	25. a

True/false

1. T	9. F (ten)	17. T
2. F (accumulation)	10. F (thirty)	18. F (horn)
3. T	11. F (meters)	19. T
4. F (lateral)	12. T	20. F (Cirques)
5. T	13. F (U-shaped)	21. T
6. F (fifty)	14. F (drumlins)	22. T
7. T	15. T	23. T
8. T	16. T	

Word choice

1. till	6. erratic	11. deflation
2. flour	7. calving	12. is not
3. mechanical	8. sorted	13. lack
4. interior	9. few	
5. Fiords	10. subdue	

Written questions

1. 1) moraines: layers or ridges of till; 2) drumlins: streamlined, asymmetrical hills composed of till; 3) eskers: ridges of sand and gravel deposited by streams flowing beneath the ice, near the

glacier's terminus; 4) kames: steep-sided hills formed when glacial meltwater washes sediment into openings and depressions in the stagnant, wasting terminus of a glacier (answers may vary)

2. 1) glacial trough: a U-shaped glaciated valley; 2) horn: a pyramid-shaped peak; 3) cirque: a bowl-shaped depression at the head of a valley glacier where snow accumulation and ice formation occur (answers may vary)

3. 1) plant and animal migrations; 2) changes in stream and river courses; 3) worldwide change in sea level that accompanied each advance and retreat of the ice sheets (answers may vary)

4. Periodic slides of sand down the slip face of a dune cause the slow migration of the dune in the direction of air movement.

ANSWER KEY

CHAPTER SIX

Vocabulary Review

1. earthquake
2. seismic sea wave (tsunami)
3. foreshock
4. focus
5. elastic rebound
6. surface wave
7. outer core

8. epicenter
9. asthenosphere
10. fault
11. crust
12. secondary (S) wave
13. seismogram
14. Mohorovicic discontinuity (Moho)

15. magnitude
16. inner core
17. primary (P) wave
18. lithosphere
19. Richter scale
20. shadow zone
21. mantle
22. seismograph

Comprehensive Review

1. 1) surface waves: travel around the outer layer of Earth; 2) primary (P) waves: travel through the body of Earth with a push-pull motion in the direction the wave is traveling; 3) secondary (S) waves: travel through Earth by "shaking" the rock particles at right angles to their direction of travel. P waves travel through all materials, whereas S waves are transmitted only through solids. Further, in all types of rock, P waves travel faster than S waves.

2. a) A; b) B

3. a) C; b) B; c) A

4. Primary (P) waves have a greater velocity than secondary (S) waves. P waves can travel through solids, liquids, and gasses; while S waves are only capable of being transmitted through solids.

5. By determining the difference in arrival times between the first P wave and first S wave and using a travel-time graph, three different recording stations determine how far they each are from the epicenter. Next, for each station, a circle is drawn on a globe corresponding to the station's distance from the epicenter. The point where the three circles intersect is the epicenter of the quake.

6. 1) magnitude of the earthquake; 2) proximity to a populated area

7. surrounding the Pacific Ocean basin and along the mid-ocean ridge (answers may vary); Most earthquake epicenters are closely correlated with plate boundaries.

8. a) E (including D and part of C); b) G; c) A; d) B; e) D; f) F; g) C

9. a) Denser rocks like peridotite are thought to make up the mantle and provide the lava for oceanic eruptions. b) The molten outer core is thought to be mainly iron and nickel.

Practice Test ━━

Multiple choice

1. e	5. a	9. b	13. b	17. c
2. c	6. a	10. e	14. c	18. b
3. e	7. c	11. e	15. d	19. a
4. b	8. d	12. a	16. b	20. d

True/false

1. T	9. T	17. T
2. F (aftershocks)	10. F (can)	18. T
3. F (two)	11. F (before)	19. T
4. T	12. F (Mohorovicic)	20. F (less)
5. T	13. T	21. T
6. T	14. F (above)	22. T
7. F (vertical)	15. T	
8. F (seismology)	16. T	

Word choice

1. S	9. asthenosphere
2. epicenter	10. seismographs: seismograms
3. surface	11. on land
4. rebound	12. flexible
5. Moho	13. Inertia
6. seismic waves	14. lithosphere
7. faults	15. stationary
8. Richter	

Written questions

1. An earthquake is the vibration of Earth produced by the rapid release of energy, usually along a fault. This energy radiates in all directions from the earthquake's source, called the focus, in the form of waves.

2. P waves are "push-pull" waves that compress and expand the rock in the direction that they are traveling; S waves, on the other hand, "shake" the materials at right angles to the direction of travel. P waves travel through all materials, whereas S waves are transmitted only through solids. Further, in all types of rock, P waves travel faster than S waves.

3. Oceanic crust is basaltic in composition, while the continental crust has an average composition similar to that of granite. The composition of the mantle is thought to be similar to the rock peridotite, which contains iron and magnesium-rich silicate minerals. Both the inner and outer cores are thought to be enriched in iron and nickel, with lesser amounts of other heavy elements.

ANSWER KEY

CHAPTER SEVEN

Vocabulary Review

1. continental drift
2. plate
3. Seafloor spreading
4. convergent boundary
5. paleomagnetism
6. plate tectonics

7. normal polarity
8. divergent boundary
9. hot spot
10. Pangaea
11. subduction zone
12. rift (rift valley)

13. transform boundary
14. Polar wandering
15. reverse polarity
16. island arc
17. deep-ocean trench

Comprehensive Review

1. 1) fit of the continents; 2) similar fossils on different landmasses; 3) similar rock types and structures on different landmasses; 4) similar ancient climates on different landmasses

2. One of the main objections to Wegener's continental drift hypothesis was his inability to provide a mechanism that was capable of moving the continents across the globe.

3. a) C; b) A; c) B

4. The theory of plate tectonics holds that Earth's rigid outer shell consists of about twenty rigid slabs called plates that are in continuous slow motion relative to each other. These plates interact in various ways and thereby produce earthquakes, volcanoes, mountains, and the crust itself.

5. 1) oceanic-continental convergence; 2) oceanic-oceanic convergence; 3) continental-continental convergence

6. a) Paleomagnetism is used to show that the continents have moved through time and, using the patterns of magnetic reversals found on the ocean floors, to explain seafloor spreading. b) Earthquake patterns are used to show plate boundaries and to trace the descent of slabs into the mantle. c) Ocean drilling has shown that the ocean basins are geologically youthful and confirms seafloor spreading by the fact that the age of the deepest ocean sediment increases with increasing distance from the ridge.

7. a) D; b) B; c) C; d) F

8. Mountains of comparable age and structure to the Appalachians are found in the British Isles and Scandinavia. When the landmasses are placed in their predrift locations, the mountain chains form a nearly continuous belt. These folded mountain belts formed roughly 300 million years ago as the landmasses collided during the formation of the supercontinent of Pangaea.

9. 1) convection current hypothesis: large convection currents within the mantle drive plate motion; 2) slab-push and slab-pull hypothesis: as new ocean crust moves away from the ridge it cools, increases in density, and eventually descends, pulling the trailing lithosphere along; 3) hot plumes hypothesis: hot plumes within the mantle reach the lithosphere and spread laterally, facilitating the plate motion away from the zone of upwelling

10. a) F; b) B; c) C; d) E; e) A; f) D; g) G

11. Using satellites, scientists are now able to confirm the fact that plates shift in relation to one another by directly measuring their relative motions using timed laser pulses that are bounced off the satellites.

12. The two sub-continents of Pangaea are the northern continent of Laurasia and the southern land-mass of Gondwanaland. Gondwanaland included the present-day continents of South America, Africa, Antarctica, Australia, and the sub-continent of India.

Practice Test

Multiple choice

1. b	7. e	13. b	19. b	25. b
2. b	8. b	14. c	20. e	26. d
3. b	9. a	15. b	21. b	27. c
4. e	10. b	16. a	22. d	
5. a	11. e	17. a	23. a	
6. d	12. a	18. c	24. d	

True/false

1. T	10. T	19. T
2. T	11. F (trenches)	20. T
3. F (centimeters)	12. T	21. T
4. F (divergent)	13. T	22. T
5. T	14. F (mantle)	23. F (thinnest)
6. F (mantle)	15. T	24. T
7. F (younger)	16. F (normal)	25. T
8. T	17. T	
9. T	18. F (youngest)	

Word choice

1. lithosphere	11. created	21. will not
2. continental shelf	12. South Pole	22. parallel
3. consumed	13. heat	23. shallow
4. also	14. thicker	24. rifts
5. oceanic	15. cooler; more	25. volcanic arcs
6. rigid	16. subduction zones	26. after
7. 200	17. rifts	27. slab-pull
8. tropical	18. rate	28. Baja Peninsula
9. subduction	19. mantle plumes	
10. more than	20. increases	

Written questions

1. Hot spots are relatively stationary plumes of molten rock rising from Earth's mantle. According to the plate tectonics theory, as a plate moves over a hot spot, magma often penetrates the surface, thereby generating a volcanic structure. In the case of the Hawaiian Islands, as the Pacific plate moved over a hot spot, the associated igneous activity produced a chain of major volcanoes. Currently, because the rising plume of mantle material is located below it, the only active island in the chain is Hawaii. Kauai, the oldest island in the chain, formed approximately five million years ago when it was positioned over the hot spot and has now moved northwest as a result of plate motion.

2. 1) divergent boundaries—where the plates are moving apart; 2) convergent boundaries—where the plates are moving toward each other; 3) transform boundaries—where the plates are sliding past each other along faults

3. The lines of evidence used to support plate tectonics include 1) paleomagnetism (polar wandering and magnetic reversals), 2) earthquake patterns, 3) the age and distribution of ocean sediments, and 4) evidence from ocean floor volcanoes that are, or were, located over hot spots.

CHAPTER EIGHT

Vocabulary Review

1. pyroclastics
2. caldera
3. aa flow
4. volcanic neck
5. dike
6. vent
7. cinder cone
8. batholith
9. shield volcano
10. flood basalt
11. pahoehoe flow
12. hot spot
13. composite cone (stratovolcano)
14. volcano
15. laccolith
16. fissure eruption
17. sill
18. nuée ardente
19. partial melting
20. viscosity
21. lahar

Comprehensive Review

1. 1) the magma's composition; 2) the magma's temperature; 3) the amount of dissolved gases in the magma

2. a) B (Mauna Loa): Shield volcanoes have recurring eruptions over a long span of time and are composed of large amounts of basaltic lava. b) A (Paricutin): The eruption history of a cinder cone is short and produces a relatively small cone composed of loose pyroclastic material. c) C (Fujiyama): A composite cone may extrude viscous lava for a long period and then, suddenly, the eruptive style changes and the volcano violently ejects pyroclastic material. Composite cones represent the most violent type of volcanic activity.

3. water vapor, carbon dioxide, nitrogen, sulfur gases

4. a) the igneous body cuts across existing sedimentary beds; b) the igneous body is parallel to existing sedimentary beds

5. The viscosity of magma is affected by its temperature, silica content (the more silica, the greater the viscosity), and quantity of dissolved gases.

6. a) C; b) E; c) F; d) D

7. a) One source of heat to melt rock is from the decay of radioactive elements in the mantle and crust. The gradual increase in temperature with increasing depth melts subducting oceanic crust and produces basaltic magma. The hot magma body formed could migrate upward to the base of the crust and intrude granitic rocks. b) Reducing confining pressure lowers a rock's melting temperature sufficiently to trigger melting. c) Partial melting produces most, if not all, magma. A consequence of partial melting is the production of a melt with a higher silica content than the parent rock. Magma generated by the partial melting of basalt is more granitic than the parent material.

8. When compared to granitic magma, basaltic magma contains less silica, is less viscous, and, in a near surface environment, basaltic rocks melt at a higher temperature.

9. 1) Along the oceanic ridge system, as the rigid lithosphere pulls apart, pressure is lessened and the partial melting of the underlying mantle rocks produces large quantities of basaltic magma. 2) Adjacent to ocean trenches, magma is generated by the partial melting of descending slabs of oceanic

crust. The melt slowly rises upward toward the surface because it is less dense than the surrounding rock. 3) Intraplate volcanism may be the result of rising plumes of hot mantle material which produce hot spots, volcanic regions a few hundred kilometers across.

Practice Test

Multiple choice

1. b	6. e	11. b	16. e	21. d
2. b	7. a	12. b	17. a	22. b
3. b	8. e	13. a	18. a	23. c
4. c	9. d	14. e	19. b	24. a
5. b	10. c	15. a	20. d	25. a

True/false

1. F (convergent)	8. T	15. T
2. T	9. F (crater)	16. T
3. T	10. T	17. T
4. T	11. T	18. F (sills)
5. F (cinder)	12. F (increase)	19. T
6. T	13. F (silica)	20. T
7. F (less)	14. F (composite)	

Word choice

1. Decreasing	6. oceanic ridge system	11. cinder
2. summit; magma chamber	7. upper mantle rock; basaltic	12. decrease; pressure
3. sulfur	8. gases; upper	13. basaltic
4. high; large	9. increase; temperature	14. subduction zone; water
5. higher	10. Increasing	15. reduced

Written questions

1. Magma and lava both refer to the material from which igneous rocks form. However, magma is molten rock below Earth's surface, including dissolved gases and crystals, while lava refers to molten rock that reaches Earth's surface.

2. 1) Shield cones are among the largest on Earth. These gently sloping domes are associated with relatively quiet eruptions of fluid basaltic lava. 2) Cinder cones are composed almost exclusively of pyroclastic material, are steep-sided, and are the smallest of the volcanoes. 3) Composite cones, as the name suggests, are composed of alternating layers of lava and pyroclastic debris. Their slopes are steeper than those of a shield volcano, but more gentle than those of a cinder cone.

3. 1) Dikes are tabular, discordant plutons. 2) Sills are tabular, concordant plutons. 3) Batholiths, by far the largest of all intrusive features, are massive, discordant plutons. (answers may vary)

ANSWER KEY

CHAPTER NINE

Vocabulary Review

1. strike-slip fault
2. horst
3. Isostasy
4. fold
5. basin
6. Orogenesis
7. fault
8. thrust fault
9. graben
10. folded mountains (complex mountains)
11. anticline
12. syncline
13. oblique-slip fault
14. fault-block mountains
15. dip-slip fault
16. upwarped mountains
17. normal fault
18. isostatic adjustment
19. joint
20. accretionary wedge
21. terrane
22. dome
23. reverse fault

Comprehensive Review

1. The presence of fossilized shells of aquatic organisms at high elevations is evidence that the sedimentary rocks were once below sea level.

2. The two reasons that the crust beneath the oceans is at lower elevation than the continental crust are 1) the crust beneath the oceans is thinner than that of the continents, and 2) oceanic rocks have a greater density than continental rocks.

3. Isostasy refers to the fact that Earth's less dense crust is "floating" on top of the denser and more easily deformed rocks of the mantle.

4. Changes in rock resulting from elastic deformation are reversible; that is, the rock will return to nearly its original size and shape when the stress is removed. Plastic deformation results in permanent changes; that is, the size and shape of a rock unit are permanently altered through folding and flowing.

5. a) An anticline is most commonly formed by the upfolding, or arching, of rock layers (letter A); b) Synclines, usually found in association with anticlines, are downfolds, or troughs (letter B).

6. Diagram A: dome; Diagram B: basin

7. a) hanging wall moves up relative to the footwall (diagram B); b) dominant displacement is along the trend, or strike, of the fault (diagram D); c) hanging wall moves down relative to the footwall (diagram A); d) hanging wall moves up relative to the footwall at a low horizontal angle (diagram C)

8. a) B; b) A

9. normal faults

10. reverse faults and thrust faults

11. a) Where oceanic and continental crusts converge, the first stage in the development of a mountain belt is the formation of a subduction zone. Deposition of sediments occurs along the continental margin. Convergence of the continental block and the subducting oceanic plate leads to deformation and metamorphism of the continental margin. A volcanic island arc forms. Sediment derived from the land as well as that scraped from the subducting plate becomes plastered

against the landward side of the trench, forming an accretionary wedge consisting of folded, faulted, and metamorphosed sediments and volcanic debris. b) Where continental crusts converge, the continental lithosphere is too buoyant to undergo any appreciable subduction and a collision between continental fragments eventually results. The result of the collision is the formation of a mountain range.

12. a) C; b) A; c) E; d) D; e) B

13. Any two of the following: Himalaya Mountains, Ural Mountains, Alps, and Appalachians.

14. An oil trap is a geologic environment that allows for economically significant amounts of oil and gas to accumulate. The two basic conditions that all oil traps have in common are 1) a porous, permeable reservoir rock that will yield petroleum and natural gas in sufficient quantities, and 2) a cap rock that keeps the upwardly mobile oil and gas from escaping at the surface.

15. Because of the great length and complexity of the San Andreas fault, it is more appropriately referred to as a "fault system." This major fault system consists primarily of the San Andreas fault and several major branches including the Hayward and Calaveras faults of central California.

Practice Test

Multiple choice

1. d	6. a	11. e	16. c	21. c
2. c	7. c	12. a	17. d	22. a
3. e	8. d	13. e	18. a	23. c
4. b	9. b	14. d	19. b	24. b
5. a	10. e	15. c	20. c	25. c

True/false

1. F (raises)	10. T	19. F (folded)
2. T	11. F (youngest)	20. F (synclines)
3. T	12. F (elastically)	21. F (downward)
4. F (Faults)	13. F (vertical)	22. T
5. T	14. T	23. T
6. T	15. F (folds)	24. F (convergent)
7. T	16. T	25. T
8. F (compressional)	17. T	26. T
9. T	18. T	27. F (reservoir)

Word choice

1. plastic	6. compressional	12. divergence
2. tensional	7. isostatic	13. less
3. impermeable; above; permeable	8. tensional	14. expose
	9. subducting	15. igneous
4. isostasy	10. synclines	16. is
5. dip-slip	11. margins	

Written questions

1. In time, extensive erosion often levels the hills on the surface, leaving only the upfolded rocks below.

2. An accretionary wedge is a chaotic accumulation of metamorphic rocks, sedimentary rocks derived from the land as well as sediment scraped from a subducting plate, and occasional pieces of ocean crust that is plastered against the landward side of a trench.

3. Most of Earth's major mountain systems have formed along convergent plate boundaries. Convergence can occur between one oceanic and one continental plate, between two oceanic plates, or between continental plates. In the first two, subduction zones often develop and the sediments and volcanics are squeezed, faulted, and deformed into mountain ranges as the plates converge. When continents collide, no subduction zone forms and the collision between the continental fragments deforms the continental material into mountains.

ANSWER KEY

CHAPTER TEN

Vocabulary Review

1. uniformitarianism
2. Cenozoic era
3. absolute date
4. correlation
5. half-life
6. catastrophism
7. relative dating

8. Precambrian
9. original horizontality
10. law of superposition
11. unconformity
12. fossil
13. radioactivity
14. period

15. inclusions
16. Paleozoic era
17. index fossil
18. principle of fossil succession
19. radiometric dating
20. eon

Comprehensive Review

1. Catastrophism states that Earth's landscape was developed by great catastrophes over a short span of time. On the other hand, uniformitarianism states that the physical, chemical, and biological laws that operate today have also operated in the geologic past. Uniformitarianism implies that the forces and processes that we observe today have been at work for a very long time.

2. a) Large coal swamps flourished in North America before the extinction of dinosaurs.
   b) Dinosaurs became extinct about 66 million years ago.

3. a) In an undeformed sequence of sedimentary rocks, each bed is older than the one above it and younger than the one below. b) Most layers of sediment are deposited in a horizontal position. c) Fossil organisms succeed each other in a definite and determinable order, and therefore any time period can be recognized by its fossil content.

4. a) A; b) C; c) B

5. 1) rapid burial; 2) possession of hard parts

6. a) a; b) e

7. a) B; b) A

8. An atom's atomic number is the number of protons in the nucleus, while the mass number is determined by adding together the number of protons and neutrons in the nucleus.

9. 1) emission of alpha particles (2 protons and 2 neutrons) from the nucleus; 2) emission of beta particles, or electrons (derived from neutrons), from the nucleus; 3) electrons are captured by the nucleus and combine with protons to form additional neutrons

10. Because the rates of decay for many isotopes have been precisely measured and do not vary under the physical conditions that exist in Earth's outer layers.

11. Carbon-14 is continuously produced in the upper atmosphere as a consequence of cosmic ray bombardment. Cosmic rays shatter the nuclei of gas atoms, releasing neutrons. These neutrons are absorbed by nitrogen atoms which, in turn, emit a proton and become the element carbon-14.

12. The primary problem is assigning absolute dates to units of time of the geologic time scale is the fact that not all rocks can be dated radiometrically.

13. a) younger (principle of cross-cutting relationships); b) younger (law of superposition); c) younger; d) older; e) older

14. 100,000 years (two half-lives, each 50,000 years)

Practice Test

Multiple choice

1. a	6. c	11. b	16. b	21. b
2. b	7. c	12. b	17. b	22. d
3. d	8. a	13. c	18. a	23. d
4. e	9. a	14. b	19. a	24. e
5. e	10. d	15. c	20. e	25. b

True/false

1. F (relative)	8. F (protons)	15. T
2. F (long)	9. T	16. T
3. T	10. F (younger)	17. T
4. T	11. F (increases)	18. F (uniformitarianism)
5. F (hard)	12. T	19. T
6. T	13. T	20. T
7. F (cannot)	14. T	

Word choice

1. index	8. conformable	15. radon
2. eons	9. wide; short	16. older
3. neutrons	10. do	17. three
4. James Hutton	11. relative	18. relative
5. organic	12. fine	19. folding and erosion
6. different from	13. daughter	20. Sedimentary
7. daughter product	14. older	

Written questions

1. Radiometric dating uses the radioactive decay of particular isotopes to arrive at an absolute date, which pinpoints the time in history when something took place; for example, the extinction of the dinosaurs about 66 million years ago. Relative dating involves placing rocks or events in their proper sequence using various laws and principles. Relative dating cannot tell us how long ago something took place, only that it followed one event and preceded another.

2. 1) law of superposition; 2) principle of original horizontality; 3) principle of cross-cutting relationships

3. Fossils help researchers understand past environmental conditions, are useful as time indicators, and play a key role in correlating rocks of similar ages that are from different places.

CHAPTER ELEVEN

Vocabulary Review

1. shields
2. nebular hypothesis
3. outgassing
4. stromatolites

Comprehensive Review

1. Diagram A: The nebula, a huge rotating cloud of dust and gases, begins to contract. Diagram B: Most of the material is gravitationally swept toward the center, producing the sun. Some dust and gases remain orbiting the central body as a flattened disk. Diagram C: The planets begin to accrete from the material that was orbiting within the flattened disk. Diagram D: Most of the debris forms the nine planets and their moons, or is swept into space by the solar wind.

2. The outer planets are composed primarily of gaseous substances.

3. Shortly after Earth formed, the decay of radioactive elements, coupled with the heat released by colliding particles, produced at least some melting of the interior. Melting allowed the denser elements, principally iron and nickel, to sink to Earth's center, while the lighter rocky components floated outward, toward the surface. This sorting of materials by density has produced an interior that consists of shells or spheres of materials having different properties.

4. The source of the gases that formed Earth's original atmosphere was volcanic outgassing. The gases were probably water vapor, carbon dioxide, nitrogen, and several trace gases; very similar to those gases released in volcanic emissions today.

5. From green plants releasing oxygen during photosynthesis.

6. a) Cenozoic; b) Precambrian; c) Mesozoic; d) Paleozoic; e) Paleozoic; f) Mesozoic; g) Paleozoic; h) Cenozoic; i) Paleozoic; j) Mesozoic

7. Most Precambrian rocks are devoid of fossils, metamorphosed and deformed, extensively eroded, and obscured by younger overlying strata which makes it difficult to divide the time interval into briefer units.

8. The beginning of the Paleozoic Era is marked by the appearance of the first life forms with hard parts—shells, scales, bones, or teeth.

9. The two hypotheses most often cited for the extinction of the dinosaurs are 1) an impact of a large asteroid or comet with Earth and 2) extensive volcanism. Both could produce a dust-laden atmosphere that would reduce the amount of sunlight that penetrated to Earth's surface, causing delicate food chains to collapse.

10. The stable continental margin of eastern North America was the site of abundant sedimentation. In the West, the Rocky Mountains were eroding and a great wedge of sediment was building the Great Plains, crustal movements created the Basin and Range Province, the Rockies were re-elevated, volcanic activity was common, and the Coast Ranges and Sierra Nevada formed.

11. The development and specialization of mammals took four principal directions; 1) increase in size, 2) increase in brain capacity, 3) specialization of teeth, and 4) specialization of limbs.

12. a) Cenozoic Era: (66 million years ago through today), the "age of mammals," the "age of flowering plants," humans evolve, large mammals become extinct; b) Mesozoic Era: (245 to 66 million years ago), "age of dinosaurs," breakup of Pangaea, widespread igneous activity in the West, gymnosperms dominant, first birds, mammals evolve, mass extinction at the close; c) Paleozoic Era: (570 to 245 million years ago), first life forms with hard parts, trilobites dominant and then become extinct, first fishes, first land plants, large coal swamps, amphibians, first reptiles, Pangaea forms, mass extinction at the close.

Practice Test

Multiple choice

1. d	8. e	15. d	22. d	29. b
2. e	9. a	16. b	23. d	30. b
3. b	10. c	17. d	24. a	31. a
4. e	11. d	18. a	25. c	32. a
5. b	12. b	19. a	26. a	
6. d	13. c	. d	27. b	
7. d	14. a	21. a	28. b	

True/false

1. T	8. F (reptiles)	15. F (Precambrian)
2. F (helium)	9. T	16. T
3. T	10. T	17. T
4. T	11. T	18. F (Cenozoic)
5. T	12. T	19. F (stromatolites)
6. F (drier)	13. F (Paleozoic)	20. T
7. F (epoch)	14. T	

Word choice

1. life forms	6. near	11. Cambrian
2. iron	7. Precambrian	12. drier
3. Paleozoic	8. Tertiary	13. increases
4. molten rock	9. before	14. above
5. angiosperms	10. Paleozoic	15. western

Written questions

1. The two hypotheses most often cited for the extinction of the dinosaurs are 1) an impact of a large asteroid or comet with Earth and 2) extensive volcanism. Both could produce a dust-laden atmosphere that would reduce the amount of sunlight that penetrated to Earth's surface, causing delicate food chains to collapse.

2. So little is known about the Precambrian because the early rock record has been obscured by Earth processes such as plate tectonics, erosion, and deposition.

3. Unlike amphibians, reptiles have shell-covered eggs that can be laid on land. The watery fluid within the reptilian egg closely resembles seawater in chemical composition. Therefore, reptiles eliminated the need for a water-dwelling stage (like the tadpole stage in frogs).

4. Mammals 1) have their young born live, 2) maintain a steady body temperature, that is, they are "warm blooded," 3) have insulating body hair, and 4) have a more efficient heart and lungs.

ANSWER KEY

CHAPTER TWELVE

Vocabulary Review

1. Oceanography
2. salinity
3. deep-ocean trench
4. continental shelf
5. turbidity current
6. abyssal plain
7. mid-ocean ridge
8. outgassing

9. terrigenous sediment
10. atoll
11. halocline
12. coral reef
13. seamount
14. hydrogenous sediment
15. thermocline
16. biogenous sediment

17. continental rise
18. guyot
19. rift zone
20. continental slope
21. echo sounder
22. desalination
23. turbidite
24. manganese nodule

Comprehensive Review

1. In the Northern Hemisphere nearly 61 percent of the surface is water. About 81 percent of the Southern Hemisphere is water. The Northern Hemisphere has about twice as much land (39 percent compared to 19 percent of the surface) as the Southern Hemisphere.

2. In the low and middle latitudes, both temperature and salinity are often highest at the surface, decrease rapidly below the surface zone, and then fall off slowly to the ocean floor.

3. chlorine and sodium

4. a) D; b) C; c) G; d) F; e) B

5. Abyssal plains are incredibly flat features found adjacent to continental slopes on ocean basin floors in all oceans. They consist of thick accumulations of sediment transported by turbidity currents far out to sea. (letter E)

6. 1) weathering of rocks on the continents; 2) from Earth's interior through volcanic eruptions, a process called outgassing

7. In the dry subtropics, where evaporation is high, the ocean's surface salinity is high because water is removed by evaporation, leaving the salts behind. Conversely, in the equatorial regions heavy precipitation dilutes ocean waters and lower salinities prevail.

8. Submarine canyons that are not seaward extensions of river valleys have probably been excavated by turbidity currents as these sand- and mud-laden currents repeatedly sweep downslope.

9. 1) warm waters with an average annual temperature about 24°C (75°F); 2) clear sunlit water; 3) a depth of water no greater than 45 meters (150 feet)

10. Coral islands, called atolls, form on the flanks of sinking volcanic islands from corals that continue to build the reef complex upward.

11. 1) Terrigenous sediment consists primarily of mineral grains that were weathered from continental rocks and transported to the ocean. 2) Biogenous sediment consists of shells and skele-

tons of marine animals and plants. 3) Hydrogenous sediment consists of minerals that crystallize directly from seawater through various chemical reactions.

12. a) C; b) F; c) B; d) A; e) E

Practice Test

Multiple choice

1. c	7. e	13. a	19. e	25. d
2. a	8. a	14. b	20. c	26. b
3. c	9. c	15. c	21. e	27. c
4. b	10. c	16. b	22. d	28. c
5. c	11. b	17. d	23. b	
6. a	12. c	18. a	24. b	

True/false

1. T	7. T	13. T
2. F (Atlantic)	8. F (lower)	14. T
3. T	9. F (mud)	15. T
4. T	10. T	16. F (slow)
5. F (slope)	11. T	
6. T	12. T	

Word choice

1. Evaporation	8. lower	15. more
2. continental	9. deep-sea fans	16. living organisms
3. mixed	10. Pacific	17. mud
4. Pacific	11. Biogenous	18. biogenous
5. larger	12. water	19. divergent
6. subsidence	13. rift	
7. greater	14. mantle	

Written questions

1. Salinity refers to the proportion of dissolved salts to pure water in the ocean, generally expressed in parts-per-thousand (‰). Salinity variations in the open ocean normally range from 33‰ to 37‰.

2. 1) continental margins; 2) ocean basin floor; 3) mid-ocean ridges

3. Mid-ocean ridges are spreading centers; that is, they are divergent plate boundaries where magma from Earth's interior rises and forms new oceanic crust. Ocean trenches form where lithospheric plates plunge into the mantle. Thus, these long, narrow, linear depressions are associated with convergent plate boundaries and the destruction of crustal material.

ANSWER KEY

CHAPTER THIRTEEN

Vocabulary Review

1. spit
2. Beach drift
3. gyre
4. wave height
5. thermohaline circulation
6. surf
7. Coriolis effect
8. wave of oscillation
9. wavelength
10. barrier island
11. tidal current
12. wave of translation
13. upwelling
14. tombolo
15. wave period
16. sea stack
17. emergent coast
18. estuary
19. tidal flat
20. wave refraction
21. Fetch
22. longshore current
23. wave-cut platform
24. tide
25. groin
26. spring tide
27. tidal delta

Comprehensive Review

1. The Coriolis effect, the deflective force of Earth's rotation on all free-moving objects, causes the movement of ocean waters to be deflected to the right, forming clockwise gyres, in the Northern Hemisphere and to the left, forming counterclockwise gyres, in the Southern Hemisphere.

2. a) C; b) F; c) D; d) H; e) G; f) A; g) B; h) E

3. In addition to influencing temperatures of adjacent land areas, cold currents also transform some tropical deserts into relatively cool, damp places that are often shrouded in fog.

4. 1) the shape of the coastline; 2) the configuration of the ocean basin

5. a) B; b) A

6. a) crest; b) trough; c) wavelength; d) wave height

7. 1) wind speed; 2) length of time the wind has blown; 3) the distance that the wind has traveled across the open water, called fetch

8. In a wave of oscillation, the wave form or shape moves forward, not the water itself. A wave of translation is the turbulent advance of water created by breaking waves.

9. 1) temperature; 2) salinity

10. 1) Arctic waters; 2) Antarctic waters

11. Wave refraction in a bay causes waves to diverge and expend less energy. As a consequence of wave refraction, sediment often accumulates and forms sheltered sandy beaches.

12. Sediment is transported along a coast by 1) beach drift, which transports sediment in a zigzag pattern along the beach and 2) by longshore currents, where turbulent water moves sediment in the surf zone parallel to the shore.

13. a) F; b) B; c) C; d) A; e) E; f) D

14. 1) as spits that were subsequently severed from the mainland by wave erosion; 2) as former sand dune ridges that originated along the shore during the last glacial period, when sea level was lower

15. 1) the topography and composition of the land; 2) prevailing winds and weather patterns; 3) the configuration of the coastline and nearshore areas (answers may vary)

16. a) Emergent coasts develop either because an area experiences uplift or as a result of a drop in sea level. Feature: elevated wave-cut platform (answer may vary); b) Submergent coasts are created when sea level rises or the land adjacent to the sea subsides. Feature: estuary (answer may vary)

17. a) Semidiurnal tides have two high and two low tides each tidal day, with a relatively small difference in the high and low water heights. b) Diurnal tides have a single high and low water height each tidal day. c) Mixed tides have two high and two low waters each day and are characterized by a large inequality in high water heights, low water heights, or both.

Practice Test

Multiple choice

1. b	7. b	13. e	19. c	25. b
2. b	8. b	14. c	20. a	26. a
3. d	9. d	15. d	21. a	27. c
4. a	10. b	16. a	22. a	
5. e	11. b	17. c	23. c	
6. b	12. c	18. d	24. c	

True/false

1. T	6. T	11. T
2. F (more)	7. T	12. T
3. F (Fetch)	8. T	13. F (right)
4. T	9. F (arch)	14. F (surf)
5. T	10. T	15. T

Word choice

1. circular	9. uplift	18. current
2. increase	10. polar	19. lose
3. waves	11. translation	20. Barrier
4. clockwise; counterclockwise	12. two	21. submergent
5. rotation	13. wind	22. more
6. emergent	14. gain	23. Fetch
7. rising	15. narrow	24. refraction
8. heat	16. eight	25. winds
	17. beach	

Written questions

1. Coastal upwelling occurs where winds are blowing equatorward and parallel to the coast. Due to the Coriolis effect, the surface water moves away from the shore area and is replaced by deeper, colder water with greater concentrations of dissolved nutrients. The nutrient-enriched waters from below promote the growth of plankton, which in turn supports extensive populations of fish.

2. Deep-ocean circulation is called thermohaline circulation. It occurs when water at the surface is made colder and/or more salty and its density increases. The cold, dense water sinks toward the ocean bottom and flows away from its source, which is generally near the poles.

3. Tides result from the gravitational force exerted on Earth by the moon, and to a lesser extent, the sun. The gravitational force causes water bulges on both the side of Earth toward the moon and the side away from it. These tidal bulges remain in place while Earth rotates "through" them, producing periods of high and low water at any one location along a coast.

4. In a wave of oscillation, each water particle moves in a nearly circular path during the passage of the wave. As a wave approaches the shore, the movement of water particles at the base of the wave are interfered with by the bottom. As the speed and length of the wave diminish, wave height increases. Eventually, water particles at the top of the wave pitch forward and the wave collapses, or breaks, and forms surf. What had been a wave of oscillation now becomes a wave of translation in which the water advances up the shore.

CHAPTER FOURTEEN

Vocabulary Review

1. troposphere
2. Weather
3. Rotation
4. thermosphere
5. Climate
6. isotherm
7. Tropic of Cancer
8. element (of weather and climate)
9. Convection
10. inclination of the axis
11. autumnal equinox
12. environmental lapse rate
13. Tropic of Capricorn
14. circle of illumination
15. mesosphere
16. radiation
17. revolution
18. Conduction
19. stratosphere
20. infrared
21. winter solstice
22. ultraviolet
23. albedo
24. greenhouse effect
25. summer solstice

Comprehensive Review

1. Weather is the state of the atmosphere at a particular place over a short period of time. Climate, on the other hand, is a generalization of the weather conditions of a place over a long period of time.

2. a) C; b) F; c) E; d) G; e) D; f) A

3. 1) rotation: the spinning of Earth about its axis; 2) revolution: the movement of Earth in its orbit around the sun

4. 1) air temperature; 2) humidity; 3) type and amount of cloudiness; 4) type and amount of precipitation; 5) air pressure; 6) the speed and direction of the wind

5. 1) Dust particles act as surfaces upon which water vapor may condense. 2) Dust may absorb or reflect incoming solar radiation.

6. 1) When the sun is high in the sky, the solar rays are most concentrated. 2) The angle of the sun determines the amount of atmosphere the rays must penetrate.

7. a) spring equinox, March 21-22, equator; b) summer solstice, June 21-22, $23\frac{1}{2}°$N (Tropic of Cancer); c) autumnal equinox, September 22-23, equator; d) winter solstice, December 21-22, $23\frac{1}{2}°$S (Tropic of Capricorn)

8. nitrogen (78 percent) and oxygen (21 percent)

9. Ozone (O_3) has three oxygen atoms per molecule, while oxygen (O_2) has two atoms per molecule. Ozone in the atmosphere absorbs the potentially harmful ultraviolet (UV) radiation from the sun.

10. a) 20; b) 50; c) 22 (see textbook Figure 14.16)

11. Most of the energy that is absorbed at Earth's surface is reradiated skyward in the form of infrared radiation. This terrestrial radiation is readily absorbed by water vapor and carbon dioxide in the atmosphere, thereby heating the atmosphere from the ground up.

12. carbon dioxide and water vapor

13. 1) Conduction is the transfer of heat through matter by molecular activity. 2) Convection is the transfer of heat by the movement of a mass or substance from one place to another. 3) Radiation is the transfer of energy (heat) through space by electromagnetic waves. It is the only mechanism of heat transfer that can transmit heat through the relative emptiness of space.

14. a) The daily mean temperature is determined by adding the maximum and minimum temperatures and then dividing by two. b) The annual mean temperature is an average of the twelve monthly means. c) The annual temperature range is computed by finding the difference between the highest and lowest monthly means.

15. a) Land heats more rapidly and to higher temperatures than water. Land also cools more rapidly and to lower temperatures than water. b) Due to the mechanism by which the atmosphere is heated, temperature generally decreases with an increase in altitude. c) A windward coastal location will often experience cool summers and mild winters when compared to an inland station at the same latitude. Leeward coastal sites will tend to have a more continental temperature pattern because the winds do not carry the ocean's influence onshore. d) Many clouds have a high albedo and reflect a significant portion of the radiation that strikes them back to space, therefore less radiation reaches the surface and daytime temperatures will be lower than if the clouds were not present.

16. January and July are selected most often for analysis because, for most locations, they represent the temperature extremes.

Practice Test

Multiple choice

1. a	7. c	13. e	19. c	25. d
2. d	8. a	14. b	20. d	26. e.
3. a	9. b	15. c	21. b	27. e
4. b	10. b	16. c	22. b	28. c
5. d	11. a	17. b	23. a	29. b
6. b	12. c	18. a	24. b	30. c

True/false

1. T	10. F (temperature)	19. F (fast)
2. F (Weather)	11. T	20. T
3. F (lower)	12. F (troposphere)	21. T
4. T	13. F (illumination)	22. T
5. T	14. F (winter)	23. F (smaller)
6. F (conduction)	15. F (gradual)	24. T
7. T	16. T	25. T
8. T	17. T	
9. T	18. F (decreased)	

Word choice

1. varies	10. decreases	19. is not
2. Cancer	11. 23.5	20. shorter
3. decreases	12. latent	21. least
4. insignificant	13. ultraviolet	22. radiation
5. more	14. equator	23. decrease
6. 3.5	15. climate	24. Earth
7. smaller	16. arid	25. convection
8. heat energy	17. temperature	26. greater
9. greater	18. land	

Written questions

1. CFCs, chlorofluorocarbons, are a group of chemicals that were once commonly used as propellants for aerosol sprays and in the production of certain plastics and refrigerants. When CFCs reach the stratosphere, chlorine breaks up some of the ozone molecules. The net effect is depletion of the ozone layer and a reduction in the layer's ability to absorb harmful ultraviolet (UV) radiation.

2. The amount of solar energy reaching places on Earth's surface varies with the seasons because Earth's orientation to the sun continually changes as it travels along its orbit. This changing orientation is due to the facts that 1) Earth's axis is inclined and 2) the axis remains pointed in the same direction (toward the North Star) as Earth journeys around the sun.

3. The atmosphere allows most of the short-wave solar energy to pass through and be absorbed at Earth's surface. Earth reradiates this energy in the form of longer wavelength terrestrial radiation, which is absorbed by carbon dioxide and water vapor in the lower atmosphere. This mechanism, referred to as the greenhouse effect, explains the general drop in temperature with increasing altitude experienced in the troposphere.

ANSWER KEY

CHAPTER FIFTEEN

Vocabulary Review

1. humidity
2. condensation
3. Relative humidity
4. Orographic lifting
5. vapor pressure
6. adiabatic temperature change
7. Latent heat
8. cloud
9. sleet
10. saturation
11. wet adiabatic rate

12. Sublimation
13. stratus
14. specific humidity
15. Frontal wedging
16. condensation nuclei
17. melting
18. rain
19. dew point
20. Rime
21. deposition
22. hail
23. dry adiabatic rate

24. hygroscopic nuclei
25. evaporation
26. Radiation fog
27. fog
28. rainshadow desert
29. psychrometer
30. convergence
31. calorie
32. supercooled
33. hygrometer

Comprehensive Review

1. a) D; b) C; c) F; d) B; e) A; f) E

2. 1) adding moisture to the air; 2) lowering the air temperature

3. a) 25 percent (5 grams per kilogram/20 grams per kilogram); b) 60 percent (3 grams per kilogram/5 grams per kilogram)

4. a) 15°C (59°F); b) 14°F (-10°C)

5. The psychrometer consists of two thermometers mounted side by side: the dry-bulb, which gives the present temperature, and the wet-bulb, which has a moist, thin muslin wick tied around the end. The lower the air's relative humidity, the more evaporation (and hence cooling) there will be from the wet-bulb thermometer. Thus, the greater the difference between the wet and dry bulb temperatures, the lower the relative humidity.

6. 42 percent

7. Once the air is cooled sufficiently to reach the dew point and condensation occurs, latent heat of condensation stored in the water vapor will be liberated, thereby reducing the rate at which the air cools.

8. Stable air resists vertical movement because the environmental lapse rate is less than the wet adiabatic rate. Unstable air wants to rise because it is less dense than the surrounding air. Unstable conditions prevail when the environmental lapse rate is greater than the dry adiabatic rate.

9. a) Orographic lifting occurs when elevated terrains, such as mountain barriers, act as barriers to flowing air. b) Frontal wedging occurs when cool air acts as a barrier over which warmer, less dense air rises. c) Convergence occurs whenever air masses flow together. Because the air must go somewhere, it moves upward.

10. 1) the form of the cloud; 2) the height of the cloud

11. a) stratus; b) cirrus; c) altocumulus

12. a) B; b) A

13. a) Steam fog, a type of evaporation fog, occurs when cool air moves over warm water and moisture evaporated from the water surface condenses. b) Frontal fog occurs when warm air is lifted over colder air by frontal wedging. If the resulting clouds yield rain, and the cold air below is near the dew point, enough rain will evaporate to produce fog.

14. 1) The Bergeron process occurs when ice crystals in a cloud collect available water vapor, eventually growing large enough to fall as snowflakes. When the surface temperature is about $4°C$ (39°F) or higher, snowflakes usually melt before they reach the ground and continue their descent as rain. 2) The collision-coalescence process occurs when large droplets form on large hygroscopic condensation nuclei and, because the bigger droplets fall faster in a cloud, they collide and join with smaller water droplets. After many collisions the droplets are large enough to fall to the ground as rain.

15. a) Sleet, a wintertime phenomenon, forms when raindrops freeze as they fall through colder air that lies below warmer air. b) Hail is produced in a large cumulonimbus cloud with violent updrafts and an abundance of supercooled water. As they fall through the cloud, ice pellets grow by collecting supercooled water droplets. The process continues until the hailstone encounters a downdraft or grows too heavy to remain suspended by the thunderstorm's updraft.

Practice Test

Multiple choice

1. d	7. b	13. b	19. b	25. a
2. b	8. e	14. a	20. d	26. b
3. b	9. c	15. c	21. a	27. d
4. a	10. b	16. b	22. d	28. c
5. b	11. a	17. d	23. a	29. c
6. c	12. c	18. e	24. a	30. a

True/false

1. T	7. T	13. F (decrease)
2. F (calorie)	8. T	14. T
3. F (height)	9. F (wet-bulb)	15. F (vaporization)
4. T	10. T	16. T
5. F (cool)	11. T	17. F (fog)
6. T	12. F (wet)	18. T

Word choice

1. releases
2. collision-coalescence
3. altitude
4. relative
5. less
6. Sleet
7. More
8. unstable
9. remains the same
10. compresses; warms
11. absolute instability
12. absorbs
13. Stratus
14. ice crystals
15. decreasing; increasing
16. lower; latent; condensation
17. Bergeron
18. lower
19. convergence; upward
20. increases

Written questions

1. Clouds are classified on the basis of their form (cirrus, cumulus, and stratus) and height (high, middle, low, and clouds of vertical development).

2. As air rises, it expands because air pressure decreases with an increase in altitude. Expanding air cools because it pushes on (does work on) the surrounding air and expends energy.

3. Stable air resists vertical movement because the environmental lapse rate is less than the wet adiabatic rate. Unstable air wants to rise because it is less dense than the surrounding air. Unstable conditions prevail when the environmental lapse rate is greater than the dry adiabatic rate.

4. (answers will vary) Depending on the place and season, the process will be either orographic lifting, frontal wedging, or convergence. In either case, when air rises it will expand and cool adiabatically. If sufficient cooling takes place, the dew point may be reached and condensation may result in the formation of clouds and perhaps precipitation.

CHAPTER SIXTEEN

Vocabulary Review

1. polar front	10. cyclone	19. anticyclone
2. isobar	11. Convergence	20. subpolar low
3. monsoon	12. land breeze	21. polar easterlies
4. Coriolis effect	13. equatorial low	22. barograph
5. wind	14. chinook	23. Divergence
6. Barometric tendency	15. sea breeze	24. subtropical high
7. aneroid barometer	16. cup anemometer	25. westerlies
8. prevailing wind	17. pressure gradient	26. trade winds
9. wind vane	18. jet stream	

Comprehensive Review

1. a) The mercury barometer is an instrument used to measure air pressure. It consists of a glass tube, closed at one end, filled with mercury. The tube is inverted in a reservoir (pan) of mercury. If the air pressure decreases, the height of the column of mercury in the tube falls; if pressure increases, the column rises. b) The wind vane is used to determine wind direction by pointing into the wind. c) The aneroid barometer measures air pressure using partially evacuated metal chambers that are compressed as air pressure increases and expand as pressure decreases. d) The cup anemometer measures wind speed using several cups mounted on a shaft that is rotated by the moving wind. e) A barograph is an aneroid barometer that continuously records pressure changes.

2. 1) pressure gradient force: the amount of pressure change over a given distance; 2) Coriolis effect is the deflective force of Earth's rotation on all free-moving objects. Deflection is to the right in the Northern Hemisphere. 3) friction with Earth's surface: slows down air movement within the first few kilometers of Earth's surface and, as a consequence, alters wind direction

3. (The diagrams in Figure 16.1 should show the following pressures and surface air movements.)
   Northern Hemisphere high: highest pressure in center, air moves out (divergence), clockwise
   Southern Hemisphere high: highest pressure in center, air moves out (divergence), counterclockwise
   Northern Hemisphere low: lowest pressure in center, air moves inward (convergence), counterclockwise
   Southern Hemisphere low: lowest pressure in center, air moves inward (convergence), clockwise

4. a) inward (convergent) and counterclockwise; b) outward (divergent) and counterclockwise

5. a) B; b) A; c) B; d) A; e) B

6. In a high pressure system (anticyclone), outflow near the surface is accompanied by convergence aloft and general subsidence of the air column. Because descending air is compressed and warmed, cloud formation and precipitation are unlikely and "fair" weather can usually be expected.

7. A: subpolar low; B: subtropical high; C: equatorial low; D: subtropical high; E: subpolar low; F: polar easterlies; G: westerlies; H: trade winds; I: trade winds; J: westerlies; K: polar easterlies

8. a) subpolar low; b) subtropical high; c) equatorial low; d) polar high

9. a) warm, dry winds often found on the leeward sides of mountains, especially the eastern slopes of the Rockies; b) cooler air over the water (higher pressure) moves toward the warmer land (lower pressure)

10. a) northeast; b) west; c) south

Practice Test

Multiple choice

1. d	6. c	11. c	16. e	21. b
2. a	7. a	12. a	17. d	22. b
3. a	8. d	13. e	18. b	23. a
4. c	9. d	14. d	19. c	24. b
5. b	10. c	15. e	20. d	25. e

True/false

1. T	7. T	13. F (subtropical)
2. F (faster)	8. F (cyclone)	14. T
3. T	9. F (dry)	15. F (convergence)
4. F (right)	10. T	16. F (colder)
5. T	11. T	17. T
6. T	12. F (greater)	

Word choice

1. higher; lower	7. west-to-east	13. higher
2. rising	8. left	14. lower
3. lower	9. west-to-east	15. lower
4. weakest	10. low	16. less
5. maintains	11. does not	17. compression
6. spacing	12. westerlies	

Written questions

1. The Coriolis effect, the deflective force of Earth's rotation on all free-moving objects, causes air to be deflected to the right of its path of motion in the Northern Hemisphere and to the left in the Southern Hemisphere.

2. A drop in barometric pressure indicates the approach of a low pressure system, called a cyclone. The cyclone is associated with converging surface winds and ascending air; hence, adiabatic cooling, cloud formation, and precipitation often occur. Conversely, an anticyclone, with its high air pressure, is associated with subsidence of the air column and divergence at the surface. Because descending air is compressed and warmed, cloud formation and precipitation are unlikely in an anticyclone and "fair" weather can usually be expected.

3. In a Northern Hemisphere cyclone (low pressure), surface winds are inward (convergence) and counterclockwise. As a consequence of surface convergence, there is a slow, net upward movement of air near the center of the cyclone. Furthermore, the surface convergence found in a cyclone is balanced by divergence of the air aloft.

ANSWER KEY

CHAPTER SEVENTEEN

Vocabulary Review

1. air mass
2. hurricane
3. front
4. tropical storm
5. source region
6. continental (c) air mass
7. warm front
8. middle-latitude cyclone
9. eye wall
10. tropical (T) air mass
11. tropical depression
12. eye
13. air-mass weather
14. cold front
15. tornado
16. maritime (m) air mass
17. tornado watch
18. Doppler radar
19. thunderstorm
20. storm surge
21. tornado warning
22. polar (P) air mass
23. occluded front
24. lake-effect snow
25. stationary front

Comprehensive Review

1. Air masses are classified according to their source region using two criteria; 1) the nature of the surface (land or water) where the air mass originates is identified using a lower case letter such as c (continental) or m (maritime), and 2) the latitude where the air mass originates is identified using an upper case letter such as P (polar) or T (tropical). Therefore, a cP air mass would have a continental polar source region.

2. A: maritime polar (mP)—cool, moist; B: continental polar (cP)—cold and dry in winter; F: continental tropical (cT)—hot and dry in summer; G: maritime tropical (mT)—warm, moist

3. continental polar (cP) from central Canada and maritime tropical (mT) from the Gulf of Mexico

4. (See textbook Figures 17.3 and 17.5) 1) A warm front occurs where the surface position of a front moves so that warm air occupies territory formerly covered by cooler air. The boundary separating the cooler air from the warmer air above has a small slope (about 1:200). 2) A cold front occurs when cold air actively advances into a region occupied by warmer air. On the average, the boundary separating the cold and warm air is about twice as steep as the slope along a warm front.

5. diagram: A; Because the cold front has caught up to the warm front, forming an occluded front.

6. a) warm front; b) cold front; c) maritime tropical (mT); d) continental polar (cP); e) southwest to south; f) northwest to north; g) thunderstorms and precipitation with the passage of the cold front, followed by cooler, drier conditions as the cP air moves over the area, change in wind direction from southwest to northwest after the cold front passes

7. a) surface: inward (convergent), counterclockwise (Northern Hemisphere), rising near center; aloft: divergence; b) surface: outward (divergent), clockwise (Northern Hemisphere), subsiding near center; aloft: convergence

8. 1) rate of movement; 2) steepness of slope

9. a) All thunderstorms require warm, moist air, which, when lifted, will release sufficient latent heat to provide the buoyancy necessary to maintain its upward flight. b) Tornadoes form in association with severe thunderstorms spawned along the cold front of a middle-latitude cyclone.

These violent windstorms that take the form of a rapidly rotating column of air that extends downward from a cumulonimbus cloud are products of the interaction between strong updrafts in the thunderstorm and winds in the troposphere.

10. 1) when they move over waters that cannot supply warm, moist, tropical air; 2) when they move onto land; 3) when they reach a location where the large-scale flow aloft is unfavorable

11. 1) wind damage; 2) storm surge damage; 3) inland freshwater flooding

Practice Test

Multiple choice

1. b	7. d	13. c	19. d	25. d
2. b	8. b	14. b	20. b	26. b
3. b	9. d	15. a	21. c	27. e
4. a	10. d	16. b	22. c	28. a
5. b	11. b	17. b	23. e	
6. a	12. e	18. a	24. d	

True/false

1. T	9. T	17. T
2. F (one)	10. T	18. T
3. T	11. T	19. T
4. T	12. F (cold)	20. T
5. F (more)	13. F (twice)	21. T
6. F (eastward)	14. T	22. T
7. F (Maritime)	15. T	
8. T	16. F (central)	

Word choice

1. cyclone	9. low
2. warm; cooler	10. occluded
3. Andrew	11. counterclockwise
4. triangles	12. descends; heats; compression
5. Pacific	13. leeward
6. dry	14. spring; cold
7. cumulonimbus	15. parallel
8. clear	

Written questions

1. A continental polar (cP) air mass that originates in central Canada will most likely be cold and dry in the winter and cool and dry in the summer. A maritime tropical (mT) air mass with its source region in the Gulf of Mexico will be warm and moist.

2. As the warm front approaches, winds would blow from the east or southeast and air pressure would drop steadily. Cirrus clouds would be sighted first, followed by progressively lower clouds, and perhaps nimbostratus clouds. Gentle precipitation could occur as the nimbostratus clouds moved overhead. As the warm front passed, temperatures would rise, precipitation would cease, and winds would shift to the south or southwest. Further, the sky clears and the pressure tendency steadies.

Later, with the approach of the cold front, cumulonimbus clouds fill much of the sky and bring the likelihood of heavy precipitation and a possibility of hail and tornado activity. The passage of the cold front is accompanied by a drop in temperature, clearing sky, a wind shift to the northwest or north, and rising air pressure. Fair weather can probably be expected for the next few days.

3. A tornado watch alerts the public to the fact that conditions are right for the formation of tornadoes; whereas a tornado warning is issued when a tornado has actually been sighted in an area or is indicated by radar.

ANSWER KEY

CHAPTER EIGHTEEN

Vocabulary Review

1. climate system
2. humid subtropical climate
3. arid (or desert) climate
4. Köppen classification
5. ice cap climate
6. subarctic climate
7. marine west coast climate
8. tropical rain forest
9. rainshadow desert
10. humid continental climate
11. tropical wet and dry climate
12. tundra climate
13. semiarid (or steppe) climate
14. dry-summer subtropical climate
15. highland climate

Comprehensive Review

1. The five parts of Earth's climate system are the atmosphere, hydrosphere, solid Earth, biosphere, and cryosphere.

2. The two most important elements in a climatic description are temperature and moisture. They are important because they have the greatest influence on people and their activities and also have an important impact on the distribution of such phenomena as vegetation and soils.

3. 1) Humid tropical (A): winterless climates; all months above 18°C. 2) Dry (B) climates: climates where evaporation exceeds precipitation; there is a constant water deficiency. 3) Humid middle-latitude climates with mild winters (C climates): mild winters; average temperature of the coldest month is below 18°C but above -3°C. 4) Humid middle-latitude climates with severe winters (D climates): severe winters; the average temperature of the coldest month is below -3°C and the warmest monthly mean exceeds 10°C. 5) Polar (E) climates: summerless climates; the average of the warmest month is below 10°C.

4. a) July, 26.4°C; b) January, -0.1°C; c) 26.5°C (-0.1°C to 26.4°C); d) June; e) late spring to early summer; f) Cfa, humid subtropical (average temperature of the coldest month is under 18°C and above -3°C, criteria for w and s cannot be met, and warmest month is over 22°C with at least 4 months over 10°C)

5. The three temperature/precipitation features that characterize the wet tropics are 1) temperatures usually average 25°C or more each month, 2) the total precipitation for the year is high, often exceeding 200 centimeters, and 3) all months are usually wet. The most continuous expanses of the wet tropics lie near the equator in South America, Africa, and Indonesia.

6. The difference is primarily a matter of degree. The semiarid is a marginal more humid variant of the arid.

7. Low-latitude deserts and steppes coincide with the subtropical high-pressure belts. Here the air is subsiding, compressed, and warmed, causing clear skies, a maximum of sunshine, and drought conditions.

8. The climate is found in the south, southeastern, and eastern United States.

9. D climates are land-controlled climates, the result of broad continents in the middle latitudes. D climates are absent in the Southern Hemisphere where the middle-latitude zone is dominated by the oceans.

10. a) dry climates; b) polar climates; c) wet tropical climates; d) marine west coast climates; e) dry climates (low latitude); f) subarctic climate; g) wet tropical climates

11. In addition to having the lowest elevation in the Western Hemisphere, Death Valley is a desert. Clear skies allow a maximum of sunshine to reach the dry, barren surface. Because no energy is used to evaporate moisture, all of the energy is available to heat the ground. Also subsiding and warming air is common to the area.

12. (see textbook Table 18.C) a) 1 to 2°C higher; b) 5 to 15 percent more; c) 100 percent more

13. Water vapor and carbon dioxide are largely responsible for the greenhouse effect of the atmosphere due to their ability to absorb the longer-wavelength outgoing terrestrial radiation and alter temperatures in the lower atmosphere.

14. Humans have contributed to the buildup of atmospheric carbon dioxide by 1) the combustion of fossil fuels, and 2) by the clearing of forests.

15. Three potential weather changes that could occur are 1) warmer temperatures, 2) a higher frequency and greater intensity of hurricanes, and 3) shifts in the paths of large-scale cyclonic storms, which, in turn, would affect the distribution of precipitation.

Practice Test

Multiple choice

1. b	6. d	11. b	16. a	21. c
2. d	7. b	12. c	17. a	22. c
3. c	8. a	13. c	18. e	23. a
4. b	9. b	14. c	19. d	24. e
5. b	10. b	15. d	20. d	25. a

True/false

1. F (five)	6. T	11. T
2. F (windward)	7. F (evaporation)	12. T
3. F (low)	8. T	13. T
4. F (precipitation)	9. T	14. F (industrialization)
5. T	10. F (below)	15. T

Word choice

1. five	6. subtropical high	11. windward
2. B	7. coniferous	12. 134
3. rise	8. 200	13. precipitation
4. temperature; precipitation	9. subsiding	14. water vapor
5. Europe	10. 0.5	15. will not

Written questions

1. In North America, climates east of the 100°W meridian vary north-to-south. A Cfa (humid subtropical warm, summer) climate located along the Gulf Coast and southeastern United States gives way to D (humid middle-latitude) climates in the central and northern portions of the continent. Polar (tundra) climate is located in the extreme northern reaches. West of the 100°W meridian are middle latitude steppes and deserts, along with highland climates in the mountains. The west coast has an area of dry-summer subtropical (Mediterranean) climate from about latitude 35 to 40°N, with a marine west coast climate extending northward along the coast into Alaska.

2. Wet tropical climates are restricted to elevations below 1000 meters because of the general decrease in temperature with altitude. At higher altitudes the temperature criterion for an A climate is not met.

3. When compared to nearby rural areas, cities have more precipitation, higher temperatures, lower wind speeds, less solar radiation, lower relative humidities, and more cloudiness. (see textbook Table 18.C)

ANSWER KEY

CHAPTER NINETEEN

Vocabulary Review

1. retrograde motion
2. sidereal day
3. heliocentric
4. sidereal month
5. geocentric
6. Rotation
7. astronomical unit (AU)
8. solar eclipse
9. declination
10. perturbation
11. perihelion
12. ecliptic
13. celestial sphere
14. constellation
15. plane of the ecliptic
16. Ptolemaic system
17. Revolution
18. phases of the moon
19. right ascension
20. synodic month
21. lunar eclipse
22. equatorial system
23. aphelion
24. precession
25. mean solar day

Comprehensive Review

1. The geocentric model held by the early Greeks proposed that Earth was a sphere that stayed motionless at the center of the universe. Orbiting Earth were the moon, sun, and the known planets, Mercury through Jupiter. Beyond the planets was a transparent, hollow sphere, called the celestial sphere, on which the stars traveled daily around Earth.

2. Retrograde motion is the apparent periodic westward drift of a planet in the sky that results from a combination of the motion of Earth and the planet's own motion around the sun. In Figure 19.1, retrograde motion is occurring between points 3 and 4. To explain retrograde motion, the geocentric model of Ptolemy had the planets moving in small circles, called epicycles, as they revolved around Earth along large circles, called deferents.

3. Early Greeks rejected the idea of a rotating Earth because Earth exhibited no sense of motion and seemed too large to be movable.

4. a) Galileo Galilei; b) Nicolaus Copernicus; c) Tycho Brahe; d) Johannes Kepler; e) Nicolaus Copernicus; f) Johannes Kepler; g) Galileo Galilei; h) Galileo Galilei; i) Sir Isaac Newton; j) Aristarchus; k) Hipparchus; l) Eratosthenes

5. Stellar parallax is the apparent shift in the position of a nearby star, when observed from extreme points in Earth's orbit six months apart, with respect to the more distant stars.

6. 1) The path of each planet around the sun is an ellipse, with the sun at one focus. 2) Each planet revolves so that an imaginary line connecting it to the sun sweeps over equal areas in equal intervals of time. 3) The orbital periods of the planets and their distances to the sun are proportional.

7. 1) four moons orbiting Jupiter; 2) the planets are circular disks rather than just points of light; 3) Venus has phases just like the moon; 4) the moon's surface is not a smooth glass sphere

8. a) November through February; b) June to July; c) gravity keeps the planet from going in a straight line out to space; d) Using Kepler's third law, a planet's orbital period squared is equal to its mean solar distance cubed ($p^2 = d^3$), the planet's orbital period (p) would be 8 Earth years (the square root of 4 cubed, or 64).

9. The equatorial system divides the celestial sphere into a coordinate system very similar to the latitude-longitude system used for locations on Earth's surface. In the sky, declination, like latitude, is the angular distance north or south of celestial equator and right ascension is the angular distance measured eastward along the celestial equator from the position of the vernal equinox.

10. a) C; b) A; c) F; d) D

11. 1. rotation; 2. revolution; 3. precession; 4. movement with the entire solar system in the direction of the star Vega at 20 kilometers per second; 5. revolution around the galaxy, a trip that requires 230 million years to traverse; 6. movement as a part of the galaxy through the universe toward the Great Galaxy of Andromeda

12. The synodic cycle is the movement of the moon through its phases, which takes 29½ days. The sidereal cycle, which takes only 27⅓ days, is the true period of the moon's revolution around Earth. The reason for the difference of nearly two between the cycles is that as the moon orbits Earth, the Earth-moon system also moves in an orbit about the sun (see textbook Figure 19.24).

13. a) 5; b) 7; c) 2; d) 1; e) 6

Practice Test

Multiple choice

1. b	7. b	13. d	19. a	25. d
2. c	8. b	14. c	20. c	26. b
3. c	9. a	15. d	21. e	27. c
4. a	10. e	16. b	22. d	28. a
5. c	11. b	17. e	23. d	29. a
6. e	12. c	18. d	24. b	30. a

True/false

1. F (elliptical)	8. T	15. T
2. F (seven)	9. T	16. F (geocentric)
3. F (Hipparchus)	10. F (perturbation)	17. F (same)
4. F (smaller)	11. T	18. T
5. F (synodic)	12. T	19. F (degrees)
6. T	13. T	20. F (Johannes Kepler)
7. F (longer)	14. F (day)	

Word choice

1. sidereal	8. geocentric	15. equal to
2. eastward	9. reflected sunlight	16. are not
3. solar	10. inertia	17. axis
4. astronomical	11. lunar	18. new-moon
5. constant; varies	12. perihelion	19. Galileo
6. longer	13. circular; Earth	20. pendulum
7. Gregorian	14. Aristarchus	

Written questions

1. The apparent westward drift of planets, called retrograde motion, results from the combination of the motion of Earth and the planet's own motion around the sun. Periodically Earth overtakes a planet in its orbit, which makes it appear that for a period of time the planet moves backward in its orbit. (see textbook Figure 19.5)

2. For an eclipse to occur, the moon must be passing through the plane of the ecliptic during the new- or full-moon phase. Because the moon's orbit is inclined about 5 degrees to the plane that contains Earth and the sun, the conditions are normally met only twice a year, and the usual number of eclipses is four.

3. Four of the six motions Earth continuously experiences are 1) rotation, 2) revolution, 3) precession, and 4) movement with the entire solar system in the direction of the star Vega. (answers may vary)

4. To explain the observable motions of the planets, the geocentric model of Ptolemy had the planets revolving around Earth along large circles called deferents, while simultaneously orbiting on small circles called epicycles.

CHAPTER TWENTY

Vocabulary Review

1. coma
2. terrestrial planet
3. asteroid
4. maria
5. meteoroid

6. Jovian planet
7. Lunar regolith
8. meteor
9. escape velocity
10. stony meteorite

11. comet
12. meteorite
13. meteor shower
14. iron meteorite
15. stony-iron meteorite

Comprehensive Review

1. The solar system consists of those celestial objects that are under the direct gravitational influence of the sun. These objects include the sun, the nine planets and their satellites, asteroids, comets, and meteoroids.

2. a) hydrogen and helium; b) principally silicate minerals and metallic iron; c) ammonia, methane, carbon dioxide, and water

3. a) The terrestrial planets (Mercury, Venus, Earth, and Mars) are all similar to Earth in their physical characteristics. b) The Jovian planets (Jupiter, Saturn, Uranus, and Neptune) are all Jupiterlike in their physical characteristics.

4. a) The Jovian planets are all large compared to the terrestrial planets. b) Because the Jovian planets contain a large percentage of gases, their densities are less than those of the terrestrial planets. c) The Jovian planets have shorter periods of rotation. d) The Jovian planets are all more massive than the terrestrial planets. e) The Jovian planets have more moons. f) The periods of revolution of the Jovian planets are longer than those of the terrestrial planets. g) The terrestrial planets are mostly rocky and metallic material, with minor amounts of gases. The Jovian planets, on the other hand, contain a large percentage of gases (hydrogen and helium), with varying amounts of ices (mostly water, ammonia, and methane).

5. Chemical differentiation is the process whereby dense elements sink to the center of a molten object, whereas the lighter components rise toward the surface. As Earth was forming, and in a molten state, the denser metallic elements (iron and nickel) sank to the center to become the core and the lighter substances (silicate minerals, oxygen, hydrogen) migrated toward the surface.

6. a) A; b) B; c) D; d) C; e) C

7. Because the gravitational attraction at the lunar surface is only one-sixth of that experienced on Earth's surface.

8. Lunar maria basins are enormous impact craters that were flooded with layer-upon-layer of very fluid basaltic lava.

9. 1) The moon and Earth consolidated at the same place in the nebula, but independent of each other, from minute rock fragments and gases that orbited the protosun and accreted into planetary-sized bodies. 2) A giant asteroid collided with Earth to produce the moon.

10. a) 3; b) 1; c) 5; d) 4; e) 2

11. a) Mars; b) Venus; c) Jupiter; d) Saturn; e) Mercury; f) Saturn; g) Venus; h) Uranus; i) Jupiter;
    j) Neptune; k) Saturn; l) Mars; m) Jupiter; n) Venus; o) Neptune; p) Mercury; q) Pluto

12. The most probable origin of the Martian moons is that they are asteroids that have been captured
    by the gravity of Mars.

13. a) Jupiter; b) Mars; c) Jupiter; d) Neptune; e) Uranus; f) Pluto; g) Saturn

14. A recent proposal suggests that Pluto was a satellite of Neptune that was displaced from its orbit
    by a collision. Charon is considered evidence that the event broke the Neptunian satellite into two
    pieces and sent them into an elongated orbit around the sun.

15. 1) they might have formed from the breakup of a planet; 2) perhaps several large bodies once
    coexisted in close proximity and their collisions produced numerous smaller objects

16. a) B; b) A; c) C; d) D

17. 1) radiation pressure: pushes dust particles away from the coma; 2) solar wind: responsible for
    moving the ionized gases

18. A meteor is a brilliant streak of light produced when a meteoroid enters Earth's atmosphere and
    burns up. A meteorite is any portion of a meteoroid that survives its traverse through Earth's atmo-
    sphere and strikes the surface.

19. The three most common types of meteorites classified by their composition are 1) irons—mostly
    iron with 5-10 percent nickel; 2) stony—silicate minerals with inclusions of other minerals; and
    3) stony-irons—mixtures.

Practice Test

Multiple choice

1. b	9. a	17. e	25. e	33. a
2. e	10. b	18. d	26. e	34. a
3. d	11. b	19. b	27. b	35. c
4. a	12. c	20. c	28. b	36. d
5. d	13. d	21. d	29. b	37. b
6. e	14. c	22. d	30. c	
7. c	15. b	23. c	31. e	
8. a	16. d	24. e	32. d	

True/false

1. T	8. T	15. T
2. F (Saturn)	9. F (lava)	16. T
3. T	10. F (one)	17. T
4. T	11. F (highlands)	18. T
5. F (Mercury)	12. F (lose)	19. T
6. F (Jupiter)	13. F (sun)	20. T
7. F (low)	14. F (Jupiter)	

Word choice

1. away from	8. same	15. old
2. less; sixth	9. meteor	16. Venus
3. rift	10. polar caps	17. beyond
4. Mercury	11. micrometeorite impacts	18. cloud cover
5. absorb	12. 1	19. closer to
6. does not	13. hundred	20. slowly; quickly
7. parallel	14. older	

Written questions

1. Compared to the Jovian planets, the terrestrial planets are small, less massive, more dense, have longer rotational periods and shorter periods of revolution. The terrestrial planets are also warmer and consist of a greater percentage of rocky material and less gases and ices.

2. Meteoroid is the name given to any solid particle that travels through interplanetary space. When a meteoroid enters Earth's atmosphere it vaporizes with a brilliant flash of light called a meteor. On occasions when meteoroids reach Earth's surface, the remains are termed meteorites.

3. The two different types of lunar terrain are 1) maria—the dark, smooth, lowland areas resembling seas on Earth that formed when asteroids punctured the surface, letting basaltic lava "bleed" out; and 2) highlands—the densely cratered areas that make up most of the lunar surface and contain remnants of the oldest lunar crust.

ANSWER KEY

CHAPTER TWENTY ONE

Vocabulary Review

1. electromagnetic radiation
2. corona
3. refracting telescope
4. spectroscope
5. photosphere
6. nuclear fusion
7. continuous spectrum
8. reflecting telescope
9. photon
10. radio telescope
11. Doppler effect
12. solar wind
13. objective lens
14. proton-proton chain
15. solar flare
16. bright-line (emission) spectrum
17. chromosphere
18. prominence
19. Spectroscopy
20. sunspot
21. focus
22. dark-line (absorption) spectrum
23. Chromatic aberration
24. radiation pressure
25. focal length
26. granules
27. plage
28. eyepiece
29. radio interferometer
30. spicule
31. aurora

Comprehensive Review

1. The properties of light are explained by two models; 1) by acting as a stream of particles, called photons, and 2) by behaving like waves.

2. Three other types of electromagnetic radiation would be gamma rays, ultraviolet light, and radio waves (answers may vary). Each is similar in that their properties can be explained using the particle and wave models. They are different in that their wavelengths are different; hence, they have different energies. The shortest wavelength, gamma rays, have the greatest energy.

3. As white light passes through a prism, the color with the shortest wavelength, violet, is bent more than blue, which is bent more than green, and so forth. Consequently, as the light emerges from the prism it will be separated into its component parts, the "colors of the rainbow." (see textbook Figure 21.1)

4. 1) Continuous spectra, uninterrupted bands of color, are produced by an incandescent solid, liquid, or gas under high pressure. 2) Dark-line (absorption) spectra, which appear as continuous spectra with dark lines running through them, are produced when white light is passed through a comparatively cool gas under low pressure. The spectra of most stars of this type. 3) Bright-line (emission) spectra are produced by a hot (incandescent) gas under low pressure. (see textbook Figure 21.3)

5. Each element or compound that is in gaseous form in a star produces a unique set of spectral lines which act as "fingerprints."

6. The two aspects of stellar motion that can be determined from a star's Doppler shift are 1) the direction, toward or away, and 2) the rate, or velocity, of the relative movement.

7. The two basic types of optical telescopes are 1) refracting telescopes, which use a lens as their objective to bend or refract light to an area called the focus, and 2) reflecting telescopes, which use a mirror to gather and focus the light.

8. a) B; b) C; c) D; d) A; e) E

9. 1) Light-gathering power is the ability of a telescope to collect light from a distant object. The larger the lens (or mirror) of a telescope, the greater its light-gathering ability. 2) Resolving power refers to the ability of a telescope to separate close objects. The greater the resolving power, the sharper the image and the finer the detail. 3) Magnifying power is the ability to make images larger. It is changed by changing the eyepiece. Magnification is limited by atmospheric conditions and the resolving power of the telescope.

10. Reflecting telescopes do not have chromatic aberration (color distortion), the glass does not have to be of optical quality, and large mirrors can be made because they can be supported from behind.

11. Magnification is calculated by dividing the focal length of the objective by the focal length of the eyepiece. In this example, the magnification would be 20 times (500 mm divided by 25 mm).

12. Invisible infrared and ultraviolet radiation is detected and studied by using photographic film that is sensitive to these wavelengths. Since most of this radiation cannot penetrate our atmosphere, balloons, rockets, and satellites must transport the cameras "above" the atmosphere to record it.

13. Radio telescopes are 1) less affected by weather and 2) can "see" through interstellar dust clouds that obscure our view at visible wavelengths. (answers may vary)

14. a) E; b) B; c) D; d) A; e) F; f) C

15. During a strong solar flare, fast-moving atomic particles are ejected from the sun, causing the solar wind to intensify noticeably. In about a day, the ejected particles reach Earth and disturb the ionosphere, affecting long distance communication and producing the auroras, the northern and southern lights.

16. The source of the sun's energy is nuclear fusion. The nuclear reaction, called the proton-proton reaction, converts four hydrogen nuclei into the nucleus of a helium atom, releasing energy during the process by converting some of the matter to energy.

Practice Test

Multiple choice

1. a	6. c	11. d	16. e	21. c
2. b	7. b	12. e	17. e	22. b
3. c	8. a	13. e	18. c	23. e
4. c	9. c	14. c	19. d	24. b
5. b	10. b	15. d	20. e	25. b

True/false

1. F (corona)	8. F (spectroscopy)	15. T
2. T	9. T	16. F (Radio)
3. T	10. T	17. F (second)
4. T	11. F (radiation)	18. F (eleven)
5. T	12. T	19. T
6. T	13. T	20. F (focal length)
7. F (water)	14. T	21. F (increases)

Word choice

1. more	8. cooler	15. diameter
2. eyepiece	9. electromagnetic	16. Violet; gamma
3. colors	10. hydrogen	17. cooler
4. longer	11. reflecting	18. good
5. photosphere	12. greater	19. chromosphere
6. redder	13. average	20. smaller
7. decreases	14. straight; 300,000	

Written questions

1. The source of the sun's energy is nuclear fusion. The nuclear reaction, called the proton-proton reaction, converts four hydrogen nuclei into the nucleus of a helium atom, releasing energy as some of the matter is converted to energy.

2. Spectroscopic analysis of the light from a star can determine the star's 1) composition, 2) temperature, and 3) motion (direction and rate), among other properties.

3. The four parts of the sun include: 1) the solar interior, where, in the deep interior, nuclear fusion produces the star's energy; 2) photosphere, which is considered to be the sun's surface and radiates most of the sunlight we see; and the two layers of the solar atmosphere, 3) the chromosphere, the lowermost portion of the solar atmosphere, is a relatively thin layer of hot, incandescent gases, and 4) the corona, the very tenuous outermost portion of the solar atmosphere.

4. A reflecting telescope does not have chromatic aberration (color distortion), it is usually less expensive than a comparable sized refractor because the glass does not have to be of optical quality, and large mirrors can be made because they are opaque and can be supported from behind.

CHAPTER TWENTY TWO

Vocabulary Review

1. magnitude
2. irregular galaxy
3. white dwarf
4. Hertzsprung-Russell (H-R) diagram
5. hydrogen burning
6. Degenerate matter
7. Stellar parallax
8. nebula
9. supergiant
10. emission nebula
11. protostar
12. neutron star
13. red giant
14. supernova
15. dark nebula
16. pulsar
17. Hubble's law
18. apparent magnitude
19. elliptical galaxy
20. light-year
21. bright nebula
22. black hole
23. reflection nebula
24. spiral galaxy
25. absolute magnitude
26. interstellar dust
27. main-sequence stars
28. planetary nebula
29. Big Bang
30. Local Group
31. pulsating variables
32. galactic cluster
33. nova

Comprehensive Review

1. The nearest stars have the largest parallax angles, while those of distant stars are too slight to measure.

2. The difference between a star's apparent magnitude and its absolute magnitude is directly related to its distance.

3. 1) how big the star is; 2) how hot the star is; 3) how far away the star is

4. a) less than 3000 K; b) between 5000 and 6000 K; c) above 30,000 K

5. Binary stars orbit each other around a common point called the center of mass. If one star is more massive than the other, the center of mass will be located closer to the more massive star. Thus, by determining the sizes of their orbits, a determination of each star's mass can be made.

6. 1) luminosity (brightness); 2) temperature

7. a) E; b) B; c) D; d) A

8. letter C

9. B through C through D represents the main sequence

10. 1) Emission nebulae absorb ultraviolet radiation emitted by an embedded or nearby hot star and reradiate, or emit, this energy as visible light. 2) Reflection nebulae merely reflect the light of nearby stars.

11. Dark nebulae are produced when a dense cloud of interstellar material is not close enough to a bright star to be illuminated.

12. Hydrogen burning occurs when groups of four hydrogen nuclei are fused together into single helium nuclei in the hot (at least 10 million K) cores of stars.

13. a) white dwarfs; b) giants followed by white dwarfs, often with planetary nebulae; c) supernovae followed by either a neutron star or black hole

14. (line will go from dust and gases to protostar, to main-sequence star, to giant stage, to variable stage, to planetary-nebula stage, to white-dwarf stage, to black-dwarf stage)

15. The large quantities of interstellar matter that lie in our line of sight block a lot of visible light.

16. The Milky Way is a rather large spiral galaxy with at least three distinct spiral arms whose disk is about 100,000 light-years wide and about 10,000 light-years thick at the nucleus.

17. 1) Spiral galaxies are typically disk-shaped with a somewhat greater concentration of stars near their centers. Arms are often seen extending from the central nucleus, giving the galaxy the appearance of a fireworks pinwheel. 2) Barred spiral galaxies have the stars arranged in the shape of a bar. About 10 percent of the galaxies are of this type. 3) Elliptical galaxies, the most abundant group, are generally smaller than spiral galaxies, have an ellipsoidal shape that ranges to nearly spherical, and lack spiral arms. 4) Irregular galaxies, about 10 percent of the known galaxies, lack symmetry.

18. spiral galaxy

19. Hubble's law states that galaxies are receding from us at a speed that is proportional to their distance.

20. According to the Big Bang theory, the entire universe was at one time confined to a dense, hot, supermassive ball. About 20 billion years ago, a cataclysmic explosion marking the inception of the universe occurred. Eventually the material that was hurled in all directions from the big bang cooled and condensed, forming the stellar systems we now observe fleeing from their birthplace.

Practice Test

Multiple choice

1. b	8. c	15. b	22. c	29. c
2. d	9. a	16. e	23. b	30. b
3. e	10. d	17. c	24. d	31. b
4. b	11. b	18. a	25. e	32. e
5. e	12. b	19. a	26. e	
6. a	13. a	20. a	27. d	
7. c	14. a	21. c	28. d	

True/false

1. F (rapidly)	9. T	17. T
2. T	10. T	18. F (light-years)
3. T	11. T	19. F (Elliptical)
4. F (distance)	12. T	20. F (brighter)
5. F (Apparent)	13. F (Bright)	21. F (high)
6. T	14. F (distance)	22. F (binary)
7. T	15. T	23. T
8. T	16. T	

Word choice

1. are	8. red	15. 100
2. protostar	9. supernovae	16. nebulae
3. x-rays	10. -26.7; 5	17. masses
4. more	11. absolute magnitude	18. beginning; universe
5. 1987	12. supernova	19. apparent
6. disk	13. atmosphere	20. quickly
7. is not	14. contracting	

Written questions

1. Following the protostar phase, the star will become a main-sequence star. After about 10 billion years (90 percent of its life) the star will have depleted most of the hydrogen fuel in its core. The star then becomes a giant. After about a billion years, the giant will collapse into an Earth-sized body of great density called a white dwarf. Eventually the white dwarf becomes cooler and dimmer as it continually radiates its remaining thermal energy into space.

2. 1) A low-mass star never becomes a red giant. It will remain a main-sequence star until it consumes its fuel, collapses, and becomes a white dwarf. 2) A medium-mass star becomes a red giant. When its hydrogen and helium fuel is exhausted, the giant collapses and becomes a white dwarf. During the collapse, a medium-mass star may cast off its outer atmosphere and produce a planetary nebula. 3) A massive star terminates in a brilliant explosion called a supernova. The two possible products of a supernova event are a neutron star or a black hole.

3. According to the Big Bang theory, the entire universe was at one time confined to a dense, hot, supermassive ball. About 20 billion years ago, a cataclysmic explosion marking the inception of the universe occurred. Eventually the material that was hurled in all directions from the big bang cooled and condensed, forming the stellar systems we now observe fleeing from their birthplace. One of the main lines of evidence in support of the Big Bang theory is the fact that galaxies are moving away from each other as indicated by the red shifts in their spectrums.